Lecture Notes in Computer Science   **15511**

Founding Editors

Gerhard Goos
Juris Hartmanis

The series Lecture Notes in Computer Science (LNCS), including its subseries Lecture Notes in Artificial Intelligence (LNAI) and Lecture Notes in Bioinformatics (LNBI), has established itself as a medium for the publication of new developments in computer science and information technology research, teaching, and education.

LNCS enjoys close cooperation with the computer science R & D community, the series counts many renowned academics among its volume editors and paper authors, and collaborates with prestigious societies. Its mission is to serve this international community by providing an invaluable service, mainly focused on the publication of conference and workshop proceedings and postproceedings. LNCS commenced publication in 1973.

Richard Chbeir · Sergio Ilarri ·
Yannis Manolopoulos · Peter Z. Revesz ·
Jorge Bernardino · Carson K. Leung
Editors

Database Engineered Applications

28th International Symposium, IDEAS 2024
Bayonne, France, August 26–29, 2024
Proceedings

Springer

Editors
Richard Chbeir ⓘ
Univ de Pau et des Pays de l'Adour
Anglet, France

Sergio Ilarri
University of Zaragoza
Zaragoza, Spain

Yannis Manolopoulos ⓘ
Aristotle Univ of Thessaloniki
Thessaloniki, Greece

Peter Z. Revesz ⓘ
University of Nebraska–Lincoln
Lincoln, NE, USA

Jorge Bernardino
Polytechnic Institute of Coimbra
Coimbra, Portugal

Carson K. Leung ⓘ
University of Manitoba
Winnipeg, MB, Canada

ISSN 0302-9743   ISSN 1611-3349 (electronic)
Lecture Notes in Computer Science
ISBN 978-3-031-83471-4   ISBN 978-3-031-83472-1 (eBook)
https://doi.org/10.1007/978-3-031-83472-1

Preface

This LNCS volume includes research papers presented at the 28th International Database Engineered Applications Symposium (IDEAS) which was held during August 28–31, 2024, in Bayonne, France. It is remarkable that the IDEAS 2024 conference was co-organized with the 28th European Conference on Advances in Databases and Information Systems (ADBIS). This collocation helped the exchange of ideas with members of a complementary community and, thus, enhanced cross-fertilization.

The first IDEAS Conference was held in Montreal, Canada (1997). Since then, IDEAS has been continuously organized as an annual event. Its previous editions were held in: Heraklion, Greece (2023); Budapest, Hungary (2022); Montreal, Canada (2021); online/Seoul, South Korea (2020); Athens, Greece (2019); Villa San Giovanni, Italy (2018); Bristol, UK (2017); Montreal, Canada (2016); Yokohama, Japan (2015); Porto, Portugal (2014); Barcelona, Spain (2013); Prague, Czech Republic (2012); Lisbon, Portugal (2011); Montreal, Canada (2010); Calabria, Italy (2009); Coimbra, Portugal (2008); Banff, Canada (2007); Delhi, India (2006); Montreal, Canada (2005); Coimbra, Portugal (2004); Hong Kong, China (2003); Edmonton, Canada (2002); Grenoble, France (2001); Yokohama, Japan (2000); Montreal, Canada (1999); Cardiff, UK (1998). IDEAS is an international forum for the discussion of issues related to data engineering, which involves not only database technology but also related areas such as artificial intelligence, information retrieval, natural language processing, and human-computer interaction, among others.

The program of IDEAS 2024 included keynote talks, research papers, and tutorials. The main conference attracted 34 paper submissions. All the papers went through a process of rigorous single-blind reviewing by at least three reviewers. Eventually, the Program Committee selected 26 submissions as full contributions appearing in the present volume. Actually, the authors had the opportunity to submit their final camera-ready version after the end of the conference, taking into account the fruitful discussions that took place on site. Thus, one could say that there was an enhanced review and double improvement phase.

The selected papers span a large spectrum of topics in the broader field of data management and were organized in eight sessions: (1) Theoretical Issues; (2) Classification; (3) Text and Languages; (4) Big Data; (5) Query Processing and Applications; (6) Machine Learning and Rules; (7) Recommendation, and (8) Indexing and Event Detection. Besides, during the event we selected the following two equally ranked papers for the best paper award, taking into account the novel contributions of the papers and the quality of the presentations:

- "Fairness in Group Recommender Systems using Variational Autoencoders" authored by Muhammad Shahzaib and Kostas Stefanidis, and
- "Predicting Dyskinetic Events through Verified Multivariate Time Series Classification" authored by Giacomo Bergami, Emma Packer, Kirsty Scott, and Silvia Del Din.

Following a successful tradition, selected best papers of IDEAS 2024 will be invited for special issues of the following journals: SN Computer Science (Springer), Information (MDPI), Digital (MDPI), and Computer Science & Information Systems (ComSIS Consortium). Therefore, the PC chairs would like to express their sincere gratitude to the Editors-in-Chief of the above journals for their approval regarding these special issues.

For this IDEAS event the following keynote talks by experts in different fields of the broader area of data management were scheduled. In particular (in alphabetical order):

- Angela Bonifati (CNRS Liris research lab, Lyon 1 University, France) delivered a talk on *"The Future is Hybrid Graphs: Combining Graphs with Time Series"*.
- Christian S. Jensen (Center for Data Intensive Systems, Aalborg University, Denmark) dealt with the topic of *"Temporal and Spatio-Temporal Data Management and Analytics – A Personal Perspective"*, and
- Schahram Dustdar (TU Wien, Austria) covered the topic of *"Distributed Intelligence in the Computing Continuum"*.

Due to unexpected problems, the last keynote talk was replaced by:

- Yannis Manolopoulos (Open University of Cyprus, University of Nicosia, Cyprus) presented advancements related to "Recommenders for Scientometric Objects".

In addition, two tutorials were also included in the program, which were delivered by:

- Zheying Zhang and Kostas Stefanidis (Tampere University, Finland), who presented a tutorial entitled *"Data-Driven Analysis for Monitoring Software Evolution"*, and
- Witold Andrzejewski, Bartosz Bębel, Paweł Boiński and Robert Wrembel (Poznań University of Technology, Poland), who delivered a tutorial entitled *"On Customer Data Deduplication – Research vs. Industrial Perspective: Lessons Learned from an R&D Project in the Financial Sector"*.

In addition, the conference organizers arranged for a rich social program that included:

- a welcome reception at Moma Kitchen in Bayonne on the first day of the conference,
- a gala dinner at Maetena in the Novotel, Anglet, on the second day, featuring a traditional music group that showcased the local culture of the Basque Country, and
- a half-day excursion on the final day allowed participants to explore key cities in the Basque Country, both in France and Spain. Attendees crossed the ocean by boat from France to Spain before enjoying a guided tour of Hondarribia's downtown. A traditional lunch was provided to conclude the event.

The Program Chairs are grateful for the support of some projects that fund their research activities in relation to the topics of the organized conference, such as the project PID2020-113037RB-I00 (NEAT-AMBIENCE project), funded by MICIU/AEI/10.13039/501100011033, and the support of the Departamento de Ciencia, Universidad y Sociedad del Conocimiento del Gobierno de Aragón (Government of Aragon: Group Reference T64_23R, COSMOS research group)

Finally, we would also like to wholeheartedly thank all participants, authors, PC members, session chairs, volunteers, and co-organizers for their contributions to making

IDEAS 2024 a great success. We would also like to thank the IDEAS Steering Committee and all sponsors.

November 2024

Richard Chbeir
Sergio Ilarri
Yannis Manolopoulos
Peter Z. Revesz
Jorge Bernardino
Carson K. Leung

Organization

General Chairs

Carson K. Leung   University of Manitoba, Canada
Jorge Bernardino   Institute of Education and Sciences (ISEC),
         Portugal
Peter Revesz   University of Nebraska Lincoln, USA

Program Chairs

Richard Chbeir   University of Pau and the Adour Region, France
Sergio Ilarri   University of Zaragoza, Spain

Special Issue Chair

Yannis Manolopoulos   University of Nicosia, Cyprus

Local Organization Chairs

Akram Hakiri   University of Pau and the Adour Region, France
Laurent Gallon   University of Pau and the Adour Region, France

Proceedings Chair

Karam Bou Chaaya   Expleo Group, France

Publicity Chairs

Khouloud Salameh   American University of Ras Al Khaimah, UAE
Theodoros Tzouramanis   University of Thessaly, Greece

Webmasters

Elie Chicha	University of Pau and the Adour Region, France
Fouad Achkouty	University of Pau and the Adour Region, France

Steering Committee

Peter Revesz (Chair)	University of Nebraska, USA
Bipin Desai (Honorary Chair)	Concordia University, Canada
Jorge Bernardino	Institute of Education and Sciences (ISEC), Portugal
Rui Chen	Samsung, USA
Le Gruenwald	University of Oklahoma, USA
Irena Holubova	Charles University, Czech Republic
Hideyuki Kawashima	Kyoto University, Japan
Bart Kuijpers	Hasselt University, Belgium
Carson Leung	University of Manitoba, Canada
Yannis Manolopoulos	University of Nicosia, Cyprus
Kamran Munir	University of the West of England, UK

Program Committee

Sabri Allani	Expleo Group, France
Toshiyuki Amagasa	University of Tsukuba, Japan
Masayoshi Aritsugi	Kumamoto University, Japan
Giacomo Bergami	Newcastle University, UK
Jorge Bernardino	Institute of Education and Sciences (ISEC), Portugal
Minal Bhise	DA-IICT, India
Karam Bou Chaaya	University of Pau and the Adour Region, France
Richard Chbeir	University of Pau and the Adour Region, France
Francesco Buccafurri	Mediterranea University of Reggio Calabria, Italy
David Chiu	University of Puget Sound, USA
Antonio Corral	University of Almería, Spain
Marcos Aurélio Domingues	State University of Maringá, Brazil
Alberto Freitas	University of Porto, Portugal
Sven Groppe	University of Lübeck, Germany
Irena Holubova	Charles University, Czech Republic
Sergio Ilarri	University of Zaragoza, Spain
Pavel Koupil	Charles University, Czech Republic

Carson K. Leung	University of Manitoba, Canada
Chuan-Ming Liu	National Taipei University of Technology, Taiwan
Jose Macedo	Federal University of Ceará, Brazil
Valéria Magalhães Pequeno	INESC-ID, Portugal
Yannis Manolopoulos	University of Nicosia, Cyprus
Miguel A. Martinez-Prieto	University of Valladolid, Spain
Danilo Montesi	University of Bologna, Italy
Yiu-Kai Ng	Brigham Young University, USA
Wilfred Ng	Hong Kong University of Science and Technology, China
Paulo Jorge Oliveira	ISEP, Portugal
Jaroslav Pokorný	Charles University, Czech Republic
Giuseppe Polese	University of Salerno, Italy
Filipe Portela	University of Minho, Brazil
Peter Revesz	University of Nebraska Lincoln, USA
Miguel Rodríguez Luaces	University of Coruña, Spain
Marinette Savonnet	University of Burgundy, France
Jeffrey Ullman	Stanford University, USA
José R.R. Viqueira	University of Santiago de Compostela, Spain
Alicja Wieczorkowska	Polish-Japanese Academy of Information Technology, Poland
Ouri Wolfson	University of Illinois Chicago, USA
Roberto Yus	University of Maryland, Baltimore County, USA

External Reviewers

Drissi Amani	University of Tunis El Manar, Tunisia
Amal Beldi	University of Tunis El Manar, Tunisia
Simone Branchetti	University of Bologna, Italy
Loredana Caruccio	University of Salerno, Italy
Oliver R. Fox	Newcastle University, UK
Gianpaolo Iuliano	University of Salerno, Italy
Jason Sawin	University of St. Thomas, USA
Sana Sellami	Aix-Marseille University, France
Roberto Stanzione	University of Salerno, Italy
Joe Tekli	Lebanese American University, Lebanon

Contents

Big Data

Query Processing and Applications

Machine Learning and Rules

Theoretical Issues

Verifiable and Friendly Metadata Management: Formal Blockchain to Property Graph Mapping

Anton Dolhopolov[✉], Arnaud Castelltort, and Anne Laurent

LIRMM, University of Montpellier, CNRS, Montpellier, France
{anton.dolhopolov,arnaud.castelltort,anne.laurent}@lirmm.fr

Abstract. Blockchain technology has gained popularity as a decentralized, tamper-proof, and verifiable data structure. Apart from financial transactions and assets management, there is also a growing interest in using it as a backbone for metadata management systems. Metadata systems aim to facilitate the organization of big data and provide users with a comprehensive interface for data discovery and navigation. However, the fundamental design of the blockchain has limitations in data querying and visualization that result in poor user experience. On the other side, graph databases are widely adapted for building user-friendly metadata systems with efficient relationship discovery capabilities. In our paper, we attempt to preserve the benefits of both types of systems through the integration of blockchain and property graph technologies. First, we formalize the structures used for metadata storage and a metadata conceptual model. Second, we provide the algorithms for mapping the (meta)data from the blockchain to the property graph model. And third, we show a prototype implementation of the integrated system to validate our proposal.

Keywords: Metadata Management · Blockchain · Property Graphs

1 Introduction

The proliferation of blockchain technology within enterprise information systems introduced a new direction of research. Beyond the financial bookkeeping applications (e.g. Bitcoin), there is a growing interest in applying this immutable, verifiable, and decentralized data structure for metadata management [5,6,8–10]. In this case, the metadata (e.g. format, location, size) is stored as the on-chain information about the off-chain data assets (e.g. datasets, media artifacts). Such architectural design aims to provide better scalability, verifiable records history, and to eliminate the downsides of storing the data in the blockchain directly [11].

Generally, metadata systems are required to provide an interface to the users for enabling data discovery, querying, analytics, etc. In recent years, graph

R. Chbeir et al. (Eds.): IDEAS 2024, LNCS 15511, pp. 3–17, 2025.
https://doi.org/10.1007/978-3-031-83472-1_1

databases become widely adapted for metadata management of big data platforms, both, in industrial tools [14,15] and in the academic works [7,12,19]. The main motivation for graph-model data storage is that it naturally supports data relationship modeling which enables scalability and flexibility [1] since it is not constrained to the strict schema enforcement compared to relational databases.

A number of recent research studies have dedicated efforts to using graph technologies for enhancing blockchain capabilities. One of the directions is Directed Acyclic Graph (DAG) based ledgers [18]. However, the goal of implementing the DAG structure is to overcome the fundamental transaction bandwidth issues seen in traditional blockchains.

Another direction focuses on conducting blockchain analytics by mapping the underlying on-chain data into the graphs. The majority of these analytical tasks are related to financial market predictions, fraud prevention, or community detection and evolution. According to the literature [13], almost all these systems are based on nodes representing account addresses or smart contracts, and edges representing asset flows or communication channels (e.g. contract calls).

We highlight that there is a growing demand for further research on the integration of graph technologies with the blockchain, particularly in the context of metadata management. As part of our contribution, we propose a formal model for metadata transformation from blockchain to property graphs and present a proof-of-concept system.

The paper is organized as follows: in Sect. 2 we review the related works and provide the necessary theoretical foundation. Section 3 depicts our formal model for generic and schema-dependent blockchain transformation. Section 4 outlines our prototype that implements the proposed formal models and in Sect. 5 we summarise our contribution and provide future research perspectives.

2   Background

In this section, we describe the related works of metadata systems and graph-enhanced blockchains. Then we introduce the blockchain and property graph formal models that will be used later for our metadata mapping rules.

2.1   Related Works

Recently, there have been a growing number of proposals to use the blockchain structure as a decentralized metadata storage and management solution.

Garcia-Barriocanal et al. have researched the way of storing the metadata by using the globally accessible distributed technology of Inter-Planetary File System [8]. Kumar and Rahman [9] have considered the case of a widely utilized Hadoop Distributed File System data storage and processing solution and proposed to improve the fault-tolerance mechanisms of the coordinator nodes by distributing the metadata information with the blockchain.

Demichev et al. have developed a system for tracking and managing the provenance metadata in healthcare data-sharing scenarios [5], while Dolhopolov et al. have introduced a generalized blockchain-based metadata management system for a decentralized data mesh platform [6].

On the other side, as Sawadogo and Darmont point out [12], graph-based modeling is the most common way to implement metadata systems thanks to the advantages of automatic information enrichment and advanced querying. Indeed, such commercial products as Apache Atlas [14] and DataHub Project [15] adopt the property-graph data modeling to provide a flexible and efficient metadata system. Eichler et al. have used a Neo4j graph database for implementing a general and expandable metadata model called HANDLE [7] and Ziegler et al. have used Neo4j for metadata management in the data lake context [19].

Unsurprisingly, the interest in enhancing the blockchain with graphs has been growing. In their vision paper, Bellomarini et al. have outlined the two-way positive impact cycle between the knowledge graphs (KG) and the blockchain [3]. On one side, KG can help with the knowledge-based smart contract generation process. On the other side, blockchain can be used for knowledge verification as it represents the unmodified source of data.

Cano-Benito et al. have proposed to use the blockchain for storing the ontologies in a way that would help to build the semantic web [4] and Fluree [16] database is an industrial solution based on Resource Description Framework technology that supports immutable data structure similar to the blockchain.

The closest work to ours is a contribution made by Tsoulias et al. [17]. The authors used the Neo4J database for directly storing the blockchain data as nodes and edges. They implemented the proof-of-work and proof-of-stake protocols with Python programming language and used the property-graph model to store the data. The work also verifies the solution against different real-world situations, like a 51% Attack.

However, their proposal does not account for the integration of already running blockchain systems but rather considers the situation of a new system bootstrap. It does not provide any discussion on the data migration from existing blockchains to the graph model which is in fact addressed in our proposal by means of provided algorithms and mapping rules.

2.2 Property Graph Model

In recent years property graphs (PGs) have gained popularity as a NoSQL data model. It provides a way to efficiently represent the relationships and connections of natural phenomena in the form of nodes and edges that are enriched with properties. For instance, data assets and users can be seen as nodes while the actions performed by users over the data will be represented as edges. When a user queries an asset, we create an edge with the label *HasQueried* from the user to the asset of interest and add the timestamp when it was done as an edge property. In Fig. 1 we show the described property graph example.

We define the formal PG model as a tuple $\mathcal{PG} = (N, E, L, P, lbl, edge, prop)$:

- N is a finite set of **nodes**, E is a finite set of **edges**, L is a finite set of **labels**, and P is a finite set of **properties** such that $N \cap E \cap P \cap L = \varnothing$
- $lbl : N \cup E \to L$ is a function that assigns a single label to a node or an edge
- $edge : E \to N \times N$ is a function that assigns each edge to a pair of nodes (n, n'), where n is the source node and n' is the target node
- $prop : N \cup E \to \mathcal{P}^+(P)$ is a partial function that associates a node or an edge with a non-empty set of properties P, such that $p = (k, v), p \in P, k \in K, v \in V$

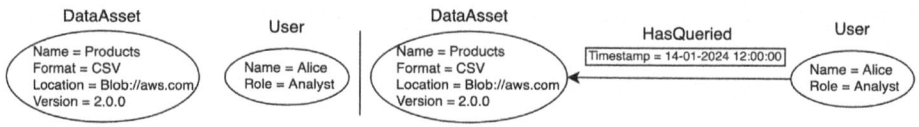

Fig. 1. Property graph example: node types (left) and edge formation (right)

As follows, K is a set of possible property keys and V is a set of possible property values. In Fig. 1 the labels *User* and *DataAsset* from set L are also assigned to nodes and offer further functionality for distinguishing node types.

In Sect. 1 we pointed out that often metadata systems are implemented with a property graph model [12]. The flexible and schema-free nature of the PG model supports the quick evolution of enterprise data assets and their relationships. Modern graph databases do not require any predefined data schema for using it (contrary to relational databases). Moreover, there is always an option to enforce the user-provided schema to guarantee data consistency and integrity.

In general, labels are used for defining the semantic meaning of the data through the user schema. In Fig. 1 *HasQueried* label describes the nature of the action relationship between the two nodes. At all times, we assume to have a PG schema that defines the structure of possible relationships between the nodes. Put it formally, schema $\mathcal{S}_{\mathcal{PG}}$ is composed of node labels L_N and edge labels L_E, such that $L_N \cup L_E = L$, $L_N \cap L_E = \varnothing$; and a function $\eta : L_N \times L_N \to L_E$ that defines allowed relationships between a given pair of nodes.

2.3 Blockchain Model

We describe a general model that can be applied to a wide range of blockchains, including Bitcoin, Ethereum, or Hypeledger Fabric that is used in our prototype. Due to the space limits, we omitted a complete formal definition and rather provided a general description thereafter.

Blockchain $\mathcal{B}$ is a sequence of cryptographically signed blocks. The blocks are made of transactions issued by user accounts and a header h that has a hash value of all transactions within that block. The first block in the blockchain is a genesis block that may contain arbitrary data. Each following block contains a header that also has a reference to the previous header and its hash value (established via function ref_h). This way the blockchain provides a verifiable and tamper-proof data structure.

Usually, transactions change the account state. They are comprised of the operations over the assets, e.g. substitution of some currency from account x and equivalent amount addition to the account y. We can write it down as a *key-value* pair (k, v) that represents the account label and balance respectively.

In the scenario of using the blockchain as a metadata system, we can apply this approach for storing information about the data assets by using (k, v) pairs for describing its metadata entries (e.g. $k = format$ and $v = CSV$).

Depending on the blockchain implementation, user transaction tx can also contain different platform-related information. In our model, we assume that each transaction is composed of a user signature (for establishing its ownership) and a set of properties P reflecting the metadata, that is $tx = (a.Sign, P)$.

Usually, each transaction is signed by a single account which establishes its ownership rights. But sometimes, it is possible to have a *multi-signature* transaction, which requires the participation of more than a single account to consider a transaction to be valid. In this case, we will use a special function μ_{ms} to denote an "ownership" of the transaction tx_i across multiple accounts.

Attention should be paid to the header reference function ref_h that always holds true, except in the case of the genesis block (absence of the block before). To overcome it, we define a self-referencing function $ref_h(h_0) = h_0$.

In this section, we have described the related works and the necessary theoretical background. The following section goes more into the detail of metadata mapping from the blockchain to the PG data structure.

3   Designing a Formal Model for Metadata Mapping

In this section, we outline the first part of our contribution. We start by describing the properties of the data transformation we want to comply with. Then we consider a running example of the metadata system and provide a generic, schema-independent conversion. As a subsequent step, we propose a conceptual metadata model and introduce a procedure for schema-dependent conversion.

Due to the space limits, we note that the complete formal rules of the mapping are provided as part of the prototype code that is discussed in Sect. 4.

3.1   Properties of Metadata Mapping

We recall three properties of the database transformation from Angles et al. [2], which are: computability, information and semantic preservation. We briefly introduce it here and we address the reader to the original paper for more details.

Computability indicates that for any two given databases $\mathcal{D}_1$ and $\mathcal{D}_2$ there exists an algorithm $\mathcal{A}$ that computes a database mapping $\mathcal{DM} : \mathcal{D}_1 \rightarrow \mathcal{D}_2$. Our goal is to provide an algorithm showing that mapping $\mathcal{DM} : \mathcal{B} \rightarrow \mathcal{PG}$ is feasible.

Information preservation means that for a computable mapping $\mathcal{DM}$ there exists an inverse translation $\mathcal{DM}^{-1} : \mathcal{D}_2 \rightarrow \mathcal{D}_1$, or that $\mathcal{D} = \mathcal{DM}^{-1}(\mathcal{DM}(\mathcal{D}))$. It is a required property for us since we are only interested in translations that don't lose the source information.

Semantic preservation indicates that the result of the mapping is a **valid** database. In this context, a valid database means a database instance $\mathcal{I}$ which is compliant with rules and constraints of the target schema $\mathcal{S}$, denoted as $\mathcal{I} \models \mathcal{S}$. To be more precise, when we discuss database mapping, we assume that any database is composed of an instance and a schema, or that $\mathcal{D} = (\mathcal{I}, \mathcal{S})$. If the schema is not defined, it means that $\mathcal{D} = (\mathcal{I}, \varnothing)$, so we will use the terms "database" and "instance" interchangeably.

3.2   Generic Transformation from Blockchain to Property Graph

As a first step, we consider an existing metadata system based on Hyperledger Fabric [6]. Our first goal is to show the feasibility of transforming the on-chain metadata from the blockchain to the property graph structure.

A running example of the blockchain-based catalog, presented in Fig. 2, has three blocks modeling different metadata entries. The first block is a genesis block with a self-referencing header. It describes a data asset in transaction $TxA1$ and its owner (displayed to the left). The second block has a similar structure, where the header points to the previous block as expected. In our case, block $TxB1$ has two signing accounts and one transaction that references another one: $TxB1 \rightarrow TxA1$, which is valid according to our formal model. Finally, the third block describes 2 data assets from different users.

To perform a generic, schema-independent metadata mapping, we introduce a generic property graph schema $\mathcal{S}_G$ that defines the structure of the generated graph. It enables us to map any blockchain-based catalog into the PG-based catalog. Our target schema consists of a set of node labels L_N and a set of edge labels L_E, where: $L_N = \{UserAccount, BlockHeader, Transaction\}$, $L_E = \{PreviousHeader, ReferencesTo, BelongsTo, Signed\}$, $L_N \cup L_E = L_G$.

Algorithm 1: Generic Metadata Mapping

Data: Sequence of blocks B
Result: Nodes N & edges E

```
 1  N ← ∅, E ← ∅;
 2  for b ∈ Sorted(B) do
 3  │    h ← b.header;
 4  │    lbl(h) ← BlockHeader;
 5  │    Push(N, h) /* appending the header h to the set of nodes N */ ;
 6  │    e ← ∅;
 7  │    lbl(e) ← PreviousHeader, edge(e) ← (h, ref(h));
 8  │    Push(E, e);
 9  │    for tx ∈ b.transactions do
10  │  │    n ← ∅, e ← ∅ a ← ∅;
11  │  │    lbl(n) ← Transaction, n.P ← tx.P;
12  │  │    Push(N, n);
13  │  │    lbl(e) ← BelongsTo, edge(e) ← (n, h);
14  │  │    Push(E, e);
15  │  │    if μ(tx) is not in N then
16  │  │  │    lbl(a) ← UserAccount, a.Sign ← tx.sign;
17  │  │  │    Push(N, a);
18  │  │    else
19  │  │  │    a ← Get(N, tx.sign) /* finding a user a by the transaction signature in N */
20  │  │    end
21  │  │    e ← ∅;
22  │  │    lbl(e) ← Signed, edge(e) ← (a, n);
23  │  │    Push(E, e);
24  │  │    for r ∈ ref(tx) do
25  │  │  │    e ← ∅;
26  │  │  │    lbl(e) ← ReferencesTo, edge(e) ← (n, r);
27  │  │  │    Push(E, e);
28  │  │    end
29  │    end
30  end
```

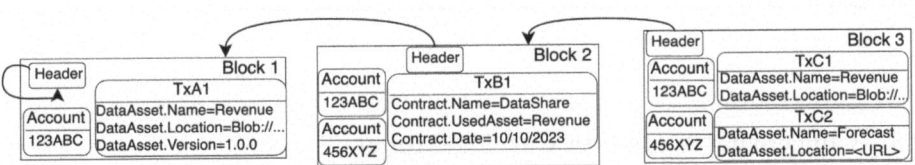

Fig. 2. A running example of the blockchain-based metadata catalog.

We provide an Algorithm 1 that guarantees the computability of the mapping $\mathcal{DM}_G : \mathcal{B} \to \mathcal{PG}$. To prove that $\mathcal{DM}_G$ is information preserving we need to have a mapping $\mathcal{DM}_G^{-1}$ such that $\mathcal{I} = \mathcal{DM}_G^{-1}(\mathcal{DM}_G(\mathcal{I}))$ for any given instance $\mathcal{I}$ that represents blockchain catalog $\mathcal{B}$. It should be noted that we consider a blockchain model as defined in Sect. 2.3. Therefore, it is sufficient to recover the initial database structure, including chain blocks, headers, transactions, accounts, and their relations. The solution to this is provided in the Algorithm 2.

The output of the generic transformation $\mathcal{DM}_G$ is a graph $\mathcal{PG} = (\mathcal{I}_G, \mathcal{S}_G)$, where $\mathcal{I}_G$ is a graph database instance as described in Sect. 2.2 and $\mathcal{S}_G$ is a generic schema created for any input blockchain catalog. It means that $\mathcal{I}_G \models \mathcal{S}_G$ by definition and that transformation $\mathcal{DM}_G$ is semantic preserving.

The result of the proposed schema-independent mapping complies with all three properties: it is computable, information and semantic preserving. One of the benefits of this mapping is that metadata stored in the form of the property-graph model gives a way to detect transaction clusters or to query all transactions created by a given user by simply following the outgoing *Signed* edges.

However, most of the time the metadata system users will define their own catalog schema which is enforced within the blockchain. Therefore, we have to have a schema-dependent mapping to generate an appropriate graph.

Algorithm 2: Inverse Metadata Mapping

Data: Nodes N & edges E
Result: Sequence of blocks B

```
 1  B ← ∅;
 2  H ← Find(BlockHeader) /* finding all nodes with BlockHeader node label */ ;
 3  for h ∈ Sorted(H) do
 4  |    b ← ∅;
 5  |    b.header ← h;
 6  |    h* ← Find(PreviousHeader, h)/*get a node by traversing a PreviousHeader edge*/;
 7  |    ref(b.header) ← h*;
 8  |    b.transactions ← ∅;
 9  |    T ← Find(BelongsTo, h);
10  |    for n ∈ T do
11  |    |    a ← Find(Signed, n);
12  |    |    tx ← (a, n.P);
13  |    |    for r ∈ Find(ReferencesTo, n) do
14  |    |    |    Push(tx.references, r)
15  |    |    end
16  |    |    Push(b.transactions, tx)
17  |    end
18  end
```

3.3   Conceptual Metadata Catalog Model

Before implementing the schema-dependent metadata mapping, we need a conceptual catalog model. Considering the vast amount of different metadata models, we adopt a simple, generic and extensible model called HANDLE [7].

We extend the HANDLE according to our previous running example and present an Entity-Relationship diagram model in Fig. 3. We omit a full definition here, but briefly state that it is seen as a tuple $\mathcal{M} = (D, M, C, U)$, where: D is a finite set of **data** assets, M is a finite set of assets **metadata**, C is a finite set of data **contracts**, U is a finite set of system **users**.

In this model, a data asset represents the raw data and it has links to the storage location, connected assets, its metadata, and owner. Metadata describes a single asset on a higher abstraction, like when, who, and how it was created. There can be many metadata entries related to the same asset but with a different context (creation, access, etc.). A data contract defines a data-sharing relation between different parties, or simply users. It also has two "ends" of a relation: the contract should be proposed by a single user. At the same time, users can sign zero or many data contracts.

Eventually, we want to check if a given blockchain catalog $\mathcal{B} = (\mathcal{I}_\mathcal{M}, \mathcal{S}_\mathcal{M})$ is valid w.r.t. the model $\mathcal{M}$, or that $\mathcal{I}_\mathcal{M} \models \mathcal{S}_\mathcal{M}$. Since we can't have explicit labels of the references in the blockchain, we fall back to considering only transaction types and the implicit link map between them. We assume that $L_\mathcal{M}$ is a set of entity type labels corresponding to model $\mathcal{M}$ and that $\lambda : tx \rightarrow L_\mathcal{M}$ is a function that returns the label of a given transaction. We say that $\mathcal{I}_\mathcal{M} \models \mathcal{S}_\mathcal{M}$, iff:

- **M1:** for any $tx \in T$, such that $\lambda(tx) = DataAsset$, there is a $tx' \in T$, where:

  - $\lambda(tx') = Metadata$ ◆ $ref(tx') = tx$ ◆ $\mu(tx) = \mu(tx')$;

- **M2:** for any $tx \in T$, such that $\lambda(tx) = Metadata$, there is a $tx' \in T$, where:

  - $\lambda(tx') = DataAsset$ ◆ $ref(tx) = tx'$;

- **M3:** for any $tx \in T$, where $\lambda(tx) = DataContract$, there is $tx' \in T$, where:

  - $\lambda(tx') = DataAsset$ ◆ $ref(tx) = tx'$ ◆ $\mu(tx) \neq \mu(tx')$;

- **M4:** for any pair $(tx, tx') \subset T$, such that:

  - $\lambda(tx) = \lambda(tx') = DataAsset$ ◆ $ref(tx) = tx'$ ◆ $\mu(tx) \neq \mu(tx')$

  ▷ there exists a transaction $\tilde{tx}$, such that:

  - $\lambda(\tilde{tx}) = DataContract$ ◆ $\mu_{ms}(\tilde{tx}) = (\mu(tx), \mu(tx'))$;

where $L_\mathcal{M} = \{DataAsset, Metadata, DataContract\}$. We intentionally omit the distinction of a $User$ entity because this information is always present as part of the considered transaction.

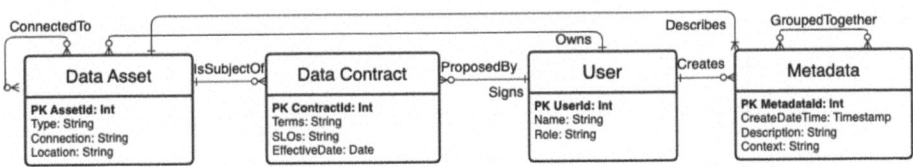

Fig. 3. Conceptual model of metadata catalog.

3.4   Schema-Dependent Metadata Catalog Mapping

As a last part of our formalization, we design a schema-dependent catalog mapping $\mathcal{DM_S} : \mathcal{B_M} \rightarrow \mathcal{PG_M}$, where: $\mathcal{B_M} = (\mathcal{I_{B_M}}, \mathcal{S_{B_M}})$ and $\mathcal{PG_M} = (\mathcal{I_{PG_M}}, \mathcal{S_{PG_M}})$ are valid databases w.r.t. previously defined model $\mathcal{M}$. The catalog conversion $\mathcal{DM_S}$ is computable and we provide a solution in the Algorithm 3.

Algorithm 3: Schema-Dependent Metadata Mapping

Data: Sequence of blocks B
Result: Nodes N & edges E

```
 1  N ← ∅, E ← ∅;
 2  for bᵢ ∈ Sorted(B) do
 3      h ← b.header, lbl(h) ← Block, h.OrderNo ← i;
 4      Push(N, h) /* appending the header h to the set of nodes N */ ;
 5      e ← ∅, lbl(e) ← Previous, edge(e) ← (h, ref(h));
 6      Push(E, e) /* appending the edge e to the set of edges E */ ;
 7      for tx ∈ b.transactions do
 8          n ← ∅, e ← ∅, a ← ∅;
 9          lbl(n) ← λ(tx), n.P ← tx.P;
10          Push(N, n);
11          lbl(e) ← BelongsTo, edge(e) ← (n, h);
12          Push(E, e);
13          if μ(tx) is not in N then
14              lbl(a) ← User, a.Sign ← tx.sign;
15              Push(N, a);
16          else
17              a ← Get(N, tx.sign) /* finding a user a by the transaction signature in N */
18          end
19          for r ∈ ref(tx) do
20              e ← ∅, eₐ ← ∅;
21              edge(e) ← (n, Get(N, r));
22              if lbl(n) is DataAsset then
23                  lbl(e) ← ConnectedTo;
24                  lbl(eₐ) ← Owns, edge(eₐ) ← (n, a);
25                  Push(E, e, eₐ);
26              else
27                  if lbl(n) is Metadata then
28                      if λ(r) is DataAsset then
29                          lbl(e) ← Describes;
30                      else
31                          lbl(e) ← GroupedTogether;
32                      end
33                      lbl(eₐ) ← Creates, edge(eₐ) ← (n, a);
34                      Push(E, e, eₐ);
35                  else
36                      lbl(e) ← IsSubjectOf ;
37                      (a₁, a₂) ← μₘₛ(tx) ;
38                      if a₂ is ∅ then
39                          lbl(eₐ) ← ProposedBy, edge(eₐ) ← (n, a);
40                          Push(E, e, eₐ);
41                      else
42                          lbl(eₐ₁) ← ProposedBy, edge(eₐ₁) ← (n, a₁);
43                          lbl(eₐ₂) ← Signs, edge(eₐ₂) ← (n, a₂);
44                          Push(E, e, eₐ₁, eₐ₂);
45                      end
46                  end
47              end
48          end
49      end
50  end
```

The obtained property graph $\mathcal{PG}_{\mathcal{M}}$ is a valid database, meaning that $\mathcal{DM}_{\mathcal{S}}$ is semantic preserving. We note that resulting schema $\mathcal{S}_{\mathcal{PG}_{\mathcal{M}}}$, as well as $\mathcal{S}_{\mathcal{B}_{\mathcal{M}}}$, is directly derived from the conceptual model $\mathcal{M}$. It is composed of node labels $L_N = \{DataAsset, Metadata, User, DataContract\}$ and edge labels $L_E = \{Describes, Owns, Creates, ProposedBy, Signs, IsSubjectOf, GropedTogether, ConnectedTo\}$. The respective entities are guaranteed by node generation code lines 8–10 and relationships are guaranteed by edge generation code lines 20–44.

To guarantee the information preservation of our mapping based on model $\mathcal{M}$, we follow a similar approach as in the generic algorithm and therefore introduce the block ordering with code lines 3–6 and 11–12.

In this section, we proposed two formal procedures for performing metadata catalog mapping from blockchain to property graph database. The schema-independent procedure allows us to obtain a generic graph that enhances the user's capacity to traverse and analyze the metadata. At the same time, the schema-dependent procedure also provides a way to systematize the metadata according to the user-defined conceptual model. In addition, both translations are computable, information and semantic preserving.

In the following section, we show a proof-of-concept implementation of our formal models alongside its performance assessment.

4 Implementing Metadata Mapping from Blockchain to Property Graph

This section provides details about the developed system used for validating the defined formal models. Then it continues with evaluation methodology, provides quantitative and qualitative results, and concludes with a results discussion.

4.1 Implementation Details

To validate our theoretical proposal with practical evidence, we have developed a proof-of-concept (PoC) system for testing the transformation of metadata information from blockchain to property graph model from two perspectives.

The first aspect includes the validation of the real-world, end-to-end transformation or mapping system, which provides a suitable interface to the users. Our PoC has a decentralized ledger deployed with Hyperledger Fabric (HLF) and a Neo4J graph database that supports the property graph model. The mapping module is implemented as a RESTful API service with Swagger support. In addition to metadata transformation (Mapping Module), this middleware service also enables the communication channels with HLF (via LedgerDriver) and Neo4J (via GraphDriver) with direct HTTP(s) requests or a graphical interface provided by Swagger. The architecture of the system is presented in Fig. 4.

We use the HLF because it naturally extends our running example from Sect. 3.2 and enables us to store the information as key-document or key-value pairs which are required by the formal models. We employ the Neo4J database

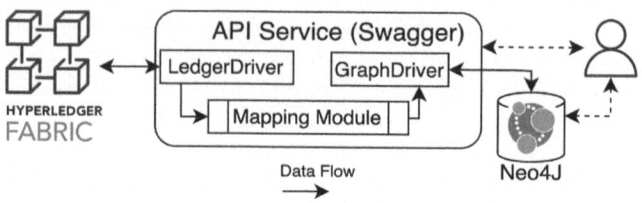

Fig. 4. Prototype system architecture.

because it is a leading, ACID-compliant graph storage and processing solution that has a lot of driver libraries.

The second aspect includes testing the performance and scalability of the proposed metadata catalog mapping. We developed two algorithms for generic and schema-dependent mappings based on the formal models in Sect. 3.

For excluding the external factors on the algorithm performance evaluation, such as database indexing or network delays, we use beforehand extracted ledger metadata information which is stored according to the blockchain model from Sect. 2. The output of each transformation algorithm is a text file containing the instructions necessary for creating required nodes, edges, and their corresponding labels and properties. The instructions are written in Cypher querying language that is supported by the Neo4J database. Both algorithms support the parallel computation model that is used for scalability assessment.

The system's code and full formal rules are available in the Seafile service[1].

4.2   Methodology and Experimental Setup

To conduct the prototype performance assessment we used the stable release of Ubuntu 22.04 (desktop version) with default post-installation parameters. The system hardware characteristics were: AMD Ryzen 5600x CPU with 6 cores (12 threads), 32 GB of RAM, 1 TB of HDD disk storage with 5400 rpm. The mapping module used in the performance benchmark was implemented as a Go binary program using version 1.20. The provided input files were represented as JSON files and the output text files contained Cypher-based instructions.

Our primary metric of the benchmark was the measurement of a transformation time from the underlying files representing metadata in the blockchain's format to the property graph nodes and edges. We employed standard library *time* package[2] to test the mapping duration with ms precision. We run the tests 10 times per transformation type per varying number of blocks.

To conduct the scalability evaluation, we tested our application with a variable number of utilized CPU cores. To achieve it on the same testing hardware, we used the *runtime.GOMAXPROCS* package function[3] that provides a way to

---

[1]   https://seafile.lirmm.fr/f/3de8b35b7cae47a9bfe0/?dl=1.

[2]   https://pkg.go.dev/time.

[3]   https://pkg.go.dev/runtime#GOMAXPROCS.

specify the maximum number of utilized processor cores by a program. We used randomly generated data to populate the Hyperledger Fabric ledger and then extracted it into the JSON.

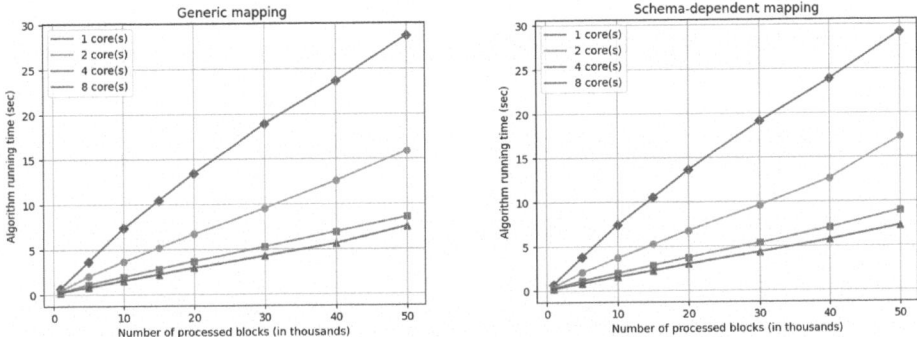

Fig. 5. Algorithm running time: generic (left) and schema-dependent method (right).

4.3   Results and Discussions

The transformation running time results are presented in Fig. 5. It reveals that our algorithms exhibit acceptable performance and scalability properties. For instance, processing 10 thousand blocks (approximately 220 Mb) with 1 processor core takes 7.5 s, while using 4 cores reduces the processing time to 2 s, on average, with the generic and schema-dependent methods. If we scale up the ledger to a larger number of blocks, we observe that the processing time increases approximately linearly for both methods. This trend is evident when handling the 50 thousand blocks (approximately 1100 Mb), where the execution runtime, on average, extends to about 29 s for 1 core setup and to about 9 s for 4 core setup. The proposed transformation algorithm passes through source blocks and transactions only once, meaning that it has the complexity of $O(n)$. This is also evident from the performance evaluation since the processing time grows linearly to the input data size.

We assume that the low improvement between 4 and 8 cores setups is due to the hardware constraint, which has only 6 full cores with hyper-threading.

The visual difference between the standard ledger visualization and graph view is presented in Fig. 6. On the left side of the figure, we can see transaction information in the form of JSON properties. The tool, called Hyperledger Explorer, automatically extracts the ledger data and stores it in the relational database. In fact, we see that such metadata presentation provides poor usability and leaves the problem of navigation and querying to the user.

On the opposite side, the built-in web interface to the graph data, provided by the Neo4J, models a lot of different entities and relationships on a single screen, which has considerably higher utility compared to the previous tool. Additionally, users can employ a Cypher querying language for efficient data querying, manipulation, and discovery. Cypher syntax is similar to SQL and it provides a way to easily define the querying graph patterns naturally.

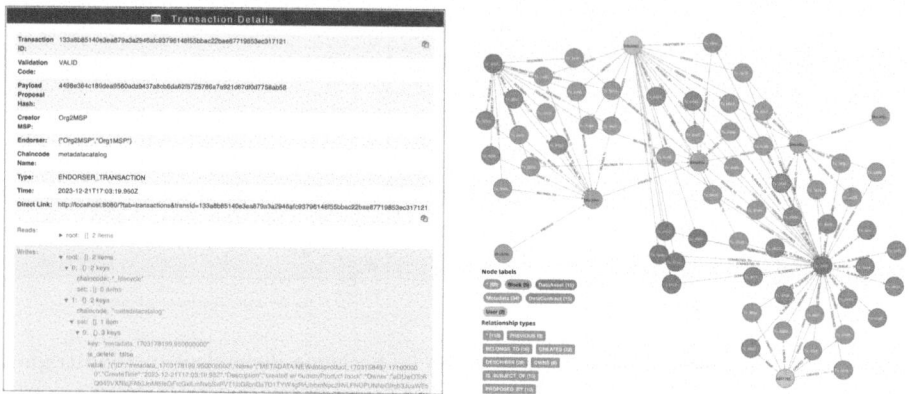

Fig. 6. Explorer transaction view (left) and Neo4J web UI sub-graph view (right).

5 Conclusions and Further Research

To conclude, this paper presents a novel way to improve data discovery and querying of the decentralized metadata management system built on top of the blockchain. It discusses the research on the integration of blockchains with graphs and highlights the missing functionalities that are satisfied by our method.

The proposed approach is based on two parts. First, we define the formal models for generic (schema-independent) and schema-dependent metadata transformation from the ledger storage to the property graph database. Furthermore, the models comply with three essential properties of database model mapping: computability, information and semantic preservation.

Second, we design and implement a prototype system that proves the feasibility and correctness of the formal models as and we provide the transformation algorithm performance and scalability evaluation. The system is based on a widely deployed permissioned blockchain - Hyperledger Fabric, and a leading transactional graph database - Neo4J.

We envision further research on several axes. We note that even though the currently used metadata model is generic and extensible, the users may still need to apply the proprietary metadata models. To achieve it, we are required to have a comprehensive metadata transformation module that processes any user-defined schema. Moreover, such a module will also benefit from incorporating the processing of a time dimension, that is from supporting the evolution of properties of entities and relationships, which is lacking as of today.

References

1. Angles, R., Gutierrez, C.: An introduction to graph data management. In: Fletcher, G., Jan, H., Larriba-Pey, J. (eds.) Graph Data Management: Fundamental Issues and Recent Developments, pp. 1–32. Springer (2018)
2. Angles, R., Thakkar, H., Tomaszuk, D.: Mapping RDF databases to property graph databases. IEEE Access **8**, 86091–86110 (2020)
3. Bellomarini, L., Nissl, M., Sallinger, E.: Blockchains as knowledge graphs-blockchains for knowledge graphs. In: Proceedings of the International Workshop on Knowledge Representation & Representation Learning, Co-located with the 24th European Conference on Artificial Intelligence (ECAI), pp. 43–51 (2020)
4. Cano-Benito, J., Cimmino, A., García-Castro, R.: Towards blockchain and semantic web. In: Proceedings of the 22nd International Workshops on Business Information Systems (BIS), Revised Papers, pp. 220–231 (2019)
5. Demichev, A., Kryukov, A., Prikhodko, N.: The approach to managing provenance metadata and data access rights in distributed storage using the hyperledger blockchain platform. In: Proceedings of the Ivannikov Ispras Open Conference (ISPRAS), pp. 131–136 (2018)
6. Dolhopolov, A., Castelltort, A., Laurent, A.: Implementing a blockchain-powered metadata catalog in data mesh architecture. In: Proceedings of the International Congress on Blockchain & Applications (BLOCKCHAIN), pp. 348–360 (2023)
7. Eichler, R., Giebler, C., Gröger, C., Schwarz, H., Mitschang, B.: Modeling metadata in data lakes-a generic model. Data Knowl. Eng. (2021)
8. García-Barriocanal, E., Sánchez-Alonso, S., Sicilia, M.A.: Deploying metadata on blockchain technologies. In: Proceedings of the 11th International Conference on Metadata & Semantic Research (MTSR), pp. 38–49 (2017)
9. Kumar, D.S., Rahman, M.A.: Simplified HDFS architecture with blockchain distribution of metadata. Int. J. Appl. Eng. Res. **12**(21), 11374–11382 (2017)
10. Liu, L., Li, X., Au, M.H., Fan, Z., Meng, X.: Metadata privacy preservation for blockchain-based healthcare systems. In: Proceedings of the 27th International ConferenceDatabase Systems for Advanced Applications (DASFAA), vol. 1, pp. 404–412 (2022)
11. Paik, H.Y., Xu, X., Bandara, H.D., Lee, S.U., Lo, S.K.: Analysis of data management in blockchain-based systems: From architecture to governance. IEEE Access **7**, 186091–186107 (2019)
12. Sawadogo, P., Darmont, J.: On data lake architectures and metadata management. J. Intell. Inf. Syst. **56**(1), 97–120 (2021)
13. Song, J., Zhang, P., Qu, Q., Bai, Y., Gu, Y., Yu, G.: Why blockchain needs graph: a survey on studies, scenarios, and solutions. J. Parallel Distrib. Comput. **180**, 104730 (2023)
14. Apache Foundation: Atlas. https://atlas.apache.org. Accessed 14 Jan 2024
15. DataHub Project: The metadata platform for the modern data stack. https://datahubproject.io/. Accessed 14 Jan 2024
16. Fluree PBC: FlureeDB. https://developers.flur.ee/. Accessed 14 Jan 2024
17. Tsoulias, K., Palaiokrassas, G., Fragkos, G., Litke, A., Varvarigou, T.A.: A graph model based blockchain implementation for increasing performance and security in decentralized ledger systems. IEEE Access **8**, 130952–130965 (2020)
18. Wang, Q., Yu, J., Chen, S., Xiang, Y.: SoK: Diving into DAG-based blockchain systems. arXiv preprint arXiv:2012.06128 (2020)
19. Ziegler, J., Reimann, P., Keller, F., Mitschang, B.: A graph-based approach to manage CAE data in a data lake. Procedia CIRP **93**, 496–501 (2020)

How to Trigger Personalized
and Context-Aware Nudges

Randi Karlsen⬤ and Anders Andersen$^{(\boxtimes)}$⬤

Department of Computer Science, UiT The Arctic University of Norway,
Tromsø, Norway
{randi.karlsen,anders.andersen}@uit.no

Abstract. A nudge represents a gentle push, intended to help people
change behavior in a desirable direction. Personalization and context-
awareness make it possible to tailor a nudge to the user's specific needs
and the current situation, and thus make it relevant and meaningful to
the user. To be effective, a·nudge must be given at a time when the
user needs to be pushed and is able to follow the nudge. This paper
presents how triggers, based on the event-condition-action (ECA) rule,
enable timely nudging by starting the nudge design process at the right
time so that a nudge is available when the user is susceptible to nudging.
The triggers are tailored to the user's situation and need for nudging, by
personalizing both event detection and the evaluation of the condition
for starting the nudge design action.

Keywords: smart nudging · digital nudging · nudge triggering ·
behavioral change · personalization and context-awareness ·
personalized triggering · adaptive nudging

1 Introduction

Nudging has become a popular technique to gently influence people to behave
according to a desirable goal, e.g., to adopt healthier or more environmentally
friendly habits. The general goal of nudging is, according to [29], to ". . . alter
people's behavior in a predictable way without forbidding any options or significantly changing their economic incentives".

Used in digital environments, nudging has the potential to reach a lot of
people (through, e.g., an application on a mobile device) and to be adapted
to the individual being nudged. Personalization and context awareness enable
nudges to be tailored to the individual user and continuously adapted to the
user's current situation and need for behavioral change.

Digital nudging has been described as a "subtle form of using design, information and interaction elements to guide user behavior in digital environments,
without restricting the individual's freedom of choice" [20]. Our focus on personalized and context-aware nudging has led to the definition of *smart nudging*,

R. Chbeir et al. (Eds.): IDEAS 2024, LNCS 15511, pp. 18–31, 2025.
https://doi.org/10.1007/978-3-031-83472-1_2

as "digital nudging, where guidance of user behavior is tailored, to be relevant to the current situation of each individual user" [16].

A smart nudging system monitors the user's activities, interests, and surroundings, and creates tailored nudges to help the user change behavior towards the nudging goal. A nudge provides the user with a suggested activity and a gentle influence to make the user follow the suggestion.

In the same way as the activity and influence of a nudge is tailored to the user, tailoring also applies to the reason for giving a nudge and the time to present it, as the need for nudging and the ability to follow a nudge vary widely from person to person. One challenge for such a smart nudging system is thus to determine when a nudge is needed.

This paper presents an approach to *personalized triggering* of smart nudges, where triggers are identified based on the user's current behavior, activity pattern, and ability to perform a suggested activity. The user's context (e.g., location and environment) and situation (e.g., available time) are also important when identifying triggers for nudging. The suggested approach is based on the well-known event-condition-action (ECA) rule [10], which we adapt to be used for personalized triggering of smart nudges.

The contribution of our work is on the adaptive triggering of nudges, where both detection of events and evaluation of conditions for executing an action are tailored to the individual user. The approach also describes a sequence of triggers, where a first trigger may cause the design of a nudge, while a second trigger executes when the nudge is scheduled to be presented to the user.

In this paper we focus on how to trigger nudging while leaving out important aspects such as nudge design, how to motivate users through nudging, privacy, and ethical issues in nudging. For this, we refer to discussions in [2,9,16,17]. In the following, we first present related work, smart nudging, and two use cases that exemplify the proposed solution. Triggering of nudges is described by detailing how the ECA rule applies to nudging. Finally, we discuss our approach and conclude.

2   Related Work

Our research focuses on the challenges of automatically designing digital nudges that are tailored to the current need of the user. The importance of tailoring in designing effective nudges has been recognized [21,24,27], and experiments [21,24] show that tailored nudges can lead to more effective nudging compared to one-size-fit-all nudges. As much work has been done on one-size-fit-all nudges [1,24], there is currently a need to target open questions concerning tailored nudging [4,6,15], and tailored triggering of nudges.

Triggering can be seen as the initial part of nudge design that starts the creation of a nudge. However, triggers are not commonly included in theoretical models for digital nudge design, such as [20,22,25,26], as these describe nudge design processes intended for designers or practitioners.

In [12] triggers are recognized as vital for persuasive products and described as a tool to help people start a behavior. Triggers are classifies into three types, where *spark* motivates behavior, a *facilitator* makes behavior easier, and a *signal* that serves as a reminder. Additionally, *timing* is about finding the opportune moment to persuade [12]. Both timing and the three trigger types are important in nudging systems.

Rule-based approaches, such as ECA, have been found useful in situation- and context-aware systems to allow push-based messages or recommendations, see e.g., [3,7,11,13,14,23]. Our work is based on findings from previous research, but differs from other approaches in that we apply a highly adaptive triggering approach, where rules are tailored to the user, and where continuous monitoring of user behavior, situation, environment, and nudging goal determine event detection and evaluation of conditions for executing an action.

3   Smart Nudging

A smart nudge is designed to support behavioral change tailored to a specific user. This means that personalization, context, and situation of the user determine when to nudge and what to nudge for. This section describes the content of smart nudges and two use cases that will illustrate the use of triggers throughout this paper.

3.1   Content of Smart Nudges

A smart nudge includes several components, i.e., the *activity* selected for the nudge, some *influence* motivating the user to do the activity, a *time frame* for when the activity is suggested done, and some *practical information* helping the user to follow the nudge [17]. All components can be tailored to the user based on monitoring the user activity, capability, preferences, situation, and environment.

A smart nudging system typically uses a mobile phone as edge device providing a nudge application to the user. A combination of back-end and edge computing determines when to nudge and what to nudge for through monitoring, selecting, and analyzing data from various sources, such as IoT sensors, third-party data sources, mobile phone sensors, and stored user data. This includes support for both static and dynamic data, e.g., activity history, user capabilities, and context (such as location, plans, and environmental conditions) [2].

Back-end processing is a continuous process allowing for a wide range of analysis, including simple data integration and analysis or more complex data mining or machine learning based analysis. *Edge processing* is done in two cases, first when the edge (a mobile device or IoT sensor) collects and analyzes data, and second when the mobile device creates and presents nudges to the user. When creating a nudge, the device has a wide range of data available, where preprocessed data from the back-end system is combined with local, possibly sensitive, and private data on the user's smartphone.

Available data from both edge and back-end is used to identify situations that trigger the creation of a nudge. These are typically situations that represent either a need for nudging (e.g., for the user to stabilize or improve behavior) or an opportunity to nudge (e.g., when a nudge is likely to be accepted).

3.2   Use Cases

The *physical activity use case* is about nudging people to be sufficiently physically active, with the motivation that regular physical activity is a known protective factor for the prevention and management of diseases [30].

A nudge presents the user with a suggestion for an activity and a time frame for doing the activity (e.g., *"go for a walk in the afternoon"* or *"take a mountain hike tomorrow"*). For the user to improve, the nudging system must challenge the user by suggesting behavior that goes beyond the user's current behavior, by, e.g., nudging the person to (i) be active more often, (ii) be active for a longer period, or (iii) engage in more challenging activities.

The *green transportation use case* is about nudging people to choose environmentally friendly transportation means (e.g., public transportation or walking). This is a reaction to urban challenges of increased traffic, congestion, air pollution, and global-scale issues of climate change and global warming [8].

Available transportation means are ranked according to environment friendliness, and a nudge suggests an alternative that is better than what the user normally prefers.

3.3   Ethics and Transparent Nudges

Ethics of nudging has been discussed and questioned by pointing to the potential danger of manipulating people and presenting unfair nudges [28,29]. Transparency, where users are made aware that nudges try to influence them, is considered an important safeguard against manipulation and nudges that undermine the users' best interests [5,18,29]. Ethical guidelines for constructing digital nudges are described in [19], and include three conditions for nudging, *transparency, ease of resistibility*, and *non-controllability*.

Transparency for smart nudges can, e.g., be achieved by stating the intention to nudge and the nudging goal in the description of a nudging application, and by warning the user each time the application starts. Despite being transparent, nudges may well be effective, as shown, e.g., [5,18].

4   Triggering a Nudge

This section describes a general approach to how nudges can be triggered using ECA rules and, for each component of ECA, demonstrates how the proposed solution can be used in the two use cases; physically activity and green transportation nudging.

4.1   Rule-Based Triggering of Nudges

To trigger nudging, we follow the ECA rule, which specifies that when an event E is detected, condition C is evaluated, and if C evaluates to true, action A is executed [10]. Evaluation of C determines if it is possible or useful to give a nudge to the user.

Contrary to the original ECA rule, we recognize only one type of action, i.e., the *nudge design* action. The action is used by every trigger in the nudging system, and, if executed, starts the design process for a nudge that is tailored to the user and the user's current situation.

An ECA rule-based approach is found to be suitable for context-aware applications [11, 23]. The approach allows for customization, flexibility, and modular definition of rules, where rules can be added, modified, and removed, and be tailored to individual users or scenarios. This supports both tailoring of nudges and tailoring of triggers to individual users of smart nudging systems.

4.2   Trigger Events

A trigger event can be described as a *significant occurrence* that initiates a nudge of the user. An occurrence represents an instance or situation where nudging is beneficial or necessary to help the user improve behavior towards the nudging goal. Trigger events are closely related to the nudging goal and must be explicitly determined for each use case.

Events are detected by monitoring the user's activities, plans, and context, and depending on the trigger type, monitoring is based on current data (e.g., the user's activity level and current conditions in the user's environment), predictions (e.g., weather or pollen level predictions), historical data (e.g., user activity patterns), and plans (e.g., traveling plans or plans for being physically active).

Generally, an event E can be a primitive or composite event, where a composite is constructed based on primitive events. In this paper, we describe primitive events and leave composite events for future work. In the following, we exemplify trigger events for the two use cases by identifying a few main trigger events, that can be further detailed as needed.

Physical Activity Nudging. To nudge users to be physically active, we identify some main trigger events, presented in Table 1. In general, an event is detected when the user is inactive or when something happens that makes it tempting or opportune for the user to be active.

A *Low Activity event* is detected by monitoring user activity and identifying when the user is not sufficiently active. The nudging system automatically sets and adjusts a personalized activity goal (denoted *ActivityGoal*) for a time period P (e.g., a week). If the user reaches the goal in P, the value of *ActivityGoal* is slightly increased for the next period, to challenge the user to improve behavior. We measure physical activity in number of steps, using a pedometer for counting steps (when walking or running), while other activities (such as swimming or climbing) are converted to steps using a step conversion factor based on time spent on the activity.

Formula 1 describes that Low Activity is true if already completed activities in P ($CompActivity(P)$) together with expected activities ($ExpectActivity(P)$) are not sufficient to reach the $ActivityGoal(P)$. Expected activities include recurring activities detected in the user's activity pattern and activities the user has committed to do during P. Testing for low activity must be done regularly so

Table 1. Trigger events for a Physical Activity nudging system.

Event	Description
Low Activity	Monitoring the user detects the need to be more physically active to maintain or improve level of activity
Required Activity	A need for doing a specific type of activity is detected, e.g., if an activity plan or habit is not followed
Situation	Events in the user's environment can represent an opportunity to be active, e.g., good conditions for skiing, or particularly nice weather.
Happening	An activity the user can join, e.g., an upcoming competition, a guided hiking tour, or an invitation from a friend to join an activity
User request	The user can request a nudge, optionally specifying activity and/or timeframe

that nudging can be triggered sufficiently often to help the user reach the nudging goal.

$$LowActivity = (ActivityGoal(P) - CompActivity(P) - ExpectActvity(P)) > 0 \quad (1)$$

A *Required Activity event* detects that the user should be nudged to perform a specific type of activity based on plans for how often certain activities should be done, or by detecting that a regular or committed activity is not performed.

A *Situation event* is about detecting environmental conditions and user situations that give opportunities for being active (e.g., low pollen activity, nice weather, good skiing conditions, or an opening in the calendar for being active).

A *Happening event* represents a specific activity and can be detected based on, e.g., information about guided tours, upcoming competitions, or an invitation from a friend to join an activity. If a nudge for the activity is accepted, it is classified as committed and a new nudge may later remind the user of the activity. The Happening event may thus result in more than one nudge.

User Request event. A nudge can be presented with alternatives or with the possibility for the user to reject the suggested activity and subsequently request a nudge for some other activity. A user request for a nudge represents a trigger event.

Green Transportation Nudging. To nudge users to choose green transportation means, we identify two main trigger events, presented in Table 2.

A *Location Change event* can automatically be detected by monitoring i) the user's activity pattern to identify recurring location changes, and ii) the user's calendar to identify appointments that require location change.

A travel is described through the following properties:

Travel={Origin, Destination, departure time, transportation means}

Table 2. Trigger events for a Green Transportation nudging system.

Event	Description
Location Change	The need to move between locations, detected through calendar information or monitoring of recurring travels
User request	To benefit from transportation suggestions and travel information provided by the nudge, the user actively requests a nudge

By analyzing the set of previous travels, the system detects recurring origin-destination pairs (e.g., between home and work), recurring time for traveling, and frequency of previously used transportation means.

As opposed to physical activity nudges, transportation nudges are time-critical. A nudge must be given in a timely manner so that the user can catch the bus or reach the destination in time for an appointment. Location changes must be detected sufficiently early so that all available transportation means are realistic options for a nudge and that the user has sufficient time to follow the nudge.

4.3 Conditions

Condition C in the ECA rule determines whether it is possible or useful to give a nudge as a result of event E, by evaluating the user's ability and opportunity to go through with an activity. An event can identify certain conditions that are used as constraints when evaluating C.

Physical Activity Nudging. Conditions for physical activity trigger events are described in Table 3. An event E can be completely open concerning what to nudge for and when to nudge, or it can be linked to a specific activity ($Activity(E)$) and/or time frame ($TimeFrame(E)$).

We use function $Capable(U, Activity(E))$ to evaluate the ability of user U to do activity $Activity(E)$ and function $Available(U, TimeFrame(E))$ to evaluate the opportunity of U to perform an activity within $TimeFrame(E)$. If C does not evaluate to true, a nudge is not given.

A *Low Activity* event means that the user is inactive and a nudge for some activity should be given as soon as possible. Condition C evaluates to false only if the user is not capable of doing any activity. However, in a case when user capability excludes every activity (e.g., because of illness or injury), nudging should be suspended until the user again is capable of being active.

The events *Required Activity*, *Situation*, and *Happening* events point to a specific activity and/or a time frame for doing an activity, and use either one or both functions to evaluate condition C. A *User Request* event will always result in a nudge, and condition C is consequently always true.

Table 3. Trigger events and conditions for Physical Activity nudging

Event	Constraints	Condition (C)
Low Activity	–	$C = true\ if$ Capable(U, <some activity>))
Required Activity	Activity(E)	$C = true\ if$ Capable(U, Activity(E))
Situation	TimeFrame(E) *Activity(E)	$C = true\ if$ (Available(U, TimeFrame(E)) AND Capable(U, Activity(E)))
Happening	TimeFrame(E) Activity(E)	$C = true\ if$ (Available(U, TimeFrame(E)) AND Capable(U,Activity(E)))
User Request	–	$C = true$ (A user request is always answered by the system)

* Included only if the event is linked to an activity

Green Transportation Nudging. For green transportation nudging, a trigger event E identifies a location pair $Loc(E) = (O, D)$ that represents origin and destination of an upcoming change of location. For a *Location Change* event, $Loc(E)$ represents a constraint when evaluating condition C. A *User Request* will also include an (O, D) pair, but this is not used as a constraint since a request always results in a nudge, and condition C is therefore always true.

Conditions for green transportation trigger events are described in Table 4. There, $LevelOfGreen(Loc(E))$ represents a measure of how environmentally friendly the user is when traveling between the locations of $Loc(E)$, while $EFgoal$ represents a threshold value for sufficient environmental friendliness and determines the usefulness of a nudge. For example, there is no need to issue a nudge if the user is already in the habit of walking to work.

$TravelModes(Loc(E))$ represents the set of available transportation means for $Loc(E)$. For a nudge to be useful, there must exist alternatives regarding transportation means.

As users may have different habits when traveling between different locations (e.g., prefers to walk to work while using the car when traveling with kids or going shopping), $LevelOfGreen$ is calculated with respect to each (O, D) pair.

We assume a set of travels for (O, D), $Travels(O, D) = \{x_1, \ldots, x_n\}$, where each $x \in Travels(O, D)$ has a value, denoted $EFvalue(x)$, representing the environmental friendliness of the transportation means used for travel x. Formula

Table 4. Trigger events and conditions for Green Transportation nudging

Event	Constraints	Condition (C)
Location Change	Loc(E)	$C = true\ if$ LevelOfGreen(Loc(E))<EFgoal AND |TravelModes(Loc(E))| >1
User Request	–	$C = true$ (A user request is always answered by the system.)

2 describes the $LevelOfGreen(O, D)$ for the set $Travels(O, D)$, as the average $EFvalue$ over every transportation means used in $Travels(O, D)$.

$$LevelOfGreen(O, D) = \frac{\sum_{x \in Travels(O,D)} EFvalue(x)}{|Travels(O, D)|} \tag{2}$$

$LevelOfGreen$, $EFvalue$, and $EFgoal$ are measured on the scale $(0, 1)$, where the higher the value the more environmentally friendly the user is. If $Loc(E)$ represents a new location change, $LevelOfGreen$ is set to the initial value of 0.

To account for behavioral change, $LevelOfGreen(O, D)$ is calculated based on $Travels(O, D)$ from a time period P, e.g., the last month or week. To challenge the user to improve behavior, the nudging system can automatically set and adjust the personalized $EFgoal$ by, over time, slightly increasing the goal.

4.4   Action

The nudge design action creates a nudge that responds to the situation detected by the trigger event. Some events specify which *activity* and/or *time frame* to include in the nudge, while other events leave this to be decided during the design process.

Selecting an *activity* involves many concerns, including that the user must be able to do the activity, the activity must represent an improvement with respect to the nudging goal, and must be possible to do in the current situation. Choosing a *time frame* is about determining when the user can perform the selected activity.

The design process is also affected by *alerts*, which are significant occurrences that represent harmful or hazardous situations (such as harmful pollution or severe weather). To avoid harmful situations, alerts will influence which activity to select for nudging.

Nudge design will, in addition to determining activity and time frame, also include an influence component and information about the activity. However, this paper does not discuss the content of these components. For a description of all components see [9, 17].

Physical Activity Nudging. In general, for events where no time frame is given, there will normally be an implicit requirement that a nudge should be given as soon as possible, to avoid a decline in activity level. For events where no activity is given, nudge design should select an activity the user is capable of doing, possibly giving some priority to preferred activities.

The *Low Activity event* is not linked to a specific activity or time frame, and nudge design should create a nudge for some activity that the user can perform immediately or in the near future.

For *Required Activity, Situation*, and *Happening* events, the nudge must satisfy the constraints linked to the event by including the specific activity and/or time frame. A *User Request* may include both activity and time frame, or be an open request, letting activity and/or time frame be determined during nudge design.

When designing a nudge where the time frame is not determined by the event, time since the last activity is important for determining when a nudge can be given. The system should not nudge if the user is currently active or recently has finished an activity. This implies that there must be a sufficiently large interval between the end of the last activity and the time frame for the next activity.

Green Transportation Nudging. The *Location Change event* identifies the (O, D) locations, and the latest arrival time for a required travel. O and D determine possible transportation means and routes, while the latest arrival time determines when the travel must take place and when to present the nudge to the user. The time frame of the nudge will depend on the selected activity (i.e., transportation means). For example, the user needs more time to walk compared to taking the bus. The nudge must also be presented to the user in time to prepare before starting the travel.

The time-critical aspect of transportation nudges implies that the time interval between nudges may well be short. The nudge frequency is determined by the frequency of Location Change events. Also, while physical activity nudging seeks to avoid interrupting the user with nudges, transportation nudges may need to interrupt the user to present a time-critical nudge.

A *User Request* for traveling between O and D may either be an open request or may specify a specific transportation means, a time frame for traveling, and/or the latest arrival time.

4.5   To Trigger a Nudging Plan

So far we have described triggers that initiate the design of a single nudge. It may additionally be beneficial to set up a nudging plan to help the user reach an *ActivityGoal* within a current period P.

A *nudging plan*, described in Formula 3, consists of a set of potential nudges, where each nudge, $n \in NudgingPlan(P)$, consists of an activity, A, a time frame, T, and a timed trigger, $designT$, for starting the design of the planned nudge.

$$NudgingPlan(P) = \{n_1, \ldots, n_m\}, \text{where } n_i = (A, T, designT) \tag{3}$$

The plan is set up based on a prediction of the user's behavior for period P, and includes actions that, if followed, will make the user reach the targeted behavioral goal. The A and T are tentative and may be changed if the user situation at nudge design time makes it necessary or beneficial.

A nudging plan may be useful in, e.g., physical activity nudging, where activities necessary to reach the nudging goal, are identified and scheduled ahead of time. Formula 4 describes a prediction (denoted $RemainSteps(P)$) for how much additional activity is needed for the user to reach the $ActivityGoal(P)$. As in

Formula 1, $CompActivity(P)$ and $ExpectActivity(P)$ represent accomplished and expected activities, respectively.

$$RemainSteps(P) = ActivityGoal(P) - CompActivity(P) - ExpectActvity(P) \tag{4}$$

A positive value for $RemainSteps(P)$ can be used as a trigger for setting up a nudging plan. Since the user may be more or less active than predicted in $ExpectActvity(P)$, $RemainSteps(P)$ must be reevaluated during P, and the nudging plan may have to be adjusted.

5 Discussion

This section discusses some observations made during our work, including differences between use cases, personalization of triggers, and the distinction between triggering nudge design and triggering nudging.

5.1 Differences Between Use Cases

A fundamental difference between our two use cases is that for physical activity nudging, the $ActivityGoal$ can be reached by adding new activities, while for green transportation nudging, the user can reach a behavioral goal only by improving the environmental friendliness of the user's travels.

In the physical activity case, lack of activity triggers nudging, while for green transportation, the user's normal activities must be intercepted, to nudge for a better transportation means than what is normally chosen. In this case, the activity itself (i.e., the travel) triggers a nudge. As demonstrated throughout this paper, the suggested generic trigger approach can serve as a basis for both use cases, and for different types of triggers.

5.2 Personalized Triggers

Triggers are personalized, both for event detection and for evaluating conditions for executing the nudge design action. In physical activity nudging, event detection for Low Activity is based on the user's current level of behavior, while Required Activity is based on the user's activity plans and habits. Conditions are evaluated based on the user's ability to do an activity and the opportunity for being active in the given time frame.

For green transportation nudging, event detection is based on the user's need to change location determined through the user's travel plans and habits, while the user's level of behavior (i.e., $LevelOfGreen$) is a component in the evaluation of the condition.

The set of trigger events can also be personalized. For example, a Situation event based on pollen detection is only relevant for people with pollen allergies, while a weather condition event may be useful only for people who find it motivating to go for a walk in nice weather. User capabilities and limitations, found in the user profile, can contribute to determine which trigger events are relevant to a specific user. Also, a user's activity pattern and reactions to previous nudges can detect if there are relations between activity and, for example, environmental conditions.

5.3   To Trigger Nudge Design vs Trigger Nudging

Nudging involves two important aspects of timing. First, the nudge has a time frame suggesting when the activity should be done, and, second, the timing with respect to when the nudge is given to the user. For example, a nudge at lunchtime can suggest going for a walk in the afternoon.

Starting the nudge design action does not imply that there immediately will be issued a nudge to the user. The nudge design can be done in advance, the prepared nudge can be stored, and a timer determines when to present the nudge to the user.

In the green transportation use case, the selected means of transportation determines when the nudge should be given. A nudge for walking will typically be given earlier in time compared to a nudge for taking the bus. Nudge design must start sufficiently early so that any of the available transportation means can be chosen, while presenting the nudge is postponed until it is (believed to be) best received by the user. For the physical activity use case, a nudge should be postponed if the user is i) currently exercising or otherwise busy, ii) has recently received a number of nudges, or iii) the time frame for the activity is way ahead.

The frequency of nudging must be balanced so that the system nudges sufficiently often to help the user improve behavior, but without flooding the user with nudges. Presenting the user with many nudges over a short period of time may be irritating or disturbing, and possibly counterproductive to reaching the nudging goal.

6   Conclusion

This paper presents an approach to personalized triggering of smart nudges, where the user's behavior, situation, and context determine both trigger events and conditions for starting a nudge design action. Our approach follows the well-known event-condition-action ECA rule, which after detecting an event, evaluates the condition, and if the condition evaluates to true, the action is executed. To trigger smart nudging, personalization applies to all three components of the ECA rule.

We describe a general approach to nudge triggering, where we, for each component of the ECA rule, first describe the approach, followed by an exemplification of how the proposed solution can be used in the two use cases; physical activity nudging and green transportation nudging. Evaluating the proposed general approach on other use cases is left for future work. Future work also includes the language needed for specifying ECA rules in a nudging system.

The specific contribution of this paper is the demonstration of how triggers for nudging can be personalized and how a general approach to triggering can be useful for use cases with considerably different characteristics. This paper describes triggers that initiate the design of a single nudge and triggers for setting up a nudging plan.

References

1. Anagnostopoulou, E., et al.: From mobility patterns to behavioural change: leveraging travel behaviour and personality profiles to nudge for sustainable transportation. J. Intell. Inf. Syst. **54**(1), 157–178 (2020)
2. Andersen, A., Karlsen, R., Yu, W.: Green transportation choices with IoT and smart nudging. In: Maheswaran, M., Badidi, E. (eds.) Handbook of Smart Cities: Software Services and Cyber Infrastructure, chap. 13, no. 1, pp. 331–354. Springer (2018)
3. Beer, T., Rasinger, J., Höpken, W., Fuchs, M., Werthner, H.: Exploiting E-C-A rules for defining and processing context-aware push messages. In: Proceedings of the International Symposium on Rules & Rule Markup Languages for the Semantic Web (RuleML), pp. 199–206 (2007)
4. Bergram, K., Djokovic, M., Bezençon, V., Holzer, A.: The digital landscape of nudging: a systematic literature review of empirical research on digital nudges. In: Proceedings of Conference on Human Factors in Computing Systems (CHI) (2022)
5. Bruns, H., Kantorowicz-Reznichenko, E., Klement, K., Jonsson, M.L., Rahali, B.: Can nudges be transparent and yet effective? J. Econ. Psychol. **65**, 41–59 (2018)
6. Caraban, A., Karapanos, E., Gonçalves, D., Campos, P.: 23 ways to nudge: a review of technology-mediated nudging in human-computer interaction. In: Proceedings of the Conference on Human Factors in Computing Systems (CHI), pp. 1–15 (2019)
7. Cimino, M.G., Lazzerini, B., Marcelloni, F., Ciaramella, A.: An adaptive rule-based approach for managing situation-awareness. Expert Syst. Appl. **39**(12), 10796–10811 (2012)
8. Comission of the European Communities: Green paper: towards a new culture for urban mobility. MEMO/07/379 (2007)
9. Dalecke, S., Karlsen, R.: Designing dynamic and personalized nudges. In: Proceedings of the 10th International Conference on Web Intelligence, Mining & Semantics (WIMS), pp. 139–148 (2020)
10. Dayal, U., Buchmann, A.P., McCarthy, D.R.: Rules are objects too: a knowledge model for an active, object-oriented database system. In: Proceedings of the 2nd International Workshop on Object-Oriented Database Systems (OODBS), pp. 129–143 (1988)
11. Dockhorn Costa, P., Ferreira Pires, L., Van Sinderen, M.: Architectural patterns for context-aware services platforms. In: Proceedings of the 2nd International Workshop on Ubiquitous Computing (IWUC), pp. 3–18 (2005)
12. Fogg, B.J.: A behavior model for persuasive design. In: Proceedings of the 4th International Conference on Persuasive Technology (PERSUASIVE), p. 7 (2009)
13. Hermoso, R., Dunkel, J., Krause, J.: Situation awareness for push-based recommendations in mobile devices. In: Proceedings of the 19th International Conference on Business Information Systems (BIS), pp. 117–129 (2016)
14. Ilarri, S., Trillo-Lado, R.: An approach for proactive mobile recommendations based on user-defined rule. Expert Syst. Appl. **242**, 122714 (2024)

15. Jesse, M., Jannach, D.: Digital nudging with recommender systems: survey and future directions. Comput. Hum. Behav. Rep. **3**, 100052 (2021)
16. Karlsen, R., Andersen, A.: Recommendations with a nudge. Technologies **7**(2), 45 (2019)
17. Karlsen, R., Andersen, A.: The impossible, the unlikely, and the probable nudges: a classification for the design of your next nudge. Technologies **10**(6), 110 (2022)
18. Loewenstein, G., Bryce, C., Hagmann, D., Rajpal, S.: Warning: you are about to be nudged. Behav. Sci. Policy **1**(1), 35–42 (2015)
19. Meske, C., Amojo, I.: Ethical guidelines for the construction of digital nudges. In: Proceedings of the Annual Hawaii International Conference on System Sciences (HICSS), pp. 3928–3937 (2020)
20. Meske, C., Potthoff, T.: The DINU-model: a process model for the design of nudges. In: Proceedings of the 25th European Conference on Information Systems (ECIS), pp. 2587–2597 (2017)
21. Mills, S.: Personalized nudging. Behav. Public Policy **6**(1), 150–159 (2022)
22. Mirsch, T., Lehrer, C., Jung, R.: Making digital nudging applicable: the digital nudge design method. In: Proceedings of the International Conference on Information Systems (ICIS), pp. 1–16 (2018)
23. Moore, P., Hu, B., Jackson, M.: Rule strategies for intelligent context-aware systems: the application of conditional relationships in decision-support. In: Proceedings of the International Conference on Complex, Intelligent & Software Intensive Systems (CISIS), pp. 9–16 (2011)
24. Peer, E., Egelman, S., Harbach, M., Malkin, N., Mathur, A., Frik, A.: Nudge me right: personalizing online security nudges to people's decision-making styles. Comput. Hum. Behav. **109** (2020)
25. Purohit, A.K., Holzer, A.: Functional digital nudges: identifying optimal timing for effective behavior change. In: Extended Abstracts of the Conference on Human Factors in Computing Systems (CHI), pp. 1,6 (2019)
26. Schneider, C., Weinmann, M., vom Brocke, J.: Digital nudging: guiding online user choices through interface design. Commun. ACM **61**(7), 67–73 (2018)
27. Schöning, C., Matt, C., Hess, T.: Personalised nudging for more data disclosure? On the adaption of data usage policies format to cognitive styles. In: Proceedings of the Annual Hawaii International Conference on System Sciences (HICSS), pp. 4395–4404 (2019)
28. Sunstein, C.R.: The ethics of nudging. Yale J. Regul. **32**, 413–450 (2015)
29. Thaler, R.H., Sunstein, C.R.: Nudge: Improving Decisions About Health, Wealth, and Happiness. Yale University Press (2008)
30. WHO: WHO Guidelines on physical activity and sedentary behaviour (2020)

Classification

Enhanced Classification of Embedded System Vulnerabilities Using Ensemble Embedding and BiLSTM Networks

Aissa Ben Yahya[✉][iD], Hicham El Akhal[iD],
and Abdelbaki El Belrhiti El Alaoui[iD]

Moulay Ismail University of Meknes, Meknes, Morocco
{ai,hi}@edu.umi.ac.ma, a.elbelrhitielalaoui@umi.ac.ma

Abstract. Critical infrastructure increasingly relies on embedded systems, making them particularly vulnerable to cyber attacks due to their complexity and interconnectivity. Unlike general-purpose systems, embedded systems need specialized security solutions tailored to their unique vulnerabilities. Accurate classification of embedded system vulnerabilities is essential for targeted analysis and mitigation. Traditional methods using pre-trained embeddings like Word2Vec, GloVe, and FastText often struggle with Out-of-Vocabulary (OOV) words, reducing their effectiveness. We address this with a novel ensemble embedding technique that combines multiple pre-trained embeddings, enhancing the classification of embedded system vulnerabilities. Our BiLSTM-based model, tested on datasets such as NVD and CNNVD, achieved 82.61% accuracy on unseen data, outperforming traditional embeddings.

Keywords: Embedded Systems · Ensemble Embedding · Vulnerability Classification · Out-of-Vocabulary Words · Cybersecurity

1 Introduction

The pervasive deployment of embedded systems in critical infrastructure sectors, including healthcare, transportation, and industrial automation, underscores the imperative of ensuring their security. The sensitive nature of the data handled by these systems, coupled with their specialized requirements, renders them uniquely susceptible to security breaches. The consequences of such breaches can be far-reaching, encompassing service outages [26], unauthorized access to sensitive information [10,29], physical harm to equipment [21], and even loss of human life [20]. One of the significant challenges in enhancing the security of embedded systems is the lack of dedicated classification techniques that distinguish embedded system vulnerabilities (ESV) from general-purpose system vulnerabilities (GPSV). Without this distinction, security experts struggle to analyze the unique characteristics of ESV and develop specialized solutions for embedded systems. To the best of our knowledge, this research is the first to

R. Chbeir et al. (Eds.): IDEAS 2024, LNCS 15511, pp. 35–48, 2025.
https://doi.org/10.1007/978-3-031-83472-1_3

propose an automatic classification framework to classify ESV from their GPSV counterpart.

Recent advancements in machine and deep learning have facilitated the development of automatic classification mechanisms, which have been successfully applied to the realm of vulnerability classification [13, 19, 28]. The current state-of-the-art in embedded systems vulnerability classification is the work in [2], which employs Term Frequency Inverse Document Frequency (TF-IDF) to generate weighted word vectors and Support Vector Machine (SVM) as a classifier to classify vulnerabilities. However, this approach has limitations, including its reliance on TF-IDF, which may not capture complex semantic relationships between words [34], and its use of SVM, which can be sensitive to feature scaling and may not generalize well to new data.

Traditional machine learning approaches, although successful in general software vulnerability classification, often encounter limitations when applied to more specialized contexts. These limitations stem from their dependence on pre-trained word embeddings, which suffer from the Out-of-Vocabulary (OOV) problem [27], where words not seen during training are not well-represented in the model, leading to poor classification performance.

To address these challenges, we propose a novel ensemble embedding technique that integrates GloVe, Word2Vec, and FastText, generating more comprehensive word representations that reduce the impact of OOV words. This technique is coupled with a Bi-directional Long Short-Term Memory (BiLSTM) model, which excels in capturing long-range dependencies in textual data.. Furthermore, a significant contribution of our work is the creation of a comprehensive benchmark dataset, collected from multiple databases, which provides a diverse range of data for model training and evaluation. The key contributions of our research include:

– Proposing a novel framework for distinguishing and classifying vulnerability reports, enabling the separation of embedded system vulnerabilities from general-purpose system vulnerabilities for dedicated analysis.
– Introducing an innovative ensemble embedding technique that integrates GloVe, Word2Vec, and FastText to generate weighted word vectors, significantly reducing the impact of out-of-vocabulary (OOV) words.
– Compiling a comprehensive benchmark dataset from multiple sources, offering a diverse and representative collection of data for robust model training and evaluation.

This paper is structured as follows: we provide an overview of relevant literature in Sect. 2, followed by a background in Sect. 3. The detailed description of our methodology is provided in Sect. 4. The evaluation results of our proposed mechanism are presented and discussed in Sect. 6. Finally, we conclude the paper in Sect. 7, where we summarize the key findings and insights derived from our research.

2   Related Work

Recently, machine learning (ML) and deep learning (DL) techniques have gained prominence in automating vulnerability classification. These approaches reduce the reliance on human expertise and can identify newly emerging or unanticipated vulnerabilities. Various studies have employed ML/DL models, such as Naive Bayes [19], TFI-DNN [13], and word embedding with CNN [28], to classify vulnerabilities into their respective types or severity levels.

Pioneering research by [32] introduced a novel model and methodology for classifying and categorizing vulnerabilities based on their security types using Natural Language Processing techniques. Similarly, [35] developed the Automatic Security Vulnerability Categorization (ASVC) framework, which utilized text mining techniques to automate vulnerability classification.

Several studies have explored the application of deep learning techniques for vulnerability classification. For instance, [12] and [17] proposed deep learning-based algorithms that utilize word embeddings and convolutional neural networks (CNN) to predict the severity of software vulnerabilities. [13] introduced the TF-IDFNN model, which combines Term Frequency-Inverse Document Frequency (TF-IDF) and information gain (IG) with a Deep Neural Network (DNN) to categorize vulnerabilities.

[16] and [5] introduced a novel weighting mechanism known as "Term Frequency-Inverse Gravity Moment" to facilitate the classification of software vulnerabilities according to their severity. [3] created a training dataset focused on extracting and categorizing vulnerabilities related to home appliance devices using Support Vector Machines.

[28] introduced a vulnerability prioritization system that uses word embedding and CNN methodologies to classify vulnerabilities into three severity levels. [33] and [31] presented algorithms that classify vulnerabilities by establishing associations between CVE entries and CWE-IDs, and using self-attention deep neural networks (SA-DNN) and text mining techniques, respectively.

[34] presented an enhanced automatic vulnerability classification algorithm, incorporating weighted word vectors and a fusion neural network to tackle the sparsity challenge in conventional vector representations.

Other studies have proposed frameworks that combine multiple sources of information to improve vulnerability classification. For example, a framework that uses a BiGRU-TextCNN model to classify vulnerabilities into weaknesses based on vulnerability descriptions has been proposed [23]. Another study has developed a deep learning-based vulnerability classifier, DeKeDVer, that combines information from both vulnerability descriptions and source code to improve classification accuracy [8]. Additionally, a machine learning-based approach that uses natural language processing techniques and various multiclass classification algorithms to automate the classification of software vulnerability vectors has been proposed [15].

Existing approaches to vulnerability classification often rely on individual word embeddings such as TF-IDF, Word2Vec, GloVe, or FastText. Our research aims to enhance the accuracy of vulnerability report classification by exploring the effectiveness of ensemble embeddings, combining multiple embedding techniques to leverage their strengths and mitigate their individual weaknesses.

3   Background

3.1   Word2vec

Word2Vec [18] is a seminal word embedding technique that has been widely adopted in natural language processing (NLP) tasks. The approach represents words as vectors in a high-dimensional space. Word2Vec employs two architectures: Continuous Bag of Words (CBOW) and Skip-Gram.

3.2   Glove

GloVe [25] is another prominent word embedding technique that has been widely used in NLP tasks. The approach represents words as vectors in a high-dimensional space, capturing semantic relationships between words such as synonymy, antonymy, and hyponymy. It's effective for general semantic tasks but may not adapt well to dynamic vocabularies or capture specific contextual nuances. GloVe employs a matrix factorization approach to learn word embeddings.

3.3   FastText

FastText [4] enhances traditional word embeddings by representing words as the sum of their character n-grams. This approach allows FastText to generate embeddings even for words that were not present in the training data. The main drawback of FastText is its increased computational complexity compared to simpler models like CBOW, as it involves additional processing to handle character n-grams. FastText employs a hierarchical softmax approach to learn word embeddings.

4   The Proposed Approach

The main method to create an embedded system vulnerability classification model is illustrated in Fig. 1.

4.1   Data Collection and Preprocessing

Data Collection. Our dataset was constructed by collecting vulnerability reports from multiple credible sources, including the National Vulnerability Database (NVD) [22], Chinese National Vulnerability Database (CNNVD) [6], China National Vulnerability Database (CNVD) [7], VarIoT [30], the Cyber Emergency Response Team (ICS-CERT) [14], Fortiguard [9], and the Zero Day

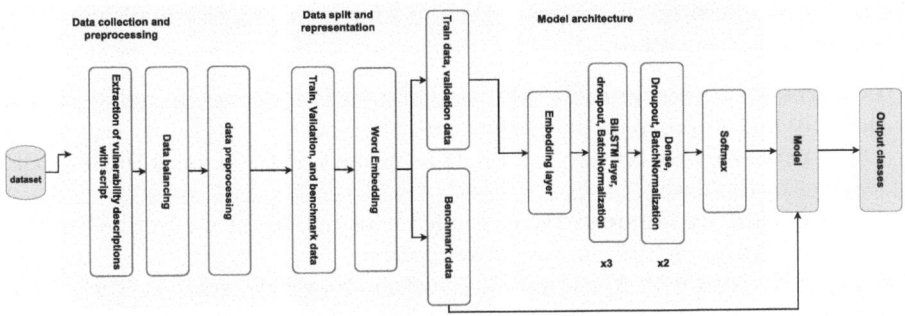

Fig. 1. Main method

Initiative (ZDI) [36]. These sources provided a comprehensive collection of software vulnerability reports, which we leveraged to develop a robust dataset[1] for embedded system vulnerability classification.

To obtain the NVD data, we downloaded the data in JSON format from the NVD feeds [1]. However, we excluded certain entries that were not suitable for model training, specifically reserved, rejected, and disputed CVEs. These entries were excluded because they do not represent confirmed vulnerabilities.

For CNNVD and CNVD, we downloaded the data in XML format from their official websites. However, the descriptions and information were in Chinese, which was not suitable for our model training. We therefore translated all entries from Chinese to English using the Google Translate API [11].

As VarIoT did not provide a downloadable version of their database, we developed a Python scraper to extract the relevant data. For ICS-CERT, Fortiguard, and ZDI, we used their data only for testing purposes, and did not contribute in our training dataset; the data collected from these databases are referred to as "Unseen" throughout this work. To ensure the highest level of data quality and relevance, we opted for manual curation of their data for our specific research purposes. The entire collected dataset is illustrated in Table 1

Table 1. Data set collected from different databases

	ESV	GPSV
NVD	13515	32387
CNNVD	12026	27535
CNVD	7457	14364
VarIoT	6775	4586
ICS-CERT, ZDI, Fortiguard (Unseen)	68	100

---

[1] https://www.kaggle.com/datasets/aissaultimate/ensemble-embedding-data.

Data Preprocessing. To prepare the data for modeling, we applied the following preprocessing steps:

- **Lowercase conversion:** We converted all text to lowercase to reduce dimensionality and improve model performance.
- **Digit removal:** We removed all digits from the text to focus on the linguistic features of the data.
- **Contraction expansion:** We expanded contractions (e.g., "you're" to "you are", "i'm" to "i am") to improve tokenization.
- **Stopword removal:** We removed common stopwords, such as "the", "and", etc., that do not add significant value to the text, to improve model accuracy.
- **Porter Stemming:** We applied Porter Stemming to reduce words to their base form, thereby reducing dimensionality and improving model performance.
- **Special character removal:** We removed special characters from the text to focus on the linguistic features of the data.
- **Base form conversion:** We converted words to their base form (e.g., "ran" to "run") to further reduce dimensionality.

4.2   Data Labeling

We employed a Python script[2] to label the data, utilizing a two-tiered keyword-based approach. The script relies on two distinct keyword lists: one for embedded systems and another for general-purpose systems. These lists were compiled through an exhaustive search of online resources and generative large language models (LLMs) to identify relevant terms associated with embedded devices, protocols, and software. A similar process was applied to create the general-purpose systems keyword list. A sample of these keyword lists is presented in Table 2.

Table 2. White and black list keywords

Embedded system terms	General-purpose terms
Asus RT	Huawei FusionServer
Huawei Quidway switches	Windows Desktop
Cisco 4000 Series	Apple OS X
Zhone GPON	Xen hypervisor
ZigBee	Google Earth
Modbus	Dell XPS 13
Embedded Linux	Skype
RTOS	HPE ProLiant
Android	Terraform

---

[2] https://github.com/aissa302/CVE-Aanlysis-Project.

Our script then analyzed each vulnerability description to determine the presence of keywords from both lists. If a description contained embedded system keywords but no general-purpose system keywords, it was assigned an embedded system vulnerability label. Conversely, if a description contained general-purpose system keywords but no embedded system keywords, it was labeled as a general-purpose system vulnerability. To minimize ambiguity during the training phase, descriptions containing keywords from both lists were excluded from the dataset.

This keyword-based approach was inspired by previous research [24], which employed a similar methodology to classify vulnerabilities. By leveraging this approach, we ensured that our dataset was accurately labeled and ready for model training.

4.3 Benchmark Dataset Creation

Current studies often use the National Vulnerability Database (NVD) for training and testing machine learning models. While convenient, this approach has a significant limitation: it fails to simulate real-world scenarios, leading to models that may overfit to the training data, achieving high accuracy but lacking true robustness.

To address this, models should be evaluated on datasets that differ in structure, content, and style, offering a more realistic assessment of their generalizability. We developed a comprehensive benchmark dataset to meet this need, incorporating vulnerability descriptions from CNNVD, CNVD, VarIoT, and a manually curated collection. These sources provide a richer and more varied set of examples, ensuring that models are tested under conditions that better reflect real-world complexities.

To create the benchmark dataset, we extracted vulnerability descriptions from each of the three sources and split them into two groups: those with CVE IDs and those without. Our benchmark dataset consist of entries without CVE ID to ensure they were unique to each database and not addressed in the NVD database.

The composition of the benchmark dataset is summarized in Table 3. The table shows the number of vulnerability descriptions from each source.

Table 3. Benchmark Dataset Composition

	ESV	GPSV
CNNVD	246	866
CNVD	1146	980
VarIoT	922	24
Unseen	68	100

5   Ensemble Embedding Technique

In natural language processing (NLP), word embedding techniques such as Word-2Vec, GloVe, and FastText are commonly employed to convert words into vector representations. These techniques, however, are prone to out-of-vocabulary (OOV) issues, where words not found in the embedding training corpus are assigned random or zero vectors. This poses challenges, particularly in domain-specific or rare-word-rich datasets, where OOV words can degrade model performance. We propose a refined ensemble embedding technique to mitigate this problem. Our approach leverages multiple word embeddings to create a more comprehensive representation, prioritizing one embedding as the primary source and integrating others as secondary fallback options. This allows for greater coverage and more accurate handling of OOV words. The methodology is as follows:

1. **Primary Embedding Selection:** One word embedding is designated as the primary embedding, chosen based on the characteristics of the dataset and the problem at hand. This embedding will serve as the primary source of word vectors.
2. **Word Vector Retrieval Process:**
    – If a word exists in the primary embedding, its vector representation is directly used.
    – If the word does not exist in the primary embedding, the algorithm searches for the word in the secondary embeddings. If found, the top 5 most similar words are retrieved.
    – The algorithm then checks if any of these similar words exist in the primary embedding. If a match is found, the vector of the most similar word in the primary embedding is assigned to the original word.
    – If no matching similar word is found, the search continues across additional embeddings until all options are exhausted.
3. **Handling OOV Words:** If the word is not found in any embedding, a randomly initialized vector is assigned to ensure it has representation in the model, even though it lacks semantic accuracy.

Algorithm 1 formalizes the procedure.

6   Evaluatioin

6.1   Experimentation Setup

A series of experiments were conducted to determine the optimal scenario for classifying vulnerability descriptions. These experiments were performed on a computer featuring a Xeon CPU running at 2.40 GHz, 32 GB of RAM, and dual NVIDIA M2000 GPUs. The operating system employed was Ubuntu 22.04 LTS. Throughout the experimentation process, we utilized TensorFlow version 2.12.0 in combination with Keras version 2.12.0. The hyperparameters applied across various scenarios are detailed in Table 4.

Algorithm 1. Ensemble Embedding Approach

Input: Set of word embeddings: $E = \{E_1, E_2, \ldots, E_n\}$ where E_1 is the primary embedding.
Input: Vocabulary V of words to be embedded.

Input: Vector dimension d.
Output: Word vector matrix M where each word $w \in V$ has a corresponding vector $v_w \in \mathbb{R}^d$.
Initialize an empty matrix M of size $|V| \times d$.
for each word $w \in V$ **do**
    Primary Embedding Search:
    if $w \in E_1$ **then**
        Set $v_w = E_1(w)$ and store in M.
    else
        Secondary Embedding Search:
        for each $E_i \in \{E_2, E_3, \ldots, E_n\}$ **do**
            if $w \in E_i$ **then**
                Retrieve top 5 most similar words $\{s_1, s_2, \ldots, s_5\}$ from E_i.
                for each $s_j \in \{s_1, s_2, \ldots, s_5\}$ **do**
                    if $s_j \in E_1$ **then**
                        Set $v_w = E_1(s_j)$, store in M, and **break**.
                    end if
                end for
            end if
        end for
        Handling Unresolved Words:
        if v_w is not assigned **then**
            Initialize a random vector $v_w \in \mathbb{R}^d$ and store in M.
        end if
    end if
end for
Return: Matrix M.

Table 4. Hyperparameters

Hyperparamters	BiLSTM
Batch-size	32
Epochs	50
Layers	19
Droupout	0.3
Loss function	sparse_categorical_crossentropy
Optimizer	Adam
Early Stop/patience	val_loss/5
Kernel Initializer	he_uniform
Kernel Regularizer	l2(0.01)
Total params	17,816,522
Trainable params	1,411,610
Non-trainable params	16,404,912

6.2   Evaluation Metrics

To comprehensively evaluate the performance of our classification models, we employed a set of standard evaluation metrics: accuracy, precision, recall, and F1-score. These metrics provide a holistic view of the models' effectiveness in classifying vulnerability descriptions.

– **Accuracy:** The proportion of correctly classified instances out of the total. It provides an overall measure of model performance.

$$\text{Accuracy} = \frac{\text{TP} + \text{TN}}{\text{Total Number of Instances}} \tag{1}$$

– **Precision:** The ratio of true positives to all predicted positives. Key for scenarios where false positives are costly.

$$\text{Precision} = \frac{\text{TP}}{\text{TP} + \text{FP}} \tag{2}$$

– **Recall:** The ratio of true positives to all actual positives. Important when missing a positive instance is critical.

$$\text{Recall} = \frac{\text{TP}}{\text{TP} + \text{FN}} \tag{3}$$

– **F1-Score:** The harmonic mean of precision and recall, useful for imbalanced datasets.

$$\text{F1-Score} = 2 \times \frac{\text{Precision} \times \text{Recall}}{\text{Precision} + \text{Recall}} \tag{4}$$

6.3   Results

The performance of the proposed algorithm was evaluated on a validation set and four benchmark datasets: CNNVD, CNVD, VarIoT, and Unseen. The results are presented in Table 5 and Fig. 2.

Table 5. Model validation results

	Accuracy	Precision	Recall	F1-score
Glove	99.14780	99.15184	99.15254	99.14780
Word2Vec	97.78961	97.85273	97.80386	97.78925
FastText	94.22103	94.64473	94.25693	94.21026
Ensemble embedding	**99.13448**	**99.13419**	**99.13472**	**99.13444**

Table 5 shows the validation results, where the Ensemble Embedding approach achieves an accuracy of 99.13%, demonstrating its effectiveness in capturing the semantic relationships between words in the vulnerability reports. The

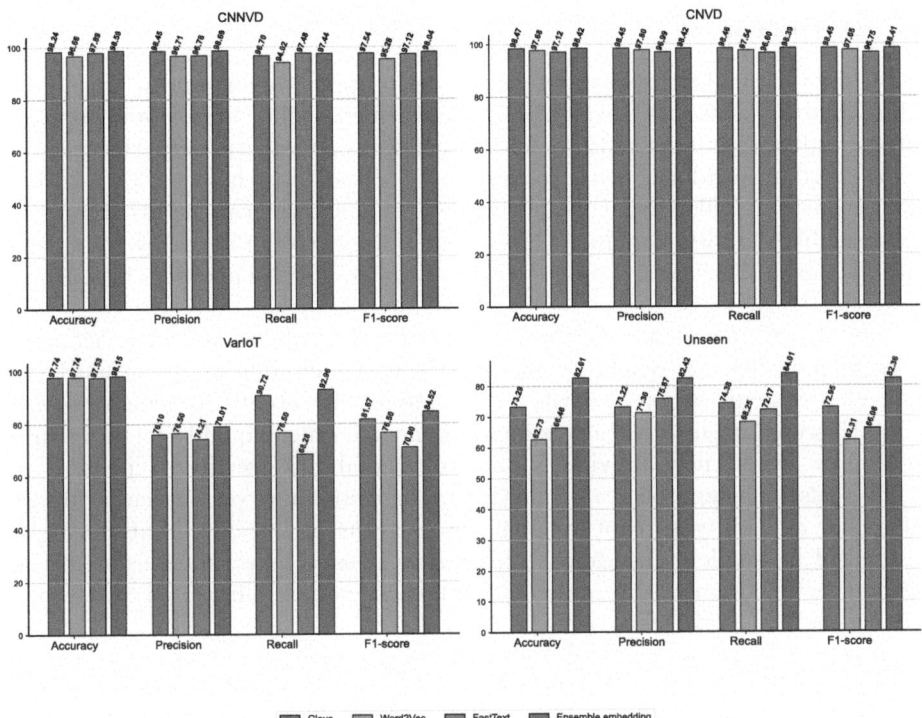

Fig. 2. Benchmark data test results

high accuracy and F1-score indicate that the proposed algorithm is capable of accurately classifying vulnerability reports, which is a crucial step in identifying and mitigating vulnerabilities in embedded systems.

Figure 2 shows that the Ensemble Embedding approach consistently outperforms individual embeddings like GloVe, Word2Vec, and FastText across all datasets, including the manually curated Unseen dataset, which contains different device/system vulnerabilities than those in the training data. This dataset was particularly challenging due to its diverse content, making it a robust test for the models.

On the Unseen dataset, the Ensemble Embedding achieved the highest scores: accuracy of 82.61%, precision of 82.42%, recall of 84.01%, and F1-score of 82.36%. This superior performance can be attributed to its ability to reduce Out-of-Vocabulary (OOV) issues. By leveraging multiple embeddings, it captures a broader range of vocabulary, making it more effective at handling diverse and previously unseen vulnerabilities compared to single embedding methods.

6.4   Discussion

The results highlight the effectiveness of the proposed algorithm in classifying vulnerability reports and distinguishing between embedded system and general-purpose vulnerabilities. The Ensemble Embedding consistently outperforms individual word embeddings, with accuracy and F1-score as high as 82.61% and 82.36% on the challenging Unseen dataset. This success can be attributed to the ensemble's ability to reduce Out-of-Vocabulary (OOV) issues by leveraging multiple embeddings, which allows it to capture complex semantic relationships more effectively.

The algorithm's strong performance across the validation and benchmark datasets indicates its reliability in identifying vulnerabilities, particularly in embedded systems, where security is paramount. Its ability to generalize well to the Unseen dataset underscores its robustness, a crucial quality for real-world applications where new, previously unseen vulnerabilities regularly emerge.

The ensemble approach's strength lies in combining various embeddings, enhancing its ability to capture nuanced patterns in the data that might be overlooked by single embeddings. This makes it especially valuable for distinguishing between embedded and general-purpose vulnerabilities, an area where existing methods have struggled.

In conclusion, the proposed algorithm offers a significant advancement in vulnerability classification, paving the way for more tailored security solutions for embedded systems, which are critical for the security and stability of modern infrastructure.

7   Conclusion

In this paper, we have presented a novel approach to automatic vulnerability classification, specifically designed to address the unique challenges of embedded systems. By leveraging ensemble embedding and BiLSTM networks, our model is able to accurately categorize vulnerabilities and distinguish between those affecting embedded systems and general-purpose systems. The creation of a comprehensive benchmark dataset, collected from multiple databases, provides a valuable resource for future research in this area. Our experimental results demonstrate the efficacy of our approach in identifying embedded system-specific vulnerabilities, paving the way for more targeted and efficient security solutions. This research contributes to the development of more effective security measures for protecting critical infrastructure and sensitive information. As the complexity and interconnectivity of embedded systems continue to grow, our approach offers a timely and innovative solution to a pressing problem in the field.

Acknowledgments. This research received no external funding.

Disclosure of Interests. The authors have no competing interests to declare that are relevant to the content of this article.

References

1. NVD feeds. https://nvd.nist.gov/vuln/data-feeds
2. Ben Yahya, A., El Akhal, H., El Alaoui, A.E.B.: Machine learning-based collection and analysis of embedded systems vulnerabilities. In: Enhancing Performance, Efficiency, and Security Through Complex Systems Control, pp. 242–261. IGI Global (2024)
3. Blinowski, G.J., Piotrowski, P.: CVE based classification of vulnerable IoT systems. In: Proceedings of the 15th International Conference on Dependability & Complex Systems (DepCoS), pp. 82–93 (2020)
4. Bojanowski, P., Grave, E., Joulin, A., Mikolov, T.: Enriching word vectors with subword information. Trans. Assoc. Comput. Linguist. **5**, 135–146 (2017)
5. Chen, J., et al.: An automatic software vulnerability classification framework using term frequency-inverse gravity moment and feature selection. J. Syst. Softw. **167**, 110616 (2020)
6. CNNVD: Chinese national vulnerability database (2024). https://www.cnnvd.org.cn/. Accessed August 2024
7. CNVD: China national vulnerability database (2024). https://www.cnvd.org.cn/. Accessed June 2024
8. Dong, Y., Tang, Y., Cheng, X., Yang, Y.: DeKeDVer: a deep learning-based multitype software vulnerability classification framework using vulnerability description and source code. Inf. Softw. Technol. **163**, 107290 (2023)
9. FortiGuard: Fortiguard (2024). https://www.fortiguard.com/. Accessed June 2024
10. Franceschi-Bicchierai, L.: How this internet of things stuffed animal can be remotely turned into a spy device (2017). https://tinyurl.com/yc4xfdpd
11. Google translate API: Google translate API (2024). https://pypi.org/project/googletrans/. Accessed July 2024
12. Han, Z., et al.: Learning to predict severity of software vulnerability using only vulnerability description. In: Proceedings of the IEEE International Conference on Software Maintenance and Evolution (ICSME), pp. 125–136 (2017)
13. Huang, G., et al.: Automatic classification method for software vulnerability based on deep neural network. IEEE Access **7**, 28291–28298 (2019)
14. ICS-CERT: Industrial control systems cyber emergency response team (2024). https://www.cisa.gov/. Accessed June 2024
15. Keküll, H., Ergen, B., Arslan, H.: A multiclass hybrid approach to estimating software vulnerability vectors and severity score. J. Inf. Secur. Appl. **63**, 103028 (2021)
16. Kudjo, P.K., et al.: Improving the accuracy of vulnerability report classification using term frequency-inverse gravity moment. In: Proceedings of the 19th IEEE International Conference on Software Quality, Reliability & Security (QRS), pp. 248–259 (2019)
17. Li, X., Chang, X., Board, J.A., Trivedi, K.S.: A novel approach for software vulnerability classification. In: Proceedings Annual Reliability and Maintainability Symposium (RAMS), pp. 1–7 (2017)
18. Mikolov, T., Chen, K., Corrado, G., Dean, J.: Efficient estimation of word representations in vector space. arXiv:1301.3781 (2013)
19. Na, S., Kim, T., Kim, H.: A study on the classification of common vulnerabilities and exposures using naïve Bayes. In: Proceedings of the 11th International Conference on Broad-Band Wireless Computing, Communication & Applications (BWCCA), pp. 657–662 (2017)

20. Newman, L.H.: A new pacemaker hack puts malware directly on the device. Wired (2018). https://www.wired.com/story/pacemaker-hack-malware-black-hat/
21. Nie, S., Liu, L., Du, Y.: Free-fall: hacking tesla from wireless to can bus. In: Black Hat conference US (2017). https://www.blackhat.com/docs/us-17/thursday/us-17-Nie-Free-Fall-Hacking-Tesla-From-Wireless-To-CAN-Bus-wp.pdf
22. NVD: National vulnerability database (2024). https://nvd.nist.gov/. Accessed June 2024
23. Pan, M., et al.: An automatic vulnerability classification framework based on BiGRU-TextCNN. In: Proceedings International Neural Network Society Workshop on Deep Learning Innovations & Applications (INNS DLIA), vol. 222, pp. 377–386 (2023)
24. Papp, D., Ma, Z., Buttyan, L.: Embedded systems security: threats, vulnerabilities, and attack taxonomy. In: Proceedings of the 13th Annual Conference on Privacy, Security & Trust (PST), pp. 145–152 (2015)
25. Pennington, J., Socher, R., Manning, C.D.: GloVe: global vectors for word representation. In: Proceedings of the Conference on Empirical Methods in Natural Language Processing (EMNLP), pp. 1532–1543 (2014)
26. Prigg, M.: How to get green lights all the way to work: hackers reveal how simple it is to control traffic lights in major cities using just a laptop. Mail Online (2014). https://tinyurl.com/vjzyc7jx
27. Rios, A., Lwowski, B.: An empirical study of the downstream reliability of pre-trained word embeddings. In: Proceedings of the 28th International Conference on Computational Linguistics (COLING) (2020)
28. Sharma, R., Sibal, R., Sabharwal, S.: Software vulnerability prioritization using vulnerability description. Int. J. Syst. Assur. Eng. Manage. 12, 58–64 (2021)
29. Stanislav, M., Beardsley, T.: Hacking IoT: a case study on baby monitor exposures and vulnerabilities. Rapid7 (2015). https://tinyurl.com/b8njhadw
30. VarIoT: VARIoT IoT vulnerabilities database (2024). https://www.variotdbs.pl/vulns/. Accessed August 2024
31. Vishnu, P., Vinod, P., Yerima, S.Y.: A deep learning approach for classifying vulnerability descriptions using self attention based neural network. J. Netw. Syst. Manage. 30, 1–27 (2022)
32. Wang, J.A., Guo, M.: Vulnerability categorization using Bayesian networks. In: Proceedings of the 6th Annual Workshop on Cyber Security & Information Intelligence Research, pp. 1–4 (2010)
33. Wang, Q., Gao, Y., Ren, J., Zhang, B.: An automatic classification algorithm for software vulnerability based on weighted word vector and fusion neural network. Comput. Secur. 103070 (2022)
34. Wang, Q., Gao, Y., Ren, J., Zhang, B.: An automatic classification algorithm for software vulnerability based on weighted word vector and fusion neural network. Comput. Secur. 126, 103070 (2023)
35. Wen, T., Zhang, Y., Wu, Q., Yang, G.: ASVC: an automatic security vulnerability categorization framework based on novel features of vulnerability data. J. Commun. 10(2), 107–116 (2015)
36. ZDI: Zero-day initiative (2024). https://www.zerodayinitiative.com/. Accessed June 2024

Predicting Dyskinetic Events Through Verified Multivariate Time Series Classification

Giacomo Bergami[1](✉)[iD], Emma Packer[2][iD], Kirsty Scott[2][iD], and Silvia Del Din[2,3][iD]

[1] School of Computing, Newcastle University, Newcastle upon Tyne, UK
`giacomo.bergami@newcastle.ac.uk`
[2] Translational and Clinical Research Institute, Faculty of Medical Sciences, Newcastle University, Newcastle upon Tyne, UK
`{e.packer,kirsty.scott-singer}@newcastle.ac.uk`
[3] National Institute for Health and Care Research (NIHR) Newcastle Biomedical Research Centre (BRC), Newcastle University and The Newcastle upon Tyne Hospitals NHS Foundation Trust, Newcastle upon Tyne, UK
`silvia.del-din@newcastle.ac.uk`

Abstract. While monitoring Parkinsonian patients with wearable sensors and tracking their drug assumption patterns, we want to differentiate the behaviours distinguishing periods of relative well-being from dyskinetic events. This requires solving a novel problem, where an entire multivariate time series (MTS) has its class label varying in time, thus leading to a generalised formulation of multivariate time series classification (MTSC). To achieve explainability, we premier the composition of data trend (DT) analysis with DECLARE*d*, a log temporal declarative language, to derive human-readable correlations across different MTS dimensions' trends. This is mediated by a novel temporal data representation, polyadic logs, supporting both MTS raw data and concurrent activity-labelled durative activities (constituents) for representing event-based classes and concurrent DTs across MTS dimensions. Our validation over a real patient dataset shows that our MTCS algorithm, EMeriTAte, outperforms state-of-the-art MTSC for a novel patient classification task.

Keywords: Multivariate Time Series Classification · Verified Temporal Artificial Intelligence · eXplainable AI · E-Health

1 Introduction

Parkinson disease (PD) is one of the most fast growing worldwide neurological conditions: by 2040, it will affect more than 40 million people and, in the UK, 1 over 500 people have PD. Motor symptoms such as dyskinetic events result from the degeneration of dopaminergic neurons: when tracking patients with Parkinsonian Disease for an entire week, periods of relative well-being are interspersed with dyskinetic events of different lengths/sizes occurring anytime.

ⓒ The Author(s), under exclusive license to Springer Nature Switzerland AG 2025
R. Chbeir et al. (Eds.): IDEAS 2024, LNCS 15511, pp. 49–62, 2025.
https://doi.org/10.1007/978-3-031-83472-1_4

As medical literature suggests that these patterns are directly correlated with how the drugs are taken [1], and given that these dyskinetic events are easily detectable by applying movement sensors to the patients, we can keep track of patients' motions in time. By translating drug assumption patterns into estimated blood concentration levels of active components given the drug dosages [6], we represent all patient information as distinct Multivariate Time Series (MTS) $T \colon \mathbb{N} \to \mathbb{R}^d$; these represent the value associated with multiple time series $X, Y, \ldots, Z$ as a d-dimensional vector $(X(t), Y(t), \ldots, Z(t))$ (where each time series functions $X \colon \mathbb{N} \to \mathbb{R}$ are a set of data points indexed by time) while also recording in one of the d dimensions whether t coincided with well-being or a diskynetic event according to the patient's diary.

Traditional Multivariate Time Series Classification (MTSC), on the other hand, assumes the entire MTS T_i from a dataset D to be associated to a class y_i, for which the classification learning task is to learn the function h minimising the classification error $|h(T_i) - y_i|$ [17], so classes are not associated to each natural-valued time stamp as we now require. All MTSC implementations from sktime[1] also assume that all MTS should have equal length/size, which is rather unlikely to happen in realistic use case scenarios, thus requiring to sample patients' time series into uniformly-long window sizes corresponding to the minimal duration time. As the key complexity of MTSC is that discriminatory key features might reside in the interactions across dimensions [14], this sampling comes to the cost of losing or splitting relevant patterns characterising dyskinetic events.

Current literature (Sect. 2.3) solves MTSC by either decomposing each MTS dimension into intervals described as mean, standard deviation, and slope [12], or by extracting time series features through convolutional kernels to be fed to a classifier [8]. Deep Neural Network refine the latter through attention mechanisms for better feature selection [17]. As e-health settings favour explainable AI for determining the most sensible model among all those with equal accuracy [11,13], none of these approaches involve the extraction of explainable properties in terms of trends and correlation of such trends across MTS dimensions (Sect. 2.2).

To overcome this limitation, this paper proposes a novel MTSC algorithm, EMeriTAte (Explainable MultivariatE coRrelatIonal Temporal Artificial inTelligence, Sect. 4), built on top of the temporal relational database KnoBAB[2] and scikit-learn classifiers, which is then validated over realistic healthcare data targeting the identification of dyskinetic events in Parkinsonian patients from [6] (Sect. 5). We extract interactions across MTSC dimensions by first (Sect. 4.1) transforming each dimension from MTS $T_{\mathfrak{E}}$ describing each patient $\mathfrak{E}$ into a sequence of Data Trends (DT) [10], discretising the stationarity, growth, and decrease patterns occurring within the time series as distinct *durative* (event) *constituents*. Different trends might start across dimensions at different times, so these constituents are then collected in single events. As no log data model currently supports such data representation, we propose polyadic logs (Sect. 3) retaining raw MTS data, DT information,

---

[1] https://github.com/sktime/sktime.

[2] https://github.com/datagram-db/knobab/releases/tag/v3.0.

and class labels across all patients (Sect. 4.2). From such logs, we then exploit off-the-shelf algorithms available from Verified Temporal Artificial Intelligence: we extract declarative temporal correlations across durative constituents using state-of-the-art specification mining algorithm such as Bolt2 [5], which are then used via *deviance learning* algorithms [4] to derive a temporal specification Φ_y distinguishing the temporal behaviour across classes and patients. Results (Sect. 5.2) show that the proposed methodology targets the temporal characterization of such dyskinetic events with an average accuracy of 98.52% and a precision of 97.57% while state-of-the-art algorithms (TapNet [17]) can solve our use case scenario problem with 46.42% accuracy and 37.22% precision, thus remarking that such algorithms cannot classify temporal behaviours by precisely selecting temporal features distinguishing dyskinetic from well-being.

2   Related Works

2.1   Temporal Data Models

We motivate the adoption of the Polyadic Temporal Data Model by remarking the inability of current data models to support both raw data information as well as supporting durative events.

Single-Aspect Temporal Data. Business Process Management (BPM) models information as a linear and finite sequence of single *pointwise* and *non-durative events* σ_j^i occurring in a well-defined instant j that can be easily represented as a sequence over a discrete timeline (*trace* σ^i). Each event is then associated with an *activity label*, determining the type of occurring activity and a *payload*, extending each event with a key-value association. The *traces*, representing a specific run of an environment of interest, are collected into (non-polyadic) *logs*.

Multi-Aspect Temporal Data. A *polyadic trace* σ^i (*k-sequence* in [16]) is an ordered list of polyadic events $\sigma_1^i, \ldots, \sigma_n^i$, each of which containing distinct *pointwise* (or non-*durative*) *constituents* (or *items*) $\sigma_{j,k}^i \in \sigma_j^i$. Sequence mining algorithms for polyadic traces such as SPADE [16] generalize traditional association rules for non-polyadic traces (*transactions*) by considering arbitrary size rules $\alpha_1 \to \cdots \to \alpha_q$ where each α_i might contain multiple distinct pointwise events that should occur at the same time. Although SPADE's representation extends the one typically assumed in the BPM field with concurrent constituents, while BPM assumes that each event represents one sole non-durative constituent of interest (i.e., $\sigma_1^i \equiv \sigma_{1,1}^i$), SPADE also does not consider events being associated with payload information and merely characterise non-durative events in terms of their associated activity label as per BPM dataless logs. Notwithstanding the former, both SPADE and BPM log models do not consider duration spans for constituents or events, thus voiding the possibility of performing temporal reasoning similarly to Allen's algebra, where we can also predicate on the event's overlap. Section 3 premiers a novel temporal data representation, where polyadic events are composed of durative concurrent constituents with a data payload, where duration is expressed through a span field. We consider all constituents with span = 1 as pointwise (or non-durative) constituents.

Template (c_l)	Non-polyadic logs interpretation $[\![c_l]\!]$
$Exists(A, n)$	Activations should occur at least n times
$Absence(A)$	Activations should occur at most n times
$Choice(A, A')$	One of the two activation conditions must appear.
$ExclChoice(A, A')$	Only one type of activation is admitted per trace.
$RespExistence(A, B)$	An activation requires at least one target.
$CoExistence(A, B)$	$[\![RespExistence(A, B)]\!] \wedge [\![RespExistence(B, A)]\!]$
$Precedence(A, C)$	No C precedes the activation.
$Response(A, B)$	An activation leads to a future target event.
$Succession(A, B)$	$[\![Precedence(A, B)]\!] \wedge [\![Response(A, B)]\!]$
$ChainPrecedence(A, B)$	An activation leads to an *immediately preceding* target.
$ChainResponse(A, B)$	An activation leads to an *immediately following* target.
$ChainSuccession(A, B)$	$[\![ChainPreced.(B, A)]\!] \wedge [\![ChainResponse(A, B)]\!]$

(a) Our DECLAREd [3] subset of interest, where A (respectively, B) denote activation (resp., target) conditions.

Template $\varsigma(A^\iota)$	Time Series Events $(A^\iota, A^{\neg\iota})$
IncreaseRapidly(A^ι)	$A^\iota, \ldots, A^\iota$
IncreaseSlowlyI(A^ι)	$A^{\neg\iota}, A^\iota, \ldots, A^\iota$
IncreaseSlowlyII(A^ι)	$A^\iota, A^{\neg\iota}, A^\iota, \ldots, A^\iota$
IncreaseSlowlyIII(A^ι)	$A^\iota, \ldots, A^\iota, A^{\neg\iota}, A^\iota$
IncreaseSlowlyIV(A^ι)	$A^\iota, \ldots, A^\iota, A^{\neg\iota}$
HighVolatilityI(A^ι)	$A^{\neg\iota}, A^\iota$
HighVolatilityII(A^ι)	$A^{\neg\iota}, A^\iota, A^{\neg\iota}, A^\iota$
HighVolatilityIII(A^ι)	$A^{\neg\iota}, A^\iota, \ldots, A^\iota, A^{\neg\iota}$
HighVolatilityIV(A^ι)	$A^\iota, A^{\neg\iota}, \ldots, A^{\neg\iota}, A^\iota$
HighVolatilityV(A^ι)	$A^\iota, A^{\neg\iota}, A^\iota, A^{\neg\iota}$
HighVolatilityVI(A^ι)	$A^\iota, A^{\neg\iota}$
DecreaseSlowlyI(A^ι)	$A^{\neg\iota}, \ldots, A^{\neg\iota}, A^\iota$
DecreaseSlowlyII(A^ι)	$A^{\neg\iota}, \ldots, A^{\neg\iota}, A^\iota, A^{\neg\iota}$
DecreaseSlowlyIII(A^ι)	$A^{\neg\iota}, A^\iota, A^{\neg\iota}, \ldots, A^{\neg\iota}$
DecreaseSlowlyIV(A^ι)	$A^\iota, A^{\neg\iota}, \ldots, A^{\neg\iota}$
DecreaseRapidly(A^ι)	$A^{\neg\iota}, \ldots, A^{\neg\iota}$

(b) DT as patterns and events [10].

Fig. 1. Declarative Languages for *(a)* non-polyadic logs or *(b)* time series.

2.2 Declarative Temporal Languages

Despite temporal Declarative languages only support single-aspect data representations, they provide a high-level and human-readable characterization of a specific subset of all the possible LTL^f temporal patterns deemed as significant. The models derived by EMeriTAte will combine these two languages.

DECLAREd [3] is a *log* temporal declarative language considering possible temporal behaviour in terms of correlations between activated events and targeted conditions, where the former refers to the necessary but not sufficient condition for satisfying a clause. If the clause has no activation, then this is trivially satisfied (*vacuously*) unless stated otherwise: e.g., Precedence might also be vacuously violated due to the presence of just target conditions [5]. DECLARE clauses can be instantiated over a specific alphabet Σ referring to activity labels associated with constituents and they can be composed to generate *conjunctive specifications* $\Phi = \{c_1, \ldots, c_l\}$; we say that a trace satisfies a conjunctive specification if it satisfies all its clauses. These conjunctive specifications can be extracted from temporal data represented as (non-polyadic) logs by efficiently exploiting a specific-to-general lattice search from which we can preventively prune the generation of specific clauses before testing them [5].

DT specifications consider a different temporal fragment expressing numerical trends for (univariate) Time Series (TS) X [10]: this requires a preliminary pre-processing step for which X is associated with durative constituents σ_{j+1}^X with activity labels X^c or $X^{\neg c}$: σ_{j+1}^X has activity label X^c if $X(j+1) > X(t)$ and $X^{\neg c}$ otherwise. Given this, we can then determine the occurrence of the pattern by simply joining X^c and $X^{\neg c}$, which can then be matched and rewritten into durative constituents from the first column of patterns $\varsigma(A^c)$ (Fig. 1b), providing the DT specification. Authors mainly use this declarative representation for TS forecasting purposes, but give no evidence for exploiting this in the context of MTS while correlating disparate DT across MTS variables, which might be expressed through DECLAREd. To achieve this, we present a methodology where both approaches are combined. Although DT discards the numerical information associated with such trends, it generalises the common shapelet approach [9] to arbitrary growth and decrease patterns expressed declaratively with human-understandable named patterns.

2.3 Multivariate Time Series Classification

We leveraged the latest survey paper on MTSC [14] for identifying orthogonal approaches to classification. This shows that none of these establish temporal correlations as per DECLAREd while considering DT for circumscribing the temporal trends and behaviour of interest for each MTS dimension.

KNN-based clustering for time series works by first identifying time series clusters depending on a distance function δ of choice and then associating to each cluster the majority class; a straightforward Euclidean metric can be used to determine the pointwise distance of the time series across dimensions (EuclideanKNN). The main drawback of this approach is that it neither considers the evolution of data trends within the time series nor allows for aligning similar trends being displaced in time, similar to dynamic time warping.

Rocket [8] exploits randomized convolutional kernels to extract relevant features from the time series, which are then fed to a linear classifier by associating such features to a class of interest. While this feature extraction approach is not guaranteed to adequately capture data trends and variations across dimensions, it also guarantees scarce explainability as information is summarised into kernel values. TapNet [17] improves the latter by exploiting attention networks for selecting the best classification features, while still heavily relying on up-front feature selection mechanisms such as dimensionality reduction, which might lose relevant information to the detriment of the classification precision and accuracy.

CANONICAL INTERVAL FOREST (CIF) [12] achieves explainability by exploiting white-box classifiers such as decision trees over *catch22* features describing a selection of time intervals describing different types of numerical variations rather than changes in trends and data patterns and their temporal correlations. SHAPELET TRANSFORM CLASSIFIER (STC) [9] characterize MTS in terms of distinctive temporal features, the *shapelets*, being frequently occurring trends occurring across all time series, for then describing each MTS in terms of their distance with each selected shapelet. Notwithstanding this approach considers numerical features that are completely discarded through DT, it fails to establish correlations across trends occurring in different dimensions, thus not establishing correlations across differently observed behaviours.

3 Data Model

Problem Formulation. Given a MTS T, we refer to its *size* $|T|$ the number of events recorded counting from 1, thus $\mathrm{dom}(T) = \{1, \ldots, |T|\}$. When $T: \mathbb{N} \to \mathbb{R}^d$

is clear from the context, given a dimension $x \leq d$, we use $x(\!|t|\!)$ as a shorthand for $T(t)(x)$. Given a time interval (i, j) and a MTS T of at least size j, we denote $T[i, \ldots, j]$ as its *projection* considering a subsequence of the events occurring within this interval: $T[i, \ldots, j](x) = T(x + i - 1)$ **if** $x + i - 1 \in \text{dom}(T)$ **and** $1 \leq x \leq j - i + 1$. Given a MTS $T_{\mathfrak{E}}$ within a training dataset D where c represents the time-wise classification and a set of maximal and non-overlapping temporal subsequences[3] (i, j) in $T_{\mathfrak{E}}$ targeting the same event time class value for dimension c, we want to learn a function h minimising the classification error $\sum_{i \leq \tau \leq j} |h(T_{\mathfrak{E}}[i, \ldots, j]) - T_{\mathfrak{E}}(\tau)(\mathsf{c})|$ for all intervals (i, j) in $T_{\mathfrak{E}}$ and from all $T_{\mathfrak{E}}$ in D. This is a more general formulation than the one posed by current literature, considering $(1, |T_{\mathfrak{E}}|)$ as the sole interval of interest and $T_{\mathfrak{E}}(i)(\mathsf{c})$ associated with one classification label throughout the timestamps $i \in \text{dom}(T_{\mathfrak{E}})$ per MTS $T_{\mathfrak{E}}$.

Polyadic Temporal Data Model. A *polyadic log* $\mathfrak{S}$ is a collection of distinct *polyadic traces* $\{\sigma^i, \ldots, \sigma^n\}$ referring to the auditing of a specific environment $\mathfrak{E}$ of interest (e.g., a patient); each trace is a list of temporally ordered *polyadic events* $[\sigma_1^i, \ldots, \sigma_o^i]$, where each of these σ_j^i is defined as a tuple of a set of *durative constituents* and a class, thus $\sigma_j^i = \langle c_j^i, \mathsf{class} \rangle$; all constituents in $c_j^i = \{\sigma_{j,1}^i, \ldots, \sigma_{j,m}^i\}$ start at time j while having possibly different activity labels and duration spans; each durative constituent $\sigma_{j,k}^i$ is a triplet $\langle \mathsf{a}, p, \mathsf{s} \rangle$ where a is the activity label, p is the payload collecting the raw data being associated to the durative constituent, and s denotes its temporal span s.t. $\mathsf{s} \leq o - j + 1$.

Poly-DECLAREd. Polyadic logs require the extension of the usual DECLAREd semantics, only dealing with pointwise constituent [3], to also consider durative ones. This extends the formal definition of DECLARE which, for mutual correlation clauses, needs to rephrase the "*The activation...*" sentences from Fig. 1 with "*For all activated durative events at a given time, there exists at least a target one such that...*". No further semantic extensions are required to support polyadic events with multiple constituents is needed.

4   Methodology

Algorithm 1 shows the training and testing steps for the EMeriTAte algorithm, through which we classify time series by extracting a human explainable data trend using specification mining algorithms. This is achieved by first discretising each time series describing each specific environment of interest (e.g., a patient) according to the occurring DTs per dimension (Sect. 4.1), for then grouping them according to the maximal time series subsequences associated to the same class, thus obtaining a distinct log per class in $\mathcal{Y}$ (Sect. 4.2). After extracting a description for each segmented log in terms of Poly-DECLAREd clauses using a specification mining algorithm, we obtain a conjunctive model Φ_i of unary/binary

---

3 More formally, each (i, j) with $1 \leq i \leq j \leq |T_{\mathfrak{E}}|$ should satisfy the following condition: $(1 < i \leq |T_{\mathfrak{E}}| \Rightarrow T_{\mathfrak{E}}(i - 1)(\mathsf{c}) \neq T_{\mathfrak{E}}(i)(\mathsf{c})) \wedge (j < |T_{\mathfrak{E}}| \Rightarrow T_{\mathfrak{E}}(k)(\mathsf{c}) \neq T_{\mathfrak{E}}(j)(\mathsf{c})) \wedge \forall i \leq \tau \leq j . T_{\mathfrak{E}}(i)(\mathsf{c}) = T_{\mathfrak{E}}(j)(\mathsf{c}) = T_{\mathfrak{E}}(\tau)(\mathsf{c})$.

Algorithm 1. Explainable MultivariatE coRrelatIonal Temporal Artificial inTElligence (EMeriTAte): training and testing phases.

```
 1: function EMERITATE_TRAIN(D = {T_{𝔈_1}, ..., T_{𝔈_n}}, 𝒴 = {1, ..., n}; θ, F)
 2:    𝔊 ← {𝒯(T_{𝔈_i})|T_{𝔈_i} ∈ D}                              ▷ MTS discretization Sect. 4.1
 3:    𝔊_1, ..., 𝔊_n ← 𝒮(𝔊, 𝒴)                                  ▷ Trace Segmentation Sect. 4.2
 4:    𝔊_1, ..., 𝔊_n ← F(𝔊_1, ..., 𝔊_n)                         ▷ Trace filtered for training
 5:    Φ_1, ..., Φ_n ← BOLT2(𝔊_1, θ), ..., BOLT2(𝔊_n, θ)        ▷ Specification Mining [5]
 6:    for all i ∈ 𝒴 and σ^j ∈ 𝔊_i and ⋆(A, B) ∈ ⋃_{i≤n} Φ_i do  ▷ Deviance Learning [4]
 7:       DataFrame[σ^j, ⋆(A, B)] := 1_{σ^j ⊨ ⋆(A,B)} − 1_{σ^j ⊭ ⋆(A,B)}   ▷ Determining clause sat.
 8:       DataFrame[σ^j, clazz] := i                           ▷ Associating each segmented trace its class
 9:    ℳ ←DecisionTree(DataFrame)                              ▷ Explainable Model Extraction
10:    h(T_i) :=EMERITATE_TEST_ℳ(T_i)
11:    return h, ℳ                      ▷ Returning the classifier h and explainable model ℳ

12: function EMERITATE_TEST_ℳ(T_i)
13:    σ ← 𝒯(T_i)
14:    for all ⋆(A, B) ∈ ℳ do
15:       Row[⋆(A, B)] := 1_{σ ⊨ ⋆(A,B)} − 1_{σ ⊭ ⋆(A,B)}
16:    return ℳ(Row)
```

clauses $\star(A, B)$ (Line 5). We use Bolt2 [5] providing model minimality guarantees while providing expedite mining running times. We carry out *deviance learning* [4] (Lines 6-9) by generating a `DataFrame`, where each row corresponds to the description of a trace σ^j from segmented logs $\mathfrak{G}_i$ in terms of clauses being (1) or not (−1) satisfied (Line 7) while associating it to the training class of interest (Line 8). We use `DataFrame` to fit a white-box classifier such as a `DecisionTree` (Line 9), from which we can extract an explainable model $\mathcal{M}$ for differentiating time series according to their DT patterns and their correlation (Fig. 2). This same model is later used at testing time to classify a time series (Line 16) after discretising it into a polyadic log (Line 13) and determining the satisfiability of all the clauses appearing within the trained white-box model (Line 15).

4.1 $\mathcal{T}$: MTS to Polyadic Traces

First (Fig. 2, **1c**), we discretize each MTS into durative events representing DTs patterns by extending the original algorithm from [10] by *(i)* considering DTs throughout the entire time series, and *(ii)* considering all the following *conditions* for each value of $x(\!|t + 1|\!)$ per each MTS variable or dimension x:

- **increase event** x^i, if we observe an increase $x(\!|t + 1|\!) > x(\!|t|\!)$, and we generate a **non-increase event** x^{-i} otherwise; this is the only pointwise event generation from [10].
- **present event** x^p, if the time series actually records the data pertaining to x at time t, and we generate a **missing-data event** x^{-p} otherwise.

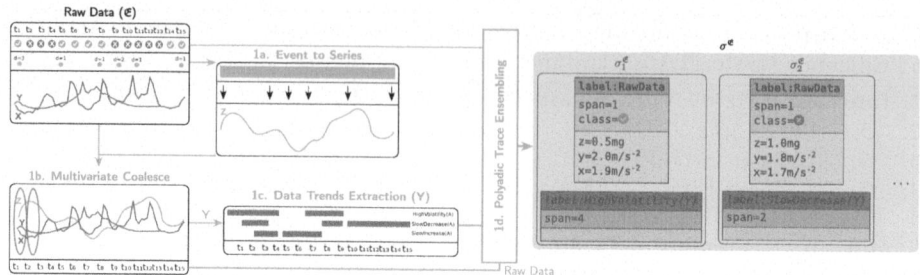

Fig. 2. Full data pipeline comprising of the data pre-processing steps required by our medical domain: we transform raw data composed by pointwise events (orange), pure time series (red/blue), and classification events (◐ and ✖) into polyadic traces in four steps (1a–1d). (Color figure online)

- **stationary event** x^s, if the time series variation is within a sensitivity ϵ, i.e. $|x(\!|t+1\!|) - x(\!|t\!|)| \leq \epsilon$, and we generate a **non-stationary event** $x^{\neg\text{s}}$ otherwise.
- **non-stimulant event** x^n, if the current value of the time series is below an absolute sensitivity ϵ pertaining to the lowest stimulus level, i.e. $|x(\!|t+1\!|)| \leq \epsilon$, and we generate a **stimulant event** $x^{\neg\text{n}}$ otherwise.

E.G., a sequence of `Total Steps`$^\neg$`i`,...,`Total Steps`$^\neg$`i` and one of `Total Steps`$^\neg$`d`,...,`Total Steps`$^\neg$`d` will generate DecreaseRapidly(`Total Steps`^{i}) and DecreaseRapidly(`Total Steps`^{d}) respectively as per specification in Fig. 1b. However, while the former will actually describe a rapid decrease in values for the variable x, the latter remarks generic absences of relevant fluctuations within the numerical data. Furthermore, *(iii)* we only keep DT referring to the longest temporal span across all the ones starting at the same time, thus removing subsuming trends.

Last (**1d**), we generate a polyadic trace per environment $\mathfrak{E}$ of interest. We represent the mined DT $\varsigma(x^c)$ starting at time j as durative events $\sigma_{j,k}^\mathfrak{E}$ with activity label $\varsigma(x^c)$, where its span s is the temporal distance between the first and the last event satisfying $\varsigma(x^c)$ (boxes in red, blue, and orange). We preserve the raw data $T_\mathfrak{E}(t)$ for each timestamp t by generating a pointwise constituent with activity label `RawData` (magenta): its payload contains property-value associations referring to its labelled dimension thus including the time-wise class information. We coalesce all the constituents starting at a specific time t_i into a single i-th polyadic event $\sigma_j^\mathfrak{E} = [\sigma_{j,1}^\mathfrak{E}, \ldots, \sigma_{j,o}^\mathfrak{E}]$ for a polyadic trace $\sigma^\mathfrak{E}$. We collect all the information into a polyadic log $\mathfrak{S}$, where $\sigma^\mathfrak{E}$ describes an environment $\mathfrak{E}$.

4.2 $\mathcal{S}$: Class-Driven Trace Segmentation and Loading

For each trace $\sigma^{\mathfrak{E}_A}$ in $\mathfrak{S}$, we scan all of its associated polyadic events and we group them by maximal contiguous sequences of polyadic events associated with the same class y, forming a new trace composed of contiguous events sharing the same desired classification outcome. By collecting all these new trace samples by class label y, we have a polyadic log $\mathfrak{S}_y$ per class y. As this transformation also preserves the Raw Data information associated with each polyadic event, this also enables atemporal event classification tasks by correlating the event's class to the Raw Data payload with the other values (e.g., specific blood concentration levels of active components) as well as reconstructing MTS projections (Sect. 5.1).

5   Experimental Evaluation

To mitigate motor symptoms caused by Parkinson's disease (PD), complex medication regimens are required. Following a period of stable response to dopaminergic medication, PD patients gradually develop progressive clinical phenomena, dyskinesias, which can be determined through patient mobility data. These require constant changes in clinical management, are intrinsically interconnected, but often require opposite medication attitudes [7]. We summarize the dataset information retrievable from [6]: we asked patients to log in a diary the period of time where dyskinesias where detected, from which we derive the event time class values; this will help us to establish correlations between movement trends within the data and these events. We are also interested in establishing temporal correlations between drug assumption patterns and fluctuations in the data series related to mobility information suggesting the insurgence of dyskinetic events. As such events harm patients' quality of life, we are interested in developing a tool for better studying these patterns. We monitor patients for an entire week, during which the dyskinetic events are recorded in the diary and interpreted as cumulative pointwise events alongside drug assumption. This information is then transformed into active principles time series estimating their blood levels.

As a pre-processing step, we transform (Fig. 2, **1a**) the desired pointwise constituents associated with the same activity label ζ and referring to the same patient/environment $\mathfrak{E}$ into a TS $Z_{\mathfrak{E}}$ according to a domain-specific transformation function. Next, (**1b**), we coalesce all univariate TS $X_{\mathfrak{E}}, Y_{\mathfrak{E}}, \ldots, Z_{\mathfrak{E}}$ referring to the same patient/environment $\mathfrak{E}$, be them generated in **1a** or collected through our motor sensors and patient diaries, as a single MTS $T_{\mathfrak{E}}(t) = (X_{\mathfrak{E}}(t), Y_{\mathfrak{E}}(t), \ldots, Z_{\mathfrak{E}}(t))$. Then, we feed the time series generated per patient to the EMERITATE_TRAIN algorithm, which is evaluated in the forthcoming subsection. Within our PD use case scenario, the first experiments evaluate the adequacy of our domain-specific transformation function, converting pointwise events concerning drug assumptions with given dosages to distinct time series referring to the predicted blood concentration levels of the active components of interest (Sect. 5.1), while the second test the adequacy of EMeriTAte of temporally characterising dyskinetic events (Sect. 5.2).

5.1   Atemporal Correlation of Estimated Blood Presence of Active Components to Occurring Dyskinetic Events

This section evaluates the adequacy of our domain-specific transformation function (**1a.**): as [1] states both *(i)* the possibility of deriving these levels by knowing (for each drug) the active components' dosage and half-life alongside their assumption times, and *(ii)* a positive correlation between levodopa blood concentration levels of and the occurrence of dyskinetic events, any strong correlation between predicted active components blood levels and the occurrence of dyskinetic events will validate our proposed estimation methodology. For this, we perform an Atemporal Event Classification over the **Raw Data** event payloads referring to both estimated levels of active components in blood and numerical values detected from motor sensors: if the trained model will only consider the blood estimated levels as features with good accuracy, we can satisfactorily progress on further validating our MTSC algorithm also using these predictions.

We extend [1] by not restricting our analysis to levodopa but by considering all the active components indicated in the drug leaflets. Thus, we can describe each drug D as a set of active components p defined as triplets $\langle \ell_p, h_p, \alpha_p \rangle$, where ℓ_p is the name of the active component, h_p is its associated half-life, and α_p refers to the quantity of active ingredient per unit of drug taken (e.g., capsule, tablet, mg, or ml). By knowing the times of the patient taking the drug $t_1^D, \ldots, t_\tau^D$ in a given dosage $q_1^D, \ldots, q_\tau^D$ we will obtain the following time series associated with the active ingredient of a single drug D: $p^D(t) = \sum_{t_i^D \geq t} \alpha_p q_i^D \cdot 2^{-\frac{t_i^D - t}{h_p}}$. Since the same active ingredient can be present in multiple drugs, we obtain the following time series defining the fluctuation of the active ingredient in the blood flow: $p(t) = \Sigma_{p \in D} p^D(t)$. Thus, we transform all drug intake events into as many time series as the distinct active ingredients across all the drugs the patient assumes.

The effectiveness of this approximation of [1] is as strong as the accuracy of predicting dyskinetic events from the expected blood levels of the active substances. Given the plausibility of the overall temporal artificial intelligence pipeline, this will also motivate the results from the subsequent section.

Results. We collect all the **Raw Data** payloads per instant of time independently from the patient of interest $\mathfrak{E}$, while retaining the class information for training an explainable model. To extract this from our data, we split the payload data set into training (70%) and testing (30%) datasets, and fit a default decision

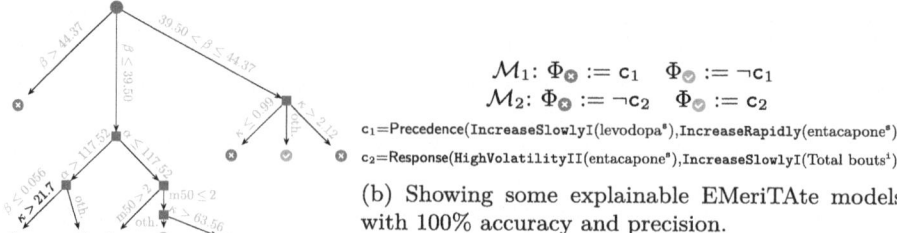

$\mathcal{M}_1$: $\Phi_\otimes := c_1$ $\Phi_\oslash := \neg c_1$
$\mathcal{M}_2$: $\Phi_\otimes := \neg c_2$ $\Phi_\oslash := c_2$

c_1=Precedence(IncreaseSlowlyI(levodopa°),IncreaseRapidly(entacapone°))
c_2=Response(HighVolatilityII(entacapone°),IncreaseSlowlyI(Total bouts¹))

(b) Showing some explainable EMeriTAte models with 100% accuracy and precision.

(a) Atemporal event classification.

Fig. 3. Explainable models extracted from the analysis over EMeriTAte: ⊗ refers to recorded dyskinetic events, while ⊘ refers to periods of patient well-being.

tree as per `scikit-learn` on the former and test over accuracy on the latter. Figure 3 shows the trained model for classifying the well-being and dyskinetic events; among all the provided variables, the entropy metric elected as the best classification predictors active components levels (α: amantadine, β: beserazide, κ: carbidopa) as well as the number of tablets from a redacted drug name (m50) with an accuracy of 94.43%. This model shows a complex interaction between different drug levels across the patients while remarking a strong correlation between estimated active principle presence in blood and dyskinetic events. This reassures us to continue with the forthcoming steps using predicted blood levels instead of drug assumption events.

5.2 Temporal Characterisation of Dyskinetic Events

Last, we select some state-of-the-art algorithms for multivariate time series classification by leveraging the best classifier across different methodological approaches from the aforementioned survey [14] and described in Sect. 2.3. While performing trace segmentation, we consider $\epsilon = 10^{-4}$ as a sensitivity parameter.

As competing algorithms do not natively support MTS changing classification labels through time, we are forced to derive our competitors' MTS from the polyadic traces resulting from the trace segmentation phase - thus being traces associated to one single class - via the values stored within the Raw Data events' payload: each numerical key appearing in any event payload is associated with a unique dimension. Notwithstanding the former transformation, these MTS will have varying sizes, which is currently not supported by the `sktime` library implementation of such algorithms [14]: we then decompose each previously generated MTS T of size k into $k - m + 1$ projected MTS $T[j, \ldots, j + m - 1]$, where m is the minimum trace size in the former transformation task. As simplistic machine learning techniques such as decision trees work by generating models whose features maximally differentiate the classes of interest, we would expect that this discretization would not impact the aforementioned models.

We consider a training/testing set split of 70%/30% for all algorithms while stratifying the samples over the classes. These experiments are repeated 20 times across datasets over the same splits; we consider the default parameters for `sktime` or, when available, the ones suggested from its documentation. Concerning EMeriTAte, we perform the stratification selection after the traces segmentation phase (Line 4), thus retaining the creamed-off ones for the testing evaluation (Line 12 via Line 11). For EMeriTAte, we also considered preliminary experiments where variations to the support threshold $\theta \in \{0, .2, .4, .6, .8, .9, 1\}$; as these did not substantially influence the average accuracy and precision results which always retained a similar average, we will consider the average for $\theta = 0$.

Results. Results from Fig. 4 clearly remark that competing approaches have an overall accuracy and precision below 50%, with the best competitor being `TapNet`, despite scoring zero precision in one experiment. These results clearly show that we cannot trivially reduce a MTSC classification task to a simple clustering algorithm via Euclidean distance similarity across the traces, as this classification methodology leads to the worst classification results. Despite our survey

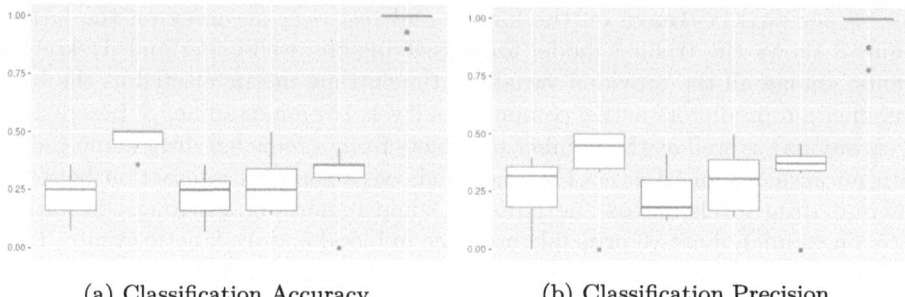

(a) Classification Accuracy. (b) Classification Precision.

Fig. 4. Box plots for 20 training experiments for MTSC over our dataset for (from left to right) Rocket, TapNet, EuclideanKNN, CIF, STC, and EMeriTAte.

of reference claims that Rocket was behaving among the best classifiers, these results remark that this classifier had the same low accuracy as a straightforward clustering approach, thus remarking that straightforward kernel approaches cannot capture the complex interactions occurring within our patient dataset. Low classification results from CIF suggest that, despite the authors also using a classification tree to discriminate across classes, the features deriving from simple average and variation statistics for classification are insufficient for discriminating across classes. Despite STC ameliorating the former by extracting shapelet information from the time series, this shows that such a way to describe the traces is insufficient to fully capture time series volatility as per our dataset, as on one occasion, it still scored zero precision and accuracy. As mentioned in the Introduction, TapNet was the best competitor, as also reported in the former survey; although the attention networks might have helped discriminate interesting temporal behaviour, this is still insufficient at ensuring high precision and accuracy in classification tasks with a small patient cohort. Given these results, our proposed approach outperformed the formers in accuracy and precision tasks: this shows no clear silver bullet at this stage can solve all the MTSC tasks. We can argue that our approach outperformed competing approaches by extracting generic data trends by exploiting DT instead of shapelets while temporally correlating across occurring DT using Poly-DECLAREd. This provided a competing advantage by not only associating a descriptive and explainable temporal behaviour to each class but also by enabling behavioural forecasting abilities across differently mined DT: as showed in Fig. 3b, these models were able to correlate entacapone, used to manage dyskinetic events of PD patients, to active principles concentrations ($\mathcal{M}_1$) and patient mobility ($\mathcal{M}_2$).

6 Conclusions

We generalise MTSC considering classification values changing in time as well as considering MTS with different sizes; the need for such a scenario is contextualised for monitoring data for PD patients being the environment $\mathfrak{E}$ of interest, for which we capture the insurgence of dyskinetic events and characterise those from well-being periods through classification labels. EMeriTAte is a new MTSC algorithm classifying MTS by correlating each class label to an explainable declarative model $\mathcal{M}$, determining the temporal correlations across DT through different MTS dimensions. This notion required extending DECLAREd's semantics to support concurrent events (Poly-DECLAREd). Our results show that our methodology outperforms current MTSC approaches over our medical scenario while providing a human-readable explanation. Notwithstanding the former, the current classification task does not consider and leverage specific Raw Data payload that might be relevant for other MTSC use cases [14]; we speculate that outperforming competing approaches over these competing scenarios will require us to extend the current specification mining algorithm further to also generate *dataful* specifications having data conditions. Future works will also consider replacing the current predictive model using binary classification trees with multi-instance learning models, which provided consolidated evidence of being better suited for numerical-based healthcare applications [2] and determining the benefit of changing the FPSpan algorithm used within the specification mining phase with SPADE, as it is better suited for mining polyadic traces [16]. We will collect extra sensor data with speech signals to increase the model prediction accuracy [15] for a larger patient cohort.

Acknowledgments. This study was supported by the Medical Research Council (MRC) Newcastle University-Confidence in Concept (CiC) 2019 (MC_PC_19047, NU-002884). We thank Professor Alison Yarnall, Professor Lynn Rochester, and Dr Lisa Alcock for their valuable comments and involvement in the project. We acknowledge the study's participants for their time and enthusiastic contribution. We thank all assessors and collaborators of the study. Silvia Del Din (SDD) was supported by the Mobilise-D project funded by Innovative Medicines Initiative 2 Joint Undertaking (JU), grant agreement No. 820820. JU is supported from European Union's Horizon 2020 research and innovation program and the European Federation of Pharmaceutical Industries and Associations (EFPIA). SDD was also supported by the Innovative Medicines Initiative 2 Joint Undertaking (IMI2 JU) project IDEA-FAST - Grant Agreement 853981. SDD was also supported by the National Institute for Health Research (NIHR) Newcastle Biomedical Research Centre (BRC) based at The Newcastle upon Tyne Hospital NHS Foundation Trust, Newcastle University and the Cumbria, Northumberland and Tyne and Wear (CNTW) NHS Foundation Trust. SDD was also supported by the NIHR/Wellcome Trust Clinical Research Facility (CRF) infrastructure at Newcastle upon Tyne Hospitals NHS Foundation Trust. SDD was supported by the UK Research and Innovation (UKRI) Engineering and Physical Sciences Research Council (EPSRC) (Grant Ref: EP/W031590/1, Grant Ref: EP/X031012/1 and Grant Ref: EP/X036146/1). All opinions are those of the authors and not the funders. The content in this publication reflects the authors' view, and neither IMI nor the European Union, EFPIA, NHS, NIHR or any associated partners are responsible for any use that may be made of the information contained herein.

Disclosure of Interests. SDD reports consultancy activity with Hoffmann-La Roche Ltd. outside of this study.

References

1. Arav, Y., Zohar, A.: Model-based optimization of controlled release formulation of levodopa for Parkinson's disease. Sci. Rep. **13**(1), 15869 (2023)
2. Avolio, M., Fuduli, A., Vocaturo, E., Zumpano, E.: On detection of diabetic retinopathy via multiple instance learning. In: Proceedings of the International Database Engineered Applications Symposium (IDEAS), pp. 170–176 (2023)
3. Bergami, G.: DECLARE*d*: a polytime LTL$_f$ fragment. Logics **2**(2), 79–111 (2024)
4. Bergami, G., et al.: Exploring business process deviance with sequential and declarative patterns. arXiv:2111.12454 (2021)
5. Bergami, G., Appleby, S., Morgan, G.: Specification mining over temporal data. Computers **12**(9), 185 (2023)
6. Debelle, H., et al.: Feasibility and usability of a digital health technology system to monitor mobility and assess medication adherence in mild-to-moderate Parkinson's disease. Front. Neurol. **14**, 1111260 (2023)
7. Del Sorbo, F., Albanese, A.: Levodopa-induced dyskinesias and their management. J. Neurol. **255**(4), 32–41 (2008)
8. Dempster, A., Petitjean, F., Webb, G.I.: ROCKET: exceptionally fast and accurate time series classification using random convolutional kernels. Data Min. Knowl. Discov. **34**(5), 1454–1495 (2020)
9. Hills, J., et al.: Classification of time series by shapelet transformation. Data Min. Knowl. Discov. **28**(4), 851–881 (2014)
10. Huo, X., et al.: A dynamic soft sensor of industrial fuzzy time series with propositional linear temporal logic. Expert Syst. Appl. **201**, 117176 (2022)
11. Leung, C.K., et al.: Explainable data analytics for disease and healthcare and informatics. In: Proceedings of the 25th International Database Engineering & Applications Symposium (IDEAS), pp. 65–74 (2021)
12. Middlehurst, M., Large, J., Bagnall, A.: The canonical interval forest (CIF) classifier for time series classification. In: Proceedings of the IEEE International Conference on Big Data (Big Data), pp. 188–195 (2020)
13. Napolitano, E.V., Fioretto, S., Masciari, E., Anniciello, A.: How pandemic affected the adoption of e-health systems. In: Proceedings of the 27th International Database Engineered Applications Symposium (IDEAS), pp. 94–98 (2023)
14. Ruiz, A.P., et al.: The great multivariate time series classification bake off: a review and experimental evaluation of recent algorithmic advances. Data Min. Knowl. Discov. **35**(2), 401–449 (2021)
15. Yuan, L., Liu, Y., Feng, H.M.: Parkinson disease prediction using machine learning-based features from speech signal. Serv. Orient. Comput. Appl. **18**(1), 101–107 (2024)
16. Zaki, M.J.: SPADE: an efficient algorithm for mining frequent sequences. Mach. Learn. **42**(1/2), 31–60 (2001)
17. Zhang, X., Gao, Y., Lin, J., Lu, C.: TapNet: multivariate time series classification with attentional prototypical network. In: Proceedings of the 34th AAAI Conference on Artificial Intelligence (AAAI), the 32nd Innovative Applications of Artificial Intelligence Conference (IAAI), the 10th AAAI Symposium on Educational Advances in Artificial Intelligence (EAAI), pp. 6845–6852 (2020)

R2-LGBM: Sales Informative Prediction System in E-commerce Application Using Ensemble Classifier

Hemn Barzan Abdalla[1,2]([✉]) [ID]

[1] Department of Computer Science, Wenzhou-Kean University, Wenzhou, China
habdalla@kean.edu
[2] Department of Computer Science and Technology, Kean University, Union, NJ, USA

Abstract. E-commerce platform renders a convenient shopping experience for remote consumers who want to buy and sell the products at their doorstep. In an e-commerce platform, the sales forecasting for commodity analysis is observed by the historical data through a time series model that makes sales inventory from a qualitative point of view. Several researchers worked on the sales prediction of e-commerce retail trade, which reported the limitations associated with interpretability, lack of data availability, large-scale dependency, time complexity, etc. A ResNet-50 Regression-based Light Gradient Boosting Machine (R2-LGBM) model is proposed to address these limitations. The R2-LGBM model enhances the ability and performance by reducing the memory requirements and gradient vanishing issues. The method forecasts the common characteristics of retail commodities and improves the reliability of prediction for sustainable development. Furthermore, the experimental research achieved the error rate of the model for the Dairy goods sale dataset is 7.57 of MAE, 3.47 of RMSE, 12.04 of MSE, and for the Superstore sale dataset is 1.40 of MAE, 2.36 of RMSE, and 5.60 of MSE for TP 80 respectively.

Keywords: Sales prediction · E-commerce · Time-series analysis · Light Gradient Boosting Machine · ResNet-50

1 Introduction

E-commerce is an effective business operation mode in which customers are connected to various business activities through browser applications on the Internet platform [1, 2]. On this platform, the customer explores online shopping, including online transactions, without face-to-face contact with sellers. In recent years, e-commerce has played a significant role in the growth of providing quality service through internet platforms, which has gained more attention among people and created huge customer purchase behavior and provider feedback for purchasing items [3, 4, 24]. Based on these customers' interests, online platforms increased the conversion ratio of purchases [5] and improved the rate of loyalty among strategic sectors [6, 7]. Regarding payment, e-commerce is classified as non-payment e-commerce and payment commerce [8]. Non-payment e-commerce performs better product delivery without transacting any online payment, which additionally has the information about information inquiry, formation of contract text, online negotiation, and so on. On the other hand, the Payment mode for e-commerce applications contains all the information resources about the purchaser along with payment options [8]. These categorizations of e-commerce platforms result in a significant development of online retail sales and an increase in the share of e-commerce sales among total retail sales [9].

With the rapid development of e-commerce, sales prediction is observed through sales and growth strategies. Based on these strategy observations, several models are used to forecast the commodity sales prediction, such as logistic regression (LR), decision tree (DT), random forest (RF), Gradient boosting decision tree (GBDT), neural network (NN), and so on [10]. In the LR [11] classifier, only the high variable relation factor is considered among the predictors by avoiding the least product factor, which leads to inaccurate outcomes. On the other hand, the CNN [10] required large-scale labeled data to perform the training process, which consumes time but provides an effective solution during the sequential data analysis [7, 12]. In the RF [13] classifier, the logical functions of the model were detected by time series analysis along with highly categorized data. During this analysis, the model suffered from unbalanced data and huge variable interpretation, affecting the hyper-parameters and reducing classification accuracy. While applying DT in sales forecasting prediction, the data quality suffers from various factors such as inaccurate information, human error, inconsistent definitions, and so on. The model suffered from data imbalance issues in the GBDT classifier [14] due to its complexity and huge memory requirements [10]. Because of these limitations of the conventional methods, a model was proposed to avoid the above-mentioned limitations and acquire effective prediction results.

The research is mainly developed to forecast sales in e-commerce applications. To provide an effective prediction outcome, an R2-LGBM model is proposed, which reduces the computational complexity and enhances the extraction capability with increased robustness. The model evaluates various marketing strategies for appropriately analyzing retail resources improving business performance by impacting operational efficiency. Based on these analyses, the R2-LGBM model possessed effective sales prediction significantly.

- **Recursive Feature Elimination with Cross-validation (RFECV):** The RFECV method eliminates the redundant and weak attribute features and organizes the informative feature subset to perform further processes.
- **ResNet-50 Regression-based Light Gradient Boosting Machine (R2-LGBM):** To evaluate retail resources in an e-commerce platform, an R2-LGBM model that possesses various marketing strategies to improve business performance is proposed. The regression analysis uses statistical concepts to determine the correlation between dependent and independent variables. Based on these correlation results, a prediction is made. The model possessed deep analysis with the help of the ResNet-50 model, which extracts the essential features to represent spatial and channel information from the second layer. Thus, in the proposed R2-LGBM, the execution speed is increased, and a highly accurate outcome is attained for the prediction.

The following sections organize the remaining part of the experimental research. Section 2 briefly elaborates on the existing methods utilized in predicting sales under e-commerce platforms. Section 3 encompasses the developed methodology along with its detailed explanation. Section 4 intercepts the achieved results obtained from the developed model. Finally, Sect. 5 provides the research conclusion along with its future directions.

2 Literature Review

In this section, the sales prediction in E-commerce platform was evaluated by several conventional methods, which are briefly elaborated below.

[15] developed a model with the combination of both GRU and Light GBM. In the method, the GRU model effectively captured the timing features and attained the ability of the Light GBM model to solve multivariable problems. The model was accessed with various sales strategies to perform the prediction process, resulting in increased robustness and high efficiency. On the other hand, the network was affected by some features' intercept ability, which made the model parameter weak. Additionally, the model suffered from the limited availability of comprehensive dataestablished a Hidden Markov Model for predicting sales based on time series under the analysis of historical sales data. The model possessed improved reliability under qualitative prediction, but it evaluates only the possible influence of external factors. For large-scale dependency, the model does not generate finite outcomes, which leads to increased computational complexity and overfitting problems.

[10] introduced a CNN model for predicting the sales commodities. The developed CNN model extracts the essential features of the data with high availability under the utilization of a real-time e-commerce dataset. The volume of sales commodity was predicted with high accuracy along with attribute information, but the model required some manual intervention to perform the extraction process. These manual mechanisms reduced the ability and increased the computation cost for performing the sales prediction function.

[12] employed a machine learning (ML) technique to forecast sales prediction in e-platforms. The ML-based methods include GBM and RF, which inherit the classification and regression process to predict the weak product based on time series form. The RF classifier provides individual class prediction for large data dependencies, which result in increased memory and computational requirements. To address these complexity issues, additional attributes were developed in the model to generate appropriate results.

[9] utilized an integrated framework for forecasting the retail e-commerce trade, including both the statistical and neural network methods. This integrated framework method pursues improved accuracy and flexibility. On the other hand, the model could not make an appropriate decision, which would impact policymakers and retail sellers.

2.1 Challenges

The remarkable limitations obtained from the existing methods are given as follows,

- Considering the various external factors associated with correlation, economic development, and commodity sales volume are affected, which badly impacts the prediction performance [9].
- The method presented in [12] performed better for minimal data during the sales commodity forecasting, but for large-scale data, the model suffered due to high time consumption and required expertise-intensive tasks for performing the prediction process.
- In the ML-based technique, the time-series analysis cannot be done using the differential pattern identity, affecting the model's sales forecast [12]. The layer parameter in the network is weak in the model because of the feature interpretability, which also leads to complex computation and limited availability of data [15].

This research proposes a correlation-based model using hybridized Resnet and Light-GBM to solve the issues above. Let us take a deep insight into the proposed R2LGBM model for sales prediction in the online sales platform.

3   Sales Prediction in E-commerce Application with Ensemble Classifier Model

The research aims to develop the R2-LGBM model to predict retail trade sales in e-commerce applications. The historical data from the Dairy goods sale dataset [16] and the Superstore sale dataset [17], is fed as the input data for performing the prediction analysis. Then the input data is fed into the pre-processing stage that eliminates the redundant and unstructured data by performing the data normalization technique. The process which efficiently organizing data in the dataset so that the redundant data or the irrelevant data are eliminated by the data normalization technique. By using this data normalization technique the duplicate data or unstructured data is eliminated. The normalized data is subjected to the feature extraction phase, where the feature elimination method is carried out, and the remaining redundant features are eliminated by performing the RFECV method. The outcome of the RFECV method is obtained as an effective subset feature as the input to the R2LGBM model to predict the desired outcome, in which the density block and identity block reduces the complexity requirements and Gradient vanishing problem. The R2-LGBM model predicts the sales by considering the historical and sales features. LGBM is a distributed and incredibly effective technique. The basic block diagram of the R2-LGBM model is illustrated in Fig. 1.

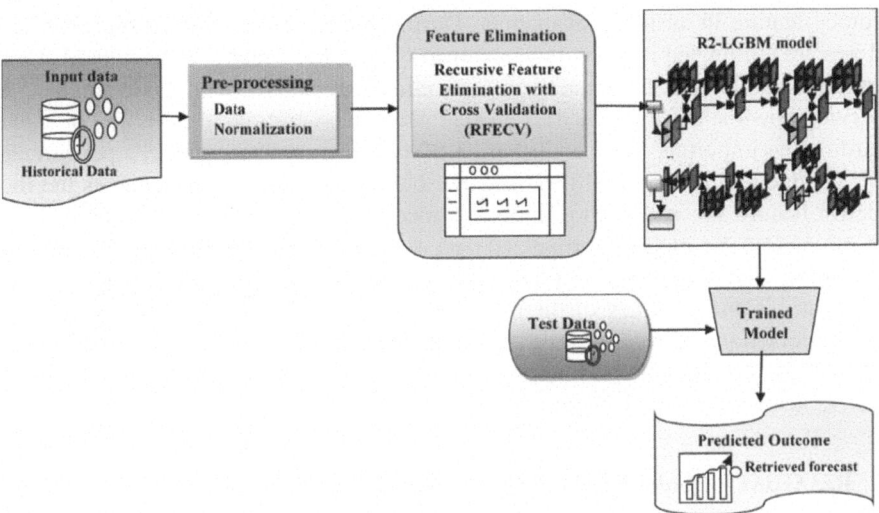

Fig. 1. Block diagram for predicting the sales of retail

3.1 Input

The input for this experiment is obtained from the Dairy Good sales dataset [16] and the Superstore Sales dataset [17]. These datasets provide vast amounts of product information, including manufacturing information, sales channel information, stock quantities, and reorder quantities. Based on this information, the obtained data input is mathematically represented as:

$$P = \{e_1, e_2, \ldots\ldots e_a, \ldots\ldots e_s\} \tag{1}$$

where P indicates the input database, and $\{e_1, e_2, \ldots, e_a, \ldots e_S\}$ defines the data attributes.

3.2 Pre-processing

The obtained raw input data is generally degraded by removed irrelevant data, and the quality is enhanced by performing the data normalization. By using the data normalization technique, the redundant and unstructured data are removed, and the data source is presented in an organized manner to perform prediction. The pre-processed data is referred to as R.

3.3 Feature Elimination with RFECV Method

The pre-processed data R is then allowed in the feature elimination phase, during which the redundant features are eliminated by the RFECV method [15]. The RFECV method combines both the recursive feature elimination and cross-validation function to ensure robustness and detect the optimal features that provide maximum model performance [18]. The RFECV classifies every feature score and eliminates the low classification accuracy feature in an iteration manner. During the recursive elimination process, as features are eliminated in a recursive manner, the accuracy metric is calculated at each iteration for evaluating the impact of feature elimination on the model performance. By observing how the accuracy metrics change with each iteration, insights can be gained regarding the importance and contribution of each feature to the model's performance [18]. The method removes all the weak attributes and features and achieves the most effective feature subset, which is allowed under the cross-validation (CV) process. For performing CV, the attained normalized data is validated by two phases such as training and testing. Under the training phase, the unseen data estimation is carried out, and the robustness and accuracy of the model are evaluated iteratively. The CV process's standard deviation and average accuracy metric measure the model's performance. Based on these RFE execution and CV tuning, high-accuracy features are selected [19] and the effective feature outcome is mentioned E.

3.4   R2LGBM for Sales Prediction in E-commerce Platform

The achieved effective feature subset E, forms the input for the R2-LGBM model that predicts the sales in e-commerce applications. The R2-LGBM model is trained in an

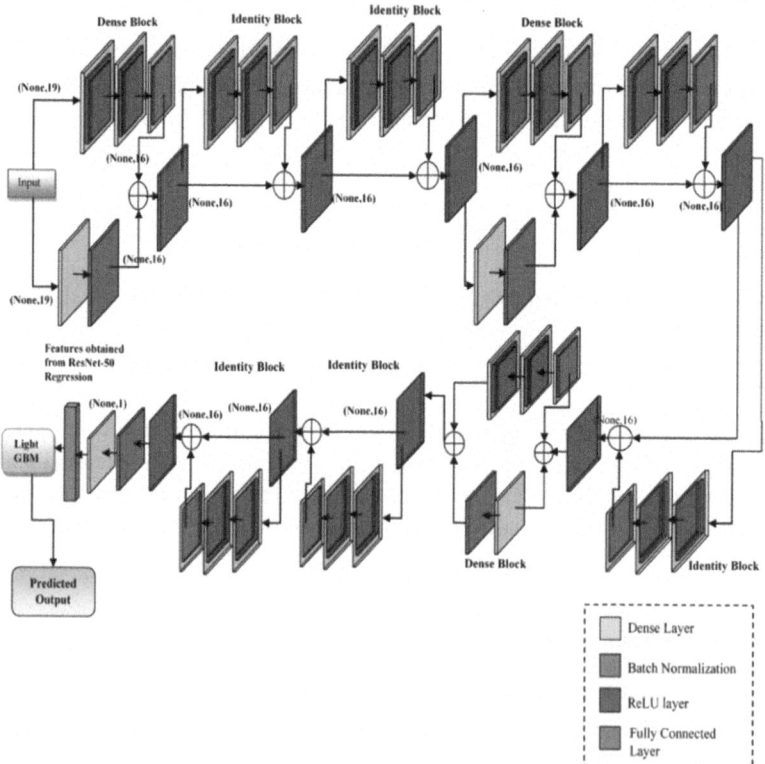

Fig. 2. Architecture of proposed R2-LGBM model

iterative manner in which the input for the model is obtained from the feature elimination RFECV method. The input dimension is *(None, 19)*, which is parallelly subjected to both the dense layer in the dense block. Along with the dense layer, the dense block contains four batch normalization layers with two activation (ReLU) layers. The outcome dimension of the dense block is *(None, 16)* concatenated with the ReLU layer and allowed into the identity block, which contains three dense and batch normalization layers along with two activation layers. Then, the outcome of the identity block is fed into the dense block iteratively. Following the identity-dense blocks, the prediction layer uses the channel and spatial representation acquired through the identity-dense blocks for making an effective sales prediction. To perform the prediction process, the R2-LGBM model concerns both the historical and sales features, which are desensitized with temporal features such as sales price and sales volume features. Thus R2-LGBM model predicts the commodity sales effectively and the basic architecture for R2-LGBM is illustrated in Fig. 2.

Dense Block
In the traditional residual network, convolutional and pooling layers are used, but in the proposed R2-LGBM model, both layers are replaced by dense layers, which perform the execution function [20]. In the dense block, the dense layer, batch normalization layer, and activation layer are organized, where the dense layer reduces the spatial dimension of the feature input, the batch normalization layer eliminates the need for a dropout layer and acts as a regularizer for the network, and the activation layer is used to reduce the heavy computation [21] and addresses the local minima issue. The input dimensions are subjected into the dense block *(None, 19)*, which is extracted and generated *(None, 16)*, as the outcome dimensions.

Identity Block
In the R2-LGBM model, the identity block contains three dense layers, three batch normalization layers, and two activation layers. In the identity block, the input activation dimension is the same as the output activation dimensions, reducing the vanishing gradient problem and improving the model performance.

Prediction Module
The prediction module is a distributed effective method that comprises Exclusive Feature Bundling (EFB), and gradient-based one-side sampling (GOSS) techniques [20]. While evaluating large-scale data, small gradient back-propagation is eliminated by the GOSS technique, and the overfitting problem is reduced by the EFB technique. The predictive module increased the execution speed and provided highly accurate predictive outcomes.

4 Results

The section encompasses the achieved outcome for sales prediction in e-commerce applications based on the R2-LGBM model, which explores the comparative evaluation with prior methods.

4.1 Experimental Setup

The experimental research is implemented in Pycharm software in the configuration of Windows 11 with the available memory of 16 GB RAM and 128GB ROM.

4.2 Dataset Description

The research predicts the sales forecasting in e-commerce applications based on the input database such as the Dairy goods sale dataset [16], and Superstore Sales dataset [17], which are briefly mentioned as follows.

Dairy Goods Sale Dataset [16] : The dataset provides an elaborate form of retail trade of dairy products, inventory management, and dairy farms. Additionally, the data encompasses product information such as farm area, location, and size, product details about date, storage conditions, sale information, sales channels, customer locations, Expiration dates, stock quantities, and reorder quantities. This dataset includes data from the period between 2019 and 2022, and it specifically focuses on the selected dairy brands operating in specific states and union territories of India.

Superstore Sales Dataset [17] : The dataset offers global superstore retail information details based on time-series form to extract the patterns. The dataset is used to assess various non-stationary applications such as weather data, stock pricing, economic data, and retail sales forecasting. The dataset is self-explanatory. It performs EDA and predicts the sales of the next 7 days from the last date of the training dataset.

4.3 Comparative Metrics

The comparative assessment of the R2-LGBM model for sales prediction with historical data input is obtained from the Dairy Goods sale dataset and Superstore Sales dataset that evaluates various evaluation metrics such as mean absolute error (MAE), Root mean square error (RMSE) and Mean square error (MSE). The error evaluation is measured by the following formula, which is included in Table 1.

Table 1. Expression of Comparative Metrics

Metrics	Formula
MAE	$\frac{1}{m} \sum\limits_{v=1}^{m} \lvert z - z^* \rvert$
RMSE	$\sqrt{\frac{1}{m} \sum\limits_{v=1}^{m} (z - z^*)^2}$
MSE	$\frac{1}{m} \sum\limits_{v=1}^{m} (z - z^*)^2$

4.4   Comparative Evaluation

The performance evaluation of the R2-LGBM model is compared with several conventional methods, such as support vector regression (SVR) [22], LGBM [23], GRU [12], and Autoregressive Integrated Moving Average (ARIMA) [22] techniques.

Comparative Assessment for Dairy Goods Sale Dataset
The R2-LGBM model is assessed using Mean Absolute error (MAE), Mean Square Error (MSE), and Root Mean Square Error (RMSE) using the Dairy Good sale dataset. At the maximum training percentage (TP) of 80%, the attained MAE error for the SVR, LGBM, GRU, and ARIMA is 23.81, 22.72, 20.64, and 16.27, while for the proposed R2LGBM, the MAE is 7.56. For RMSE, the achieved error value of R2-LGBM is 3.47, while for the existing SVR, LGBM, GRU, and ARIMA models; the RMSE is 5.30, 4.68, 4.57, and 3.97. The Proposed R2-LGBM achieves very low RMSE when compared to the existing methods. This occurs due to the addition of both Resnet 50 and the Regression. The regression and the resnet 50 is works well in the loss function. The resent 50 helps in reduce the loss by improving the training process, which also enhance the performance. On the other hand, the MSE for the proposed R2-LGBM is 12.07, while for existing SVR, LGBM, GRU, and ARIMA are 28.12, 21.99, 20.95, and 15.74. The assessment of the proposed R2-LGBM model for TP analysis is illustrated in Fig. 3 and Table 2 shows the results of K-fold analysis.

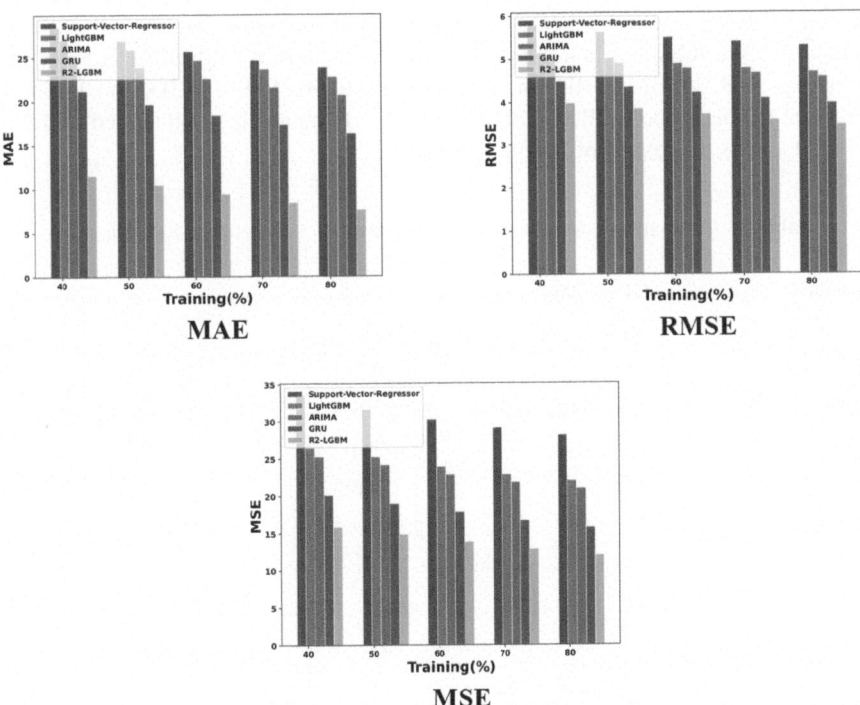

Fig. 3. Comparative Assessment of TP for Dairy Goods Sale Dataset

Table 2. Comparative Assessment of KF Analysis for Dairy Goods Sale Dataset

Methods	KF = 10		
	MAE	RMSE	MSE
SVR	24.13708	5.340398	28.51985
LGBM	14.44584	4.634112	21.47499
ARIMA	11.61324	4.52409	20.46739
GRU	7.59	3.871475	14.98832
R2-LGBM	5.714943	3.321747	11.034

Comparative Assessment for Superstore Sale Dataset
The R2-LGBM model is assessed using Mean Absolute error (MAE), Mean Square Error (MSE), and Root Mean Square Error (RMSE) using the Superstore Goods sale dataset. At the maximum training percentage (TP) of 80%, the attained MAE error for the SVR, LGBM, GRU, and ARIMA is 23.00, 13.53, 9.56, and 6.65, while for the proposed R2LGBM, the MAE is 1.40. For RMSE, the achieved error value of R2-LGBM is 2.37, while for the existing SVR, LGBM, GRU, and ARIMA models; the RMSE is 5.20, 3.99, 3.74, and 3.24. A very low RMSE is achieved for the proposed R2-LGBM when compared with the other existing methods. This improvement attains due to the inclusion or Regression and Resent 50. The regression and resnet 50 acts well in the error loss function. On the other hand, the MSE for the proposed R2-LGBM is 5.60, while for existing SVR, LGBM, GRU, and ARIMA are 27.06, 15.99, 14.03, and 10.54. The assessment of the proposed R2-LGBM model for TP analysis is illustrated in Table 3 and Fig. 4 shows the results of K-fold analysis.

Table 3. Comparative Assessment of KF analysis for Superstore Sale Dataset

Methods	KF = 10		
	MAE	RMSE	MSE
SVR	24.13708	5.321822	28.32179
LGBM	14.44584	4.059297	16.47789
ARIMA	11.61324	3.636307	13.22273
GRU	7.59	3.425493	11.734
R2-LGBM	5.714943	2.822863	7.968553

4.5 Comparative Discussion

The performance evaluation of the R2-LGBM model is compared with several conventional methods, such as SVR, LGBM, GRU, and ARIMA techniques. The kernel function parameter in the SVR model does not determine accurate results, which affects the model's performance and causes it to suffer from a loss function. Additionally, the method cannot utilize the optimal parameters to determine the model accuracy [23]. Several challenges were obtained in the light GBM model, such as overfitting problems, data interpretability, and difficulty identifying optimal parameters [22]. In the GRU technique, the method does not utilize large-scale dependencies that require high computational and memory requirements to perform specific scenarios [15]. While performing the prediction task in an ARIMA technique does not consider the external factors that may influence the model's capability for generating appropriate results [22]. The R2-LGBM model is developed to overcome these issues and predicts the desired outcome by evaluating sales retail trade in E-commerce applications. The comparative assessment for the R2-LGBM model with the respective conventional method is represented in Table 4 and Table 5 for the corresponding dataset.

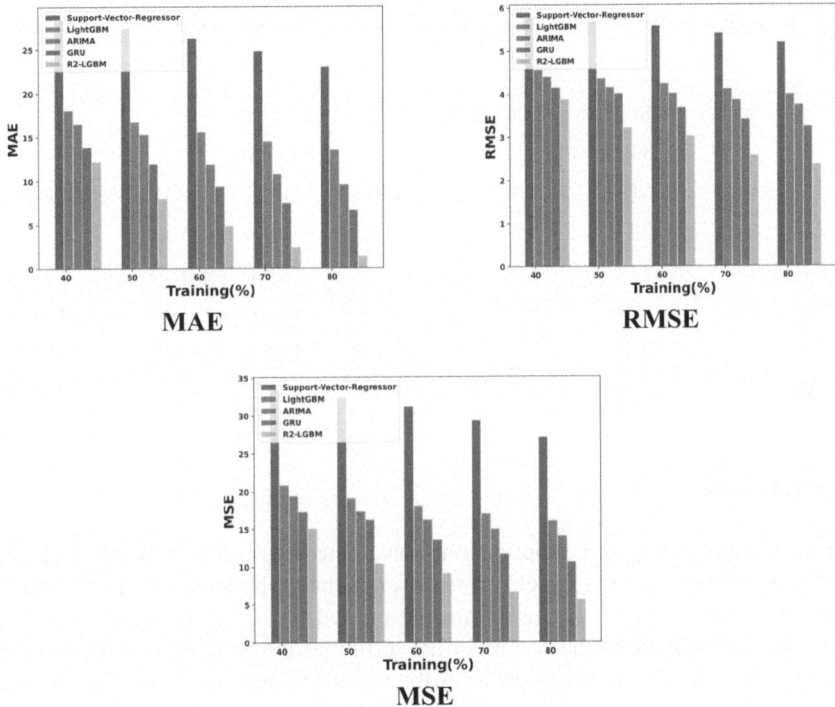

Fig. 4. Comparative Assessment of TP for Superstore Sale Dataset

Table 4. Comparative Assessment for R2-LGBM model with other prior methods (Dairy Goods Sale dataset)

Analysis/Methods	Dairy Goods sale dataset					
	TP (80%)			KF (10)		
	MAE	RMSE	MSE	MAE	RMSE	MSE
SVR	23.92	5.30	28.12	23.40	5.30	28.12
Light GBM	22.81	4.71	22.16	21.90	4.63	21.47
ARIMA	20.64	4.58	20.99	20.59	4.50	20.27
GRU	16.48	3.99	15.95	15.35	3.82	14.59
R2-LGBM	7.57	3.47	12.04	6.69	3.32	11.03

Table 5. Comparative Assessment for R2-LGBM model with other prior methods (Superstore Sale dataset)

Analysis/Methods	Superstore sale dataset					
	TP (80%)			KF (10)		
	MAE	RMSE	MSE	MAE	RMSE	MSE
SVR	22.80	5.20	27.06	24.13	5.37	28.92
Light GBM	13.53	3.99	15.99	14.44	4.06	16.47
ARIMA	9.56	3.74	14.03	11.61	3.66	13.42
GRU	6.45	3.27	10.73	7.59	3.42	11.73
R2-LGBM	1.40	2.36	5.60	6.11	2.93	8.56

5   Conclusion

Sales prediction in e-commerce applications is evaluated by the R2-LGBM model, which extracts the features from the structured time series data. The input data is organized in a structured form by utilizing data normalization and RFECV techniques. The model predicts the sales volume of the commodities, attributes information, and the total sales volume of commodities. Based on these attributes information, the model predicts retail sales effectively and produces the desired output. In the model, the computational complexity, Gradient vanishing problem, and overfitting issues are reduced, improving the model's performance and ability. By these effective performances, the R2-LGBM model predicts the sales goods automatically from the organized data and achieves effective error evaluation metrics such as MAE, RMSE, and MSE. The error rate values achieved for these evaluation metrics under TP (80%) for the dairy goods sale dataset are 7.57, 3.47, and 12.04. For the Superstore dataset, the values obtained are 1.40, 2.36, and 5.60. However, it faces challenges that only large dataset can be used in this model. In the future, more advancement will be made based on the predictive deep and hybrid models.

Acknowledgments. The authors gratefully acknowledge the financial support from Wenzhou-Kean University.

Funding. This work was supported by Leading Talents of Provincial Colleges and Universities, Zhejiang-China (Grant No: KY20220214000024) and General Program - Education Department of Zhejiang Province (Grant No. Y202045131).

Disclosure of Interests. The author declare that they have no competing interests.

References

1. Lin, J., Luo, Z., Cheng, X., Li, L.: Understanding the interplay of social commerce affordances and swift guanxi: an empirical study. Inf. Manage. **56**(2), 213–224 (2019)
2. Abdalla, H.B., Zhen, L., Yuantu, Z.: A new approach of e-commerce web design for accessibility based on game accessibility in Chinese market. Int. J. Adv. Comput. Sci. Appl. **12**(8) (2021)
3. Abdalla, H.B., Wang, L.: Evaluating Chinese potential e-commerce websites based on analytic hierarchy process. Int. J. Adv. Comput. Sci. Appl. **12**(9) (2021)
4. Zhang, Z.: Sales prediction based on ARIMA time series and multifactorial linear model. Highlights Sci. Eng. Technol. **38**, 1–8 (2023)
5. Hernandez, S., Alvarez, P., Fabra, J., Ezpeleta, J.: Analysis of users' behavior in structured e-commerce websites. IEEE Access **5**, 11941–11958 (2017)
6. Anadiotis, G.: Alibaba: building a retail ecosystem on data science, machine learning, and cloud (2017). https://www.zdnet.com/arti-cle/alibaba-building-a-retailecosystem-on-data-science-artificial-intelligence-and-cloud/
7. Zhao, K., Wang, C.: Sales forecast in e-commerce using convolutional neural network. arXiv: 1708.07946 (2017)
8. Xu, S.X., Huang, G.Q.: Efficient multi-attribute multi-unit auctions for B2B e-commerce logistics. Prod. Oper. Manag. **26**(2), 292–304 (2017)
9. Tudor, C.: Integrated framework to assess the extent of the pandemic impact on the size and structure of the e-commerce retail sales sector and forecast retail trade e-commerce. Electronics **11**(19), 3194 (2022)
10. Pan, H., Zhou, H.: Study on convolutional neural network and its application in data mining and sales forecasting for E-commerce. Electron. Commer. Res. **20**(2), 297–320 (2020)
11. Yao, Y., Wang, H.: Optimal subsampling for Softmax regression. Stat. Pap. **60**(2), 585–599 (2019)
12. Singh, K., Booma, P.M., Eaganathan, U.: E-commerce system for sale prediction using machine learning technique. In: Proceedings of the International Conference on Computational Physics in Emerging Technologies (ICCPET), Mangalore, India, p. 012042 (2020)
13. Wager, S., Athey,: Estimation and inference of heterogeneous treatment effects using random forests. J. Am. Stat. Assoc. **113**(523) (2018)
14. Rao, H., et al.: Feature selection based on artificial bee colony and gradient boosting decision tree. Appl. Soft Comput. **74**, 634–642 (2019)
15. Chen, Y., et al.: Development of a time series e-commerce sales prediction method for short-shelf-life products using GRU-LightGBM. Appl. Sci. **14**(2), 866 (2024)
16. Dairy Goods Sales Dataset. https://www.kaggle.com/datasets/suraj520/dairy-goods-sales-dataset

17. Superstore sales Dataset. https://www.kaggle.com/datasets/rohitsahoo/sales-forecasting
18. Awad, M., Fraihat, S.: Recursive feature elimination with cross-validation with decision tree: feature selection method for machine learning-based intrusion detection systems. J. Sens. Actuat. Netw. **12**(5), 67 (2023)
19. Akkaya, B.: The effect of recursive feature elimination with cross-validation method on classification performance with different sizes of datasets. In: Proceedings of the 4th International Conference on Data Science & Applications, Istanbul, Turkey (2021)
20. Wang, G.: Sales forecast of retail commodity on the basis of LightGBM and Xgboost. In: Proceedings of the International Conference on Financial Innovation, FinTech & Information Technology (FFIT), Shenzhen, China (2022)
21. Chen, D., Hu, F., Nian, G., Yang, T.: Deep residual learning for nonlinear regression. Entropy **22**(2), 193 (2020)
22. Tao, W., Wu, C., Wu, T., Chen, F.: Research on the optimization of pricing and the replenishment decision-making problem based on LightGBM and dynamic programming. Axioms **13**(4), 257 (2024)
23. Sharma, G.K., Patil, S.: Big data analysis for revenue and sales prediction using support vector regression with auto-regressive integrated moving average. SAMRIDDHI: J. Phys. Sci. Eng. Technol. (Part-1) **15**(01), 1–8 (2023)
24. Abdalla, H.B., Gheisari, M., Awlla, A.H.: Hybrid self-attention BiLSTM and incentive learning-based collaborative filtering for e-commerce recommendation systems. Electron. Commer. Res. (2024)

Text and Languages

Comparing Related Languages with a Fuzzy Morphism Matching Algorithm

Daniel Schaefer[ID] and Peter Z. Revesz[(✉)][ID]

University of Nebraska – Lincoln, Lincoln, NE 68588, USA
Dschaefer2@huskers.unl.edu, peter.revesz@unl.edu

Abstract. This paper proposes a fuzzy morphism matching algorithm for discovering similarities within related languages. The fuzzy morphism matching algorithm takes as input a novel representation of the linguistic structures of the two languages that are compared. This representation is a type of Markov model that is built from an abstract representation of the basic set of words in the languages where the abstraction is based on combinations of six phoneme categories and three positions of those phonemes within the basic sets of words. The limited number of nodes in these Markov models allows efficient calculations of partial subgraph isomorphism matchings between them, and the degree of matching leads to a natural similarity measure that depends not on the number of cognate words but only on the phonetic structure of the languages, which have greater stability. This allows the detection of a strong similarity between closely related languages such as English and German as well as a weaker similarity between more distantly related languages like English and Hungarian.

Keywords: Fuzzy · Glottochronology · Markov model · Isomorphism

1 Introduction

The identification of the degree of relationship among distantly related languages is a challenging problem in linguistics because there is no objective and reliable metric to measure the relatedness among distantly related languages, even languages that are traditionally considered to belong to distinct language families. Within closely related languages, one may count the number of cognate word pairs, but distantly related languages may contain few cognate word pairs. The phonetic structure of languages is considered more resilient to change than the set of individual words. Hence, a robust and reliable metric of language relatedness would need to consider the phonetic structure of the languages.

We propose a fuzzy morphism matching algorithm for discovering similarities among even distantly related languages based on the graphical representation their phonetic structure. We represent the set of basic words in languages using a special type of Markov model.

Graph-similarity metrics such as the graph edit distance may suggest that two languages are related but will not necessarily suggest avenues for translation. Conversely,

© The Author(s) 2025
R. Chbeir et al. (Eds.): IDEAS 2024, LNCS 15511, pp. 79–91, 2025.
https://doi.org/10.1007/978-3-031-83472-1_6

if an algorithm identifies certain isomorphisms within the Markov graph structures of two languages, those isomorphisms may suggest common structural elements between the languages. However, one limitation of isomorphism matching is that it must be exact. That is, one subgraph must be a perfect isomorphism of another subgraph. An important feature of languages is that there are many exceptions to most linguistic rules, which make exact isomorphisms unlikely. This paper introduces the fuzzy-morphism: an inexact morphism between two graphs based on a node-distance metric.

The rest of this paper is organized as follows. Section 2 reviews some basic concepts about the subgraph isomorphism problem. Section 3 describes a Markov model representation of the phonetic structure of languages. Section 4 describes the new fuzzy subgraph morphism algorithm that is designed to be applicable to these graphs. Section 5 presents the experimental results. Section 6 provides a discussion of the results. Finally, Sect. 7 presents some conclusions and directions for further research.

2 Review of the Subgraph Isomorphism Problem

The subgraph isomorphism problem (Cook, 1971) asks to find a matching M between the nodes of two graphs G1 and G2 such that the matching forms an isomorphism. If G1 has n nodes and G2 has m nodes, then a matching M is an $n \times m$ matrix where M[i, j] = 1 represents a correspondence between node i in G1 and node j in G2 and M[i, j] = 0 represents no correspondence between those nodes. Since node i in G1 can correspond with only one node in G2, there can be at most a single 1 in any given row or column of M.

The subgraph isomorphism problem is NP-complete (Ullman, 1976). Hence it is tempting to find a solution using a brute force method that generates all possible matchings and tests each one-by-one to determine if it represents an isomorphism. To generate all possible matchings, it is necessary to compute all possible permutations of 1s in the matrix M. The runtime of the brute-force algorithm grows exponentially with respect to the sizes of n and m.

Cordella et al. (1998) proposed a search algorithm that represents the possible state-space of M as a tree. This algorithm prunes the tree by eliminating possible substates of M that cannot lead to a solution. The algorithm evaluates whether a given substate of M forms an isomorphism between G1 and G2 before adding a possible new node pair to M.

This paper uses an alternative algorithm first proposed by Ullmann (1976). Ullmann's algorithm uses a set of plausibility rules to identify node pairs in M that are most likely to form an isomorphism. For example, Ullmann suggests that if the degree of a node in G1 is greater than the degree of a node in G2, than the two nodes cannot be isomorphic. Once the algorithm identifies all likely pairs, the algorithm generates their permutations, and each state is tested one-by-one.

3   A Markov Model for Representing the Phonetic Structure of Languages

Swadesh lists (are word lists in various languages that represent 210 of the most basic concepts that are thought to be culturally universal (Swadesh, 1955). Hence, we can talk about an English Swadesh list, a German Swadesh list, and so on. Swadesh lists are frequently used to study the glottochronology of the languages because the Swadesh list words tend to be relatively older in origin. For example, English and German are closely related with many cognate pairs of words in their Swadesh lists (Russ 1994). This matches our intuition because English and German are both Indo-European languages. We hypothesized that besides the cognate words, the phonetic structures of these languages are also similar, and would be different from that of Hungarian, which is a Uralic language (Bakró-Nagy et al., 2022). Hence, we primarily aimed to represent the phonetic structure of these three languages, but our approach can be extended to represent the phonetic structure of other languages too. Our approach is similar to that of Rao et al., (2009), which used a Markov model to analyze the Indus Valley Script. However, Rao et al. (2009) based their Markov model on but the sequences of signs in the various Indus Valley Script inscriptions instead of sequences of phonemes.

After collecting the *International Phonetic Alphabet* (IPA) representation of each English, German, and Hungarian word in the Swadesh list, several diacritic marks that indicate certain features of a word, like the emphasized syllable, whether a vowel is long, or other connectives between sounds were removed to simplify the Markov state space. These eliminations help to simplify the underlying problem by reducing the number of nodes in the resulting graph while still maintaining many important features of the structure of the word. After these simplifications, we were left with the set of IPA phonemes that appear in the second column of Table 1.

Since this set of IPA phonemes still seemed too large, we grouped together related sounds into six phonetic groups as shown in Table 1.

Table 1. The phonetic groups created during preprocessing of the input words.

Group	IPA phonemes
Fricative	'ɦ', 'θ', 'ʁ', 'β', 'χ', 'v', 'ʋ', 'f', 'h', 'ð', 'ɣ', 'x', 'ʍ', 'ç', 'ʒ', 'ʃ', 's', 'z', 'ʧ', 'j'
Plosives	'ʔ', 'ɟ', 'g', 'k', 'p', 'b', 'ɖ', 'd', 't', 'ʔ'
Liquids	'l', 'ɫ', 'ɹ', 'ʀ', 'r', 'ɾ'
Nasals	'ɳ', 'm', 'n', 'ɱ', 'ŋ', 'ɲ'
Semivowels	'j', 'w'
Vowels	'ə', 'ɐ', 'y', 'ɔ', 'ɤ', 'ɒ', 'i', 'ø', 'ɨ', 'ɛ', 'œ', 'u', 'ɝ', 'ɪ', 'a', 'ɚ', 'ɑ', 'æ', 'e', 'ɜ', 'ɛ', 'o', 'ʊ', 'ɛ', 'ʌ'

To represent a word, we append a letter to each phonetic group to indicate its position within the words of the language. An 'I' is appended to phonetic groups in the initial position, 'M' to phonetic groups in the medial position, and 'F' to phonetic groups in the final position. For example, Table 2 shows a set of English words and their simplified representations.

Table 2. A Subset of English Words from the Swadesh List.

Word	IPA	Remove Diacritic Marks	Simplified Representation
who	[ˈhuː]	hu	Fri(I) → Vow(F)
what	[ˈwɒt]	wɒt	Sem(I) → Vow(M) → Plo(F)
where	[ˈwɛə]	wɛə	Sem(I) → Vow(M) → Vow(F)
when	[ˈwɛn]	wɛn	Sem(I) → Vow(M) → Nas(F)
how	[ˈhaʊ]	haʊ	Fri(I) → Vow(M) → Sem(F)

To demonstrate how to construct the Markov graph, consider the subset of English words from the Swadesh list and their IPA equivalences provided in Table 2. Given the word and the IPA equivalent, we remove emphasizing and connective symbols. Next, we group the symbols based on the categories presented in Table 1 and the position of the symbol in the word. Each weighted directed edge from one node to another node indicates a possible transition in a graph where the weights are the transition probabilities. The transition probability to any node from a given node is derived from the frequency of subsequent phonemes in the given set of words.

Figure 1 shows our Markov model for the words in Table 2. The word initial nodes are blue, the word medial nodes are red, and the word final nodes are green in Fig. 1 for better visibility of the phonetic structure that the Markov model represents. Note that there are four cases in Table 2 when a word medial vowel, which is represented by the

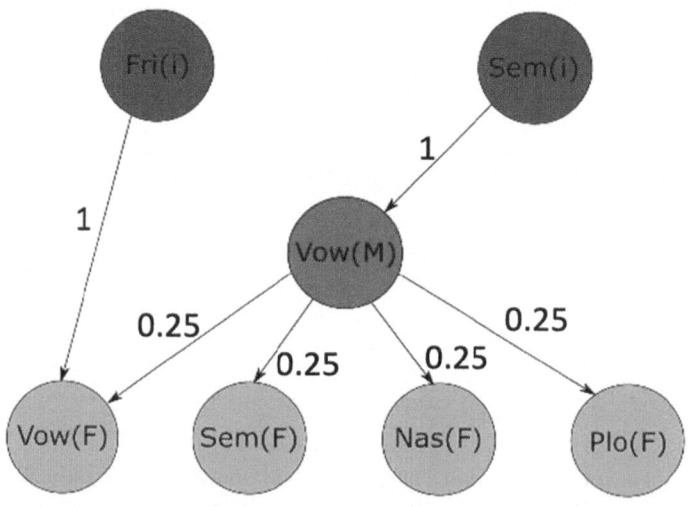

Fig. 1. The Phonetic Structure of Five English Words (Color figure online)

red node in the center of Fig. 1, is followed by a word final phoneme. Since we assume that the words have the same probability, the transition probability is 0.25 from the red node to any of the green nodes.

4   The Fuzzy Morphism Matching Algorithm

Table 3 provides the plausibility rules for the fuzzy morphism matching algorithm. Structural metrics like the average degree of predecessors or successors to the given node are especially effective at reducing the number of candidate node pairs, because they capture important information about a given node's position in the wider network.

Each of these measures is standardized: they are all either proportions or averages of probabilities of changing from one node to another node. Therefore, all measures contribute roughly equally to the total distance between each pair of nodes.

Ideally, distances between nodes representing sounds that match are very small, while distances between unmatched pairs are very large.

Table 3. List of plausibility rules used by the fuzzy morphism matching algorithm.

Plausibility Rule	Units
node in degree	proportion of possible edges
node out degree	proportion of possible edges
sound frequency	proportion of total sounds
average node in weights	average of edge weights
average node out weights	
average node predecessor in weights	
average node predecessor out weights	
average node successor in weights	
average node successor out weights	

Figure 2 shows a pseudocode of the brute force version of our fuzzy morphism algorithm. The algorithm uses several steps to search for possible fuzzy matches between the nodes of the input graphs.

Many distance metrics, defined in the *Distance* function in Step 1, are possible. We experimented with the Euclidian distance, the root mean square error, and the cosine dissimilarity measures. Eventually, we chose the cosine dissimilarity measure, although the distance measures gave similar results.

In Step 2, t is an arbitrary threshold such that nodes below the threshold are considered too distant to be plausibly the same node.

Step 4 returns the matching M_f that has the minimal total distance.

Step 1: /* Create a distance matrix D. */

```
D = CreateDistanceMatrix(G1, G2){
    D = new matrix(G1.rowcount, G2.rowcount)
    foreach N1 in G1{
        foreach N2 in G2{
            D[N1, N2] = Distance(N1, N2, G1, G2)
        }
    }
    return D
}
```

Step 2: /* Convert the distance matrix D to a transition matrix T. */

```
T = CreateTransitionMatrix(D, t) {
    for i in D {
        for j in D[i] {
            T[i,j] = 1 if D[i,j] < t else T[i,j] = 0
        }
    }
    return T
}
```

Step 3: /* From T, generate the set S of all possible matching M_1, M_2, M_3 ... M_N. */

```
S = CreateSetOfPossibleMatchings(T) {
    S = list()
    for row_index in T {
        PossibleRowsOfS = GeneratePossibleRows(T[row_index])
        S = GenerateAllCombinations(S, PossibleRowsOfS)
    }
    return S
}
```

Step 4: /* For a set of matchings S, return the matching S with the minimum total distance. */

```
Mf = GetMinimalMatching(S) {
    Mf = new matrix()
    minimum_distance = inf
    for Mi in S {
        distance = CalculateTotalDistance(M)
        if minimum_distance > distance {
            minimum = Mi
            minimum_distance = distance
        }
    }
    return Mf
}
```

Fig. 2. Pseudocode of the Fuzzy Morphism Matching Algorithm

5 Experimental Results

We considered three languages in our computer experiments: English, German and Hungarian. English and German belong to the Germanic branch of the Indo-European languages. They separated in the 5th century (Buccini and Moulton, 2024). Hungarian belongs to the Uralic language family (Bakró-Nagy et al., 2022). Some authors suppose that the Indo-European and the Uralic languages had as a common origin an ancient Eurasian language that may go back over ten thousand years (Pagel et al., 2013). This means that some of the distant similarities may be detectable by the similarity in their phonetic structures represented by our Markov models.

Figure 3 shows our Markov models of the phonetic structures of English, German and Hungarian Swadesh list words. We omitted the transition probabilities of the edges. The three graphs look superficially similar, but there are some major differences among them.

The first difference is in the set of phonemes that form the graph nodes. English and Hungarian have semivowels in the medial position, denoted as Sem(M), while German does not. In addition, Hungarian has semivowels in the final position, denoted Sem(F), while English and German do not.

The second difference is in the directed edges that connect the graph nodes. For example, German has more directed edges that end at node Nas(M), which denoted nasals in the medial position, than English and Hungarian have. German also has an unusually high number of directed edges that end at node Nas(F).

The third difference is that English and German have self-loops, that is, directed edges that start and end on the same node, for the node Vow(M), which denotes vowels in the medial position. These could be diphthongs, which Hungarian lacks. This is reflected in the Markov model in Fig. 3 (c) by a lack of such a self-loop.

Figure 4 shows the fuzzy morphism matchings that was found when the input were pairs of the English, German and Hungarian Markov models from Fig. 3. In the first matrix, the rows correspond to English phonemes and the columns correspond to German phonemes. An entry with a '1' value indicates that the fuzzy morphism matching algorithm found a match between the English and German phonemes in the corresponding row and column. These matches are arranged along the main diagonal of the topmost matrix. The other matrices are similar.

The fuzzy morphism matching algorithm found ten matches between English and German that are identical groups of phonemes in the identical positions as shown by the ten green entries in the first matrix in Fig. 4. For example, the nodes representing vowels in the initial position, denoted vowels(I), are matched in the upper left corner of the matrix.

In contrast, the fuzzy morphism matching algorithm found only seven matches between English and Hungarian and only six matches between German and Hungarian that are identical groups of phonemes as shown by the green entries in the second and third matrices in Fig. 4. The significantly lower number of green matches between Hungarian and the other two languages shows that Hungarian is more distantly related to English and German than the latter two are related to each other. Hence, this result gives some support to those linguists who suppose that these two language families have a common origin (Pagel et al., 2013).

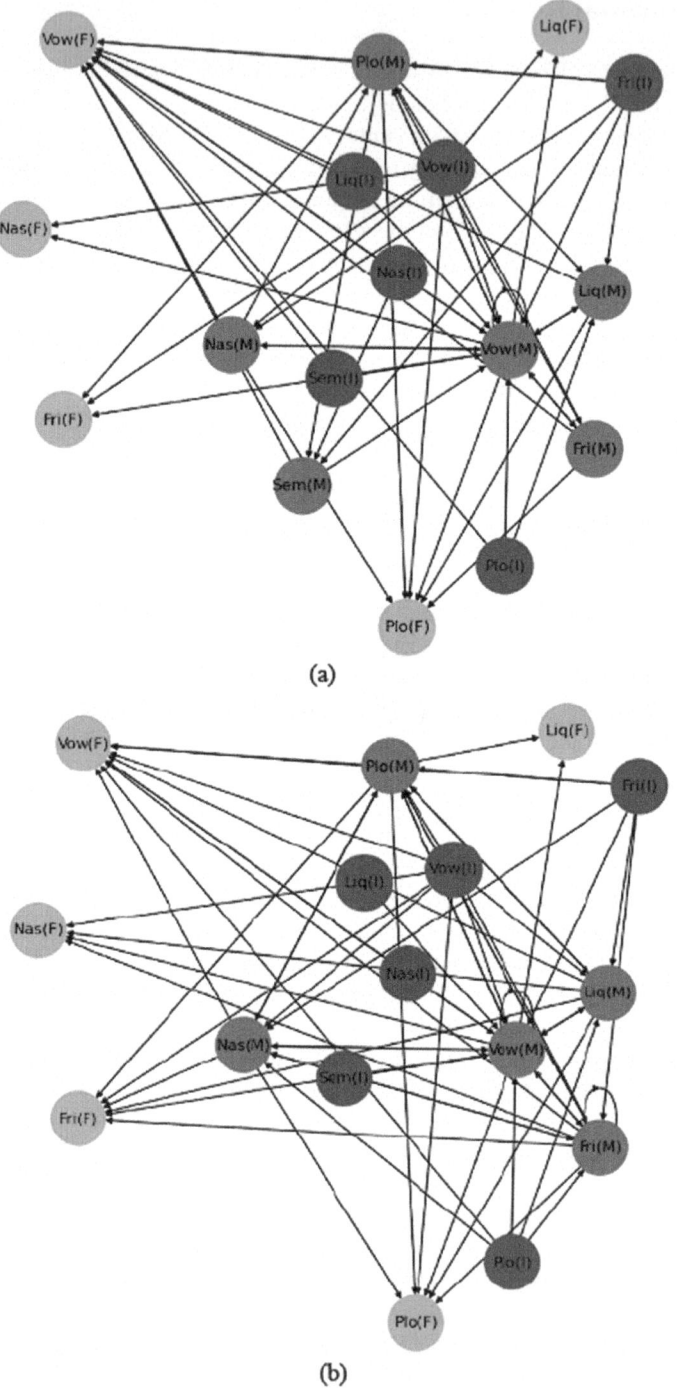

Fig. 3. The phonetic structure of (a) English, (b) German, and (c) Hungarian words.

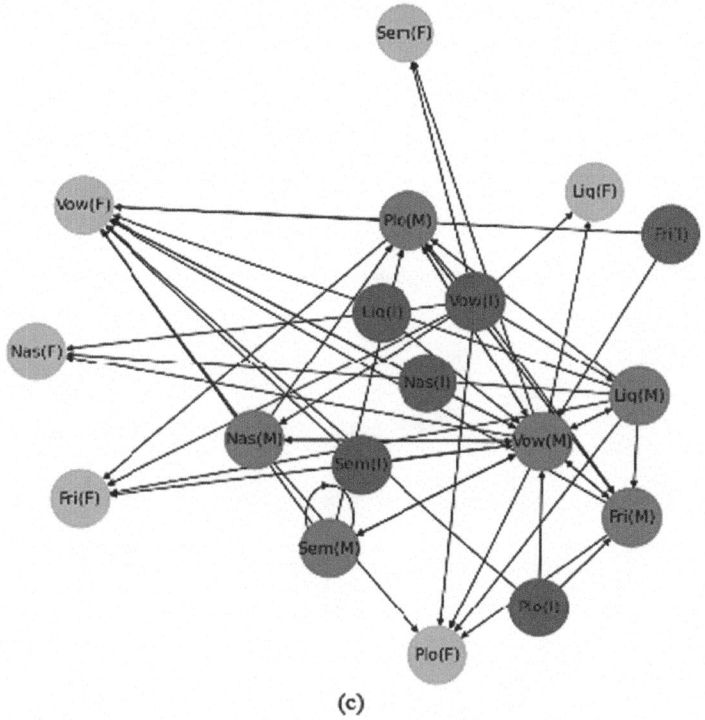

(c)

Fig. 3. (*continued*)

There are several nodes that were not matched by the fuzzy morphism matching algorithm. Therefore, some rows and columns in the matrices of Fig. 4 did not have any '1' value. The lack of matches by the fuzzy morphism matching algorithm may be due to not having enough available information.

This may be because there are not enough words in the Swadesh list of these languages to make substantial differences among some of the less frequent phonemes. This problem can be solved by extending the set of words in the input database. In another experiment, we extended our input database from the Swadesh list to the set of the 997 most frequent English words and the set of the 766 most frequent German words. This led to a more refined graph representing the phonetic structure of English as well as a more complete matrix generated by the fuzzy morphism matching algorithm, as shown in Fig. 5.

Figure 5 shows that the fuzzy morphism matching algorithm found 14 matches between the identical phonetic groups of the English and German phonemes. This is a 40 percent increase from the 10 that were found previously.

In all cases, the yellow matches between different phonemes can be also considered with a lower weight than the green matches. If we give a weight of 1 to the green matches and a weight of 0.1 to the yellow matches, then the similarities among the three languages based on the Swadesh list can be calculated to be the following:

sim(English, German) = 10 + 0.3 = 10.3

	vowels(I)'	fricative(F)'	plosives(I)'	vowels(F)'	fricative(I)'	vowels(M)'	nasals(M)'	plosives(F)'	nasals(F)'	nasals(I)'	liquids(M)'	plosives(M)'	liquids(F)'	fricative(M)'	semivowels(I)'	liquids(I)'
vowels(I)'	1	0	0	0	0	0	0	0	0	0	0	0	0	0	0	0
fricative(F)'	0	1	0	0	0	0	0	0	0	0	0	0	0	0	0	0
plosives(I)'	0	0	1	0	0	0	0	0	0	0	0	0	0	0	0	0
vowels(F)'	0	0	0	1	0	0	0	0	0	0	0	0	0	0	0	0
fricative(I)'	0	0	0	0	1	0	0	0	0	0	0	0	0	0	0	0
vowels(M)'	0	0	0	0	0	1	0	0	0	0	0	0	0	0	0	0
liquids(M)'	0	0	0	0	0	0	1	0	0	0	0	0	0	0	0	0
plosives(F)'	0	0	0	0	0	0	0	1	0	0	0	0	0	0	0	0
nasals(F)'	0	0	0	0	0	0	0	0	1	0	0	0	0	0	0	0
liquids(I)'	0	0	0	0	0	0	0	0	0	1	0	0	0	0	0	0
nasals(M)'	0	0	0	0	0	0	0	0	0	0	1	0	0	0	0	0
plosives(M)'	0	0	0	0	0	0	0	0	0	0	0	1	0	0	0	0
liquids(F)'	0	0	0	0	0	0	0	0	0	0	0	0	1	0	0	0
fricative(M)'	0	0	0	0	0	0	0	0	0	0	0	0	0	0	0	0
nasals(I)'	0	0	0	0	0	0	0	0	0	0	0	0	0	0	0	0
semivowels(M)'	0	0	0	0	0	0	0	0	0	0	0	0	0	0	0	0
semivowels(I)'	0	0	0	0	0	0	0	0	0	0	0	0	0	0	0	0

English and German

	vowels(I)'	plosives(I)'	vowels(F)'	nasals(I)'	plosives(F)'	fricative(F)'	fricative(I)'	vowels(M)'	liquids(F)'	plosives(M)'	nasals(M)'	fricative(M)'	semivowels(F)'	semivowels(M)'	semivowels(I)'	nasals(F)'	liquids(M)'	liquids(I)'
vowels(I)'	1	0	0	0	0	0	0	0	0	0	0	0	0	0	0	0	0	0
plosives(I)'	0	1	0	0	0	0	0	0	0	0	0	0	0	0	0	0	0	0
vowels(F)'	0	0	1	0	0	0	0	0	0	0	0	0	0	0	0	0	0	0
semivowels(I)'	0	0	0	1	0	0	0	0	0	0	0	0	0	0	0	0	0	0
plosives(F)'	0	0	0	0	1	0	0	0	0	0	0	0	0	0	0	0	0	0
fricative(F)'	0	0	0	0	0	1	0	0	0	0	0	0	0	0	0	0	0	0
nasals(I)'	0	0	0	0	0	0	1	0	0	0	0	0	0	0	0	0	0	0
vowels(M)'	0	0	0	0	0	0	0	1	0	0	0	0	0	0	0	0	0	0
nasals(F)'	0	0	0	0	0	0	0	0	1	0	0	0	0	0	0	0	0	0
plosives(M)'	0	0	0	0	0	0	0	0	0	1	0	0	0	0	0	0	0	0
fricative(M)'	0	0	0	0	0	0	0	0	0	0	1	0	0	0	0	0	0	0
liquids(M)'	0	0	0	0	0	0	0	0	0	0	0	1	0	0	0	0	0	0
liquids(F)'	0	0	0	0	0	0	0	0	0	0	0	0	1	0	0	0	0	0
nasals(M)'	0	0	0	0	0	0	0	0	0	0	0	0	0	1	0	0	0	0
liquids(I)'	0	0	0	0	0	0	0	0	0	0	0	0	0	0	1	0	0	0
fricative(I)	0	0	0	0	0	0	0	0	0	0	0	0	0	0	0	0	0	0
semivowels(M)'	0	0	0	0	0	0	0	0	0	0	0	0	0	0	0	0	0	0

English and Hungarian

	vowels(I)'	plosives(I)'	vowels(F)'	plosives(F)'	fricative(F)'	vowels(M)'	liquids(F)'	plosives(M)'	fricative(M)'	liquids(I)'	semivowels(F)'	semivowels(M)'	semivowels(I)'	nasals(F)'	fricative(I)'	liquids(M)'	nasals(M)'	nasals(I)'
vowels(I)'	1	0	0	0	0	0	0	0	0	0	0	0	0	0	0	0	0	0
nasals(I)'	0	1	0	0	0	0	0	0	0	0	0	0	0	0	0	0	0	0
vowels(F)'	0	0	1	0	0	0	0	0	0	0	0	0	0	0	0	0	0	0
plosives(F)'	0	0	0	1	0	0	0	0	0	0	0	0	0	0	0	0	0	0
fricative(F)'	0	0	0	0	1	0	0	0	0	0	0	0	0	0	0	0	0	0
vowels(M)'	0	0	0	0	0	1	0	0	0	0	0	0	0	0	0	0	0	0
nasals(F)'	0	0	0	0	0	0	1	0	0	0	0	0	0	0	0	0	0	0
nasals(M)'	0	0	0	0	0	0	0	1	0	0	0	0	0	0	0	0	0	0
liquids(M)'	0	0	0	0	0	0	0	0	1	0	0	0	0	0	0	0	0	0
liquids(I)'	0	0	0	0	0	0	0	0	0	1	0	0	0	0	0	0	0	0
liquids(F)'	0	0	0	0	0	0	0	0	0	0	1	0	0	0	0	0	0	0
plosives(M)'	0	0	0	0	0	0	0	0	0	0	0	0	0	0	0	0	0	0
plosives(I)'	0	0	0	0	0	0	0	0	0	0	0	0	0	0	0	0	0	0
fricative(I)'	0	0	0	0	0	0	0	0	0	0	0	0	0	0	0	0	0	0
fricative(M)'	0	0	0	0	0	0	0	0	0	0	0	0	0	0	0	0	0	0
semivowels(I)'	0	0	0	0	0	0	0	0	0	0	0	0	0	0	0	0	0	0

German and Hungarian

Fig. 4. The fuzzy morphism matching of the English and German (top), English and Hungarian (middle), and German and Hungarian (bottom) based on the Markov models in Fig. 3 that represents the phonetic structure of these three languages.

	plosives(I)	vowels(F)	fricative(M)	vowels(I)	fricative(F)	plosives(M)	plosives(F)	nasals(I)	fricative(I)	vowels(M)	nasals(F)	nasals(M)	liquids(M)	semivowels(I)	liquids(F)	semivowels(F)	liquids(I)	semivowels(M)
plosives(I)	1	0	0	0	0	0	0	0	0	0	0	0	0	0	0	0	0	0
vowels(F)	0	1	0	0	0	0	0	0	0	0	0	0	0	0	0	0	0	0
plosives(M)	0	0	1	0	0	0	0	0	0	0	0	0	0	0	0	0	0	0
vowels(I)	0	0	0	1	0	0	0	0	0	0	0	0	0	0	0	0	0	0
fricative(F)	0	0	0	0	1	0	0	0	0	0	0	0	0	0	0	0	0	0
fricative(M)	0	0	0	0	0	1	0	0	0	0	0	0	0	0	0	0	0	0
plosives(F)	0	0	0	0	0	0	1	0	0	0	0	0	0	0	0	0	0	0
nasals(I)	0	0	0	0	0	0	0	1	0	0	0	0	0	0	0	0	0	0
fricative(I)	0	0	0	0	0	0	0	0	1	0	0	0	0	0	0	0	0	0
vowels(M)	0	0	0	0	0	0	0	0	0	1	0	0	0	0	0	0	0	0
nasals(F)	0	0	0	0	0	0	0	0	0	0	1	0	0	0	0	0	0	0
semivowels(M)	0	0	0	0	0	0	0	0	0	0	0	1	0	0	0	0	0	0
liquids(M)	0	0	0	0	0	0	0	0	0	0	0	0	1	0	0	0	0	0
simivowels(I)	0	0	0	0	0	0	0	0	0	0	0	0	0	1	0	0	0	0
liquids(F)	0	0	0	0	0	0	0	0	0	0	0	0	0	0	1	0	0	0
semivowels(F)	0	0	0	0	0	0	0	0	0	0	0	0	0	0	0	1	0	0
liquids(I)	0	0	0	0	0	0	0	0	0	0	0	0	0	0	0	0	1	0
nasals(M)	0	0	0	0	0	0	0	0	0	0	0	0	0	0	0	0	0	1

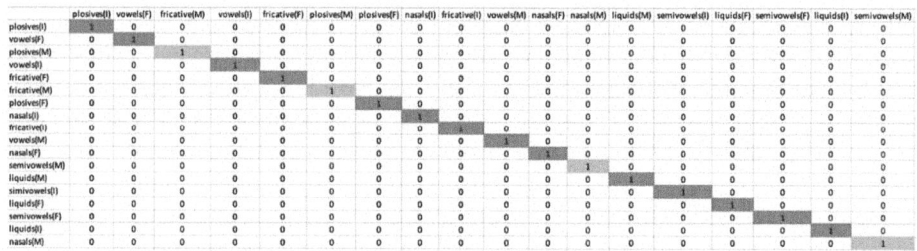

Fig. 5. The fuzzy morphism matching of the English and German phonetic structures, where the English phonetic structure was based on the 997 most frequent English words.

$$\text{sim(English, Hungarian)} = 7 + 0.8 = 7.8$$
$$\text{sim(German, Hungarian)} = 6 + 0.5 = 6.5$$

Weighted calculations could give a more refined similarity measure. However, more experimental results are needed to find the appropriate weights for the green and the yellow matches.

6 Discussion of the Results

The results of the previous section can be summarized as shown in Table 4.

Table 4. Summary of the experimental results.

	English	German	Hungarian
English		10.3	7.8
German	10.3		6.5
Hungarian	7.8	6.5	

The data of Table 4 can be used as an input to a phylogenetic tree generating algorithm such as the UPGMA algorithm (Sokal and Michener, 1958). The UPGMA algorithm would generate the tree shown in Fig. 6. The intermediate node English-German corresponds to the Germanic branch of the Indo-European language family, while the topmost node English-German-Hungarian may correspond to the Nostratic language family. Other phylogenetic algorithms also would generate the same tree in this case.

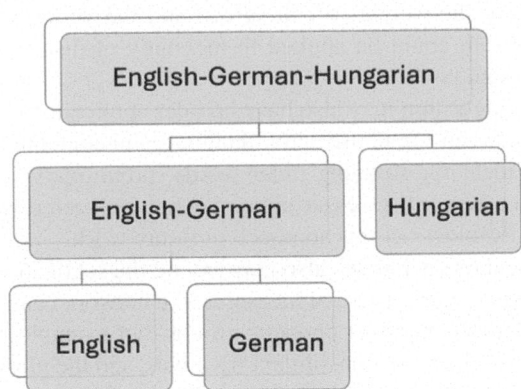

Fig. 6. The phylogenetic tree generated by the UPGMA algorithm.

An interesting aspect of the results is that the fuzzy morphism matching algorithm finds some matches between non-identical phonemes that have a regular sound change between them according to linguists. Regular sound changes are frequently used to identify related languages (Labov 2020).

For example, English plosive phonemes in the medial position, denoted by plosives(M) in Fig. 4, are matched with German fricatives in the medial position, denoted by fricative(M) in Fig. 4. This match can be considered a valid match because of a regular sound change between English /t/, which is a plosive phoneme, and German /s/, which is a fricative phoneme. Examples of this regular sound change can be found in the following pair of cognate words:

better besser

water Wasser

Given a set of cognate word pairs, it is possible to automatically identify regular sound changes between languages using associative data mining (Revesz 2019). This method helped identify regular sound changes between Sumerian and Ugric languages (Revesz 2022) and between Minoan and Hungarian (Revesz 2017, 2024).

7   Conclusions and Future Work

The fuzzy-morphism matching algorithm that was introduced in this paper has great potential to identify long-distance relationships among languages. Our method can identify connections both within single language families and among language families, thereby providing a firmer basis to the study of superfamilies that are considered impossible to solidly identify by traditional means.

A natural application of this work is to the glottochronology of languages. In the future, we plan to extend this work to other languages where the glottochronology is well understood. By studying a larger set of Indo-European languages whose relationships are well-understood, we can better evaluate the accuracy of our fuzzy morphism matching algorithm and calibrate the weights for the green and the yellow matches. After this calibration, the algorithm could be applied to the study of distant language relations across different language families.

Additionally, this algorithm may also have broader applications toward understanding the similarities between sound use across languages in contemporary use. Whereas the Swadesh list is ideal for studying older words, preliminary results suggest that expanding the number of words gives stronger matches between related languages. We plan to evaluate the usefulness of this approach in future work.

As a more technical extension, we also may extend the set of structural measures in Table 3 by incorporating other structural measures like network centrality. We also plan to refine Table 1 by introducing finer phonetic groups. For example, the fricative group could be divided into sibilant and non-sibilant fricatives, and the plosive group could be divided into bilabial, dental, and velar plosives. All these data engineering refinements could further improve the usability of the fuzzy morphism matching algorithm.

Finally, the fuzzy morphism matching algorithm uses the entire global structure of the input Markov models to discover matchings between nodes. It may also be possible to identify certain common structures or patterns within languages. These structures could be translated into graphlets (Pržulj et al., 2004) which can then be used to identify which languages those patterns belong to. We plan to explore this possible application in future work.

References

Bakró-Nagy, M., Laakso, J., Skribnik, E. (eds.): The Oxford Guide to the Uralic Languages. Oxford University Press (2022)

Buccini, A.F., Moulton, W.G.: Germanic languages. Encyclopedia Britannica (2024)

Cook, S.A.: The complexity of theorem-proving conjectures. In: Proceedings of the 3rd ACM Symposium on Theory of Computing (STOC), pp. 151–158 (1971)

Cordella, L.P., Foggia, P., Sansone, C., Tortorella, F., Vento, M.: Graph matching: a fast algorithm and its evaluation. In: Proceedings of the 14th IEEE International Conference on Pattern Recognition (ICPR), Brisbane, Australia, vol. 2, pp. 1582–1584 (1998)

Labov, W.: The regularity of regular sound change. Language **96**(1), 42–59 (2020)

Pagel, M., Atkinson, Q.D., Calude, A.S., Meade, A.: Ultraconserved words point to deep language ancestry across Eurasia. Proc. Natl. Acad. Sci. **110**(21), 8471–8476 (2013)

Pržulj, N., Corneil, D.G., Jurisica, I.: Modeling interactome: scale-free or geometric? Bioinformatics **20**(18), 3508–3515 (2004)

Rao, R.P.N., Yadav, N., Vahia, M.N., Joglekar, H., Adhikari, R., Mahadevan, I.: A Markov model of the Indus script. Proc. Natl. Acad. Sci. **106**(33), 13685–13690 (2009)

Revesz, P.Z.: Establishing the west-Ugric language family with Minoan, Hattic and Hungarian by a decipherment of linear A. WSEAS Trans. Inf. Sci. Appl. **14**, 306–335 (2017)

Revesz, P.Z.: Using data mining algorithms to discover regular sound changes among languages. In: Proceedings of the 23rd International Conference on Circuits, Systems, Communications and Computers, MATEC, vol. 292, no. 03018 (2019)

Revesz, P.Z.: Sumerian-Ugric protowords and regular sound changes, Appendix to: Parpola, S. Etymological Dictionary of the Sumerian Language, vol. 3, Winona Lake: Eisenbrauns, pp. 390–415 (2022)

Revesz, P.Z.: A tale of two sphinxes: Proof that the Potaissa Sphinx is authentic and other Aegean influences on early Hungarian inscriptions. Mediter. Archaeol. Archaeom. **24**(2), 191–216 (2024)

Russ, C.V.J.: The German Language Today: A Linguistic Introduction. Routledge (1994)

Sokal, R.R., Michener, C.D.: A statistical method for evaluating systematic relationships. Univ. Kansas Sci. Bull. **38**, 1409–1438 (1958)

Swadesh, M.: Towards greater accuracy in lexicostatistic dating. Int. J. Am. Linguist. **21**, 121–137 (1955)

Ullmann, J.R.: An algorithm for subgraph isomorphism. J. ACM **23**(1), 31–42 (1976)

Enhancing LLM Code Generation Using Natural Language Processing in the Context of Machine Learning

Jordan Nelson[1], Michalis Pavlidis[1], Andrew Fish[2], and Nikolaos Polatidis[1]([⊠])

[1] School of Architecture, Technology and Engineering,
University of Brighton, Brighton BN2 4GJ, UK
j.nelson7@uni.brighton.ac.uk, {M.Pavlidis,
N.Polatidis}@Brighton.ac.uk
[2] Department of Computer Science, University of Liverpool, Liverpool L69 3BX, UK
Andrew.Fish@Liverpool.ac.uk

Abstract. The rapid rise in popularity of Generative AI and Large Language Models (LLMs) has brought both innovation and controversy, particularly regarding plagiarism and IP law infringements. However, one underexplored concern is the generation of code by these models, which, despite their potential, often includes errors and promotes poor programming practices. This paper explores new methods to address these issues by integrating LLMs with Automated Machine Learning (AutoML). By leveraging AutoML's capabilities in hyperparameter tuning and model selection, we propose a novel approach for generating robust machine learning algorithms. This integration aims to enhance the accuracy and reliability of code generation while mitigating legal risks. Our findings include the application of Natural Language Processing (NLP) and Natural Language Understanding (NLU) techniques to interpret chatbot prompts, thereby improving the generation and customization of machine learning models. The proposed methodology demonstrates practical implementation and high prediction accuracy, offering a promising solution to the current challenges faced by LLM-based code generation. In summary the findings of the paper are as follows: A new implementation of natural language processing for natural language understanding in the context of chatbot prompts aims to serve as an initial step for feature extraction, which will be utilised by an AutoML system to generate machine learning algorithms.

Keywords: LLM · NLP · AutoML · Chatbot · Machine Learning

1 Introduction

The rise of chatbots has been well documented. However, chatbots and generative AI have faced criticism, from potential IP law infringements to issues with the accuracy of the data they provide users [1, 2]. While LLM-based code generation offers promising advantages in accelerating productivity and automating tasks for businesses, a critical analysis of its successes and failures is essential. One publication examining the correctness of synthetic

© The Author(s), under exclusive license to Springer Nature Switzerland AG 2025
R. Chbeir et al. (Eds.): IDEAS 2024, LNCS 15511, pp. 92–105, 2025.
https://doi.org/10.1007/978-3-031-83472-1_7

code identified several weaknesses and limitations in the evaluation power of the original test inputs from Human Eval [3, 4]. This publication highlighted improvements that could identify significant amounts of previously undetected code errors.

One hypothesis is to change the system entirely for chatbot-generated code. Problems often stem from the code being generated similarly to text data. Common programming practices within a chatbot's training can lead to bad programming habits being indistinguishable from good ones. To address this, we propose two concepts: firstly, that the chatbot should better distinguish what its generated code should look like, and secondly, it should improve the way it returns code to the user. While addressing these issues comprehensively is a monumental task, this paper focuses specifically on code generation for machine learning (ML) algorithms.

We aim to alleviate these issues using AutoML, which has gained popularity for making machine learning more accessible by automating processes like hyperparameter tuning and model selection. AutoML has proven popular even in academic research [5]. Therefore, we consider AutoML a superior solution for generating machine learning models, reducing code errors and the risk of IP theft compared to traditional generative approaches. To integrate an AutoML framework within a chat environment, we need to understand the user's prompt aims and gather information to assist in generating the model, such as settings and parameters. For Natural Language Processing (NLP), Convolutional Neural Networks (CNN) have been effective in various tasks [6] and are accessible via the AutoML API AutoKeras [7]. CNNs are particularly good for text classification because they can automatically capture local patterns and hierarchical features within text data, which enhances the model's ability to understand and classify text accurately.

This paper proposes an innovative approach to LLM code generation tailored for machine learning contexts. The methodology begins by suggesting an alteration to LLM techniques to discern the specific machine learning algorithm a user intends to implement. Subsequently, the system engages the user with tailored prompts to refine and personalise the code for the model or algorithm being considered. This iterative process aims to enhance user engagement and facilitate the customization of machine learning code according to the user's specific requirements and preferences.

The contributions of this work are summarised as follows:

- An application of how the LLM and generative AI code generational process can be altered to provide working, error free machine learning algorithms
- Extensive evaluation of the proposed methodology, demonstrating its practicality and effectiveness, with results showing high accuracy.

The remaining sections of this paper are structured as follows: Sect. 2 provides an overview of related work, Sect. 3 describes the dataset used, Sect. 4 details the methodology, Sect. 5 outlines the experimental evaluation steps and presents the results, while Sect. 6 concludes with insights into future research directions.

2 Related Work

Recent studies into code generation by state-of-the-art (SOTA) large language models (LLMs) such as ChatGPT [18] are limited due to the novelty of these technologies, and the full extent of their capabilities is still being explored and expanded upon. Consequently, the available research on this topic remains scarce. Some publications [8, 9] have identified several flaws and security issues in the code generated by ChatGPT. For instance, out of 21 reported use-cases, only five were initially secure, with an additional seven becoming more secure only after explicit instructions were provided by the user. Other studies have revealed that synthetic code generated by ChatGPT exhibited vulnerabilities in over a third of their use-cases, with some studies reporting vulnerabilities in approximately 12% of cases [10, 11].

Looking closer at this, a survey on LLMs for code generation [12] has identified primary challenges, such as the need for LLMs to manage complex, repository-level code generation, develop novel model architectures that better capture code structures, and curate high-quality, diverse code datasets for effective pre-training and fine-tuning. These challenges underscore the gap between academic research and practical development, emphasising the necessity for advancements to enhance LLMs' real-world applicability and performance in code generation. In terms of integrating LLMs with automated machine learning [13], the implementation of Contextual Modular Generation proves demanding, particularly for highly complex ML tasks involving intricate input and output data. Comparisons to collaborative problem-solving on platforms like Kaggle are infeasible due to the collective nature of these efforts. Nonetheless, the approach aims to reduce the effort and knowledge required for writing ML programs in everyday scenarios, achieving significant automation and performance improvements.

Research comparing deep learning models for automated source code generation and auto-completion [14] found that small language models (AWD-LSTM and AWD-QRNN) perform best with char-sized tokenization, while larger models like GPT-2 excel in source code generation even with different tokenization methods. Pre-trained models generally outperform those not pre-trained, even if initially trained on non-programming languages. However, transformer models like BERT and RoBERTa, despite their high accuracy, underperformed in specific tasks like source code auto-completion, highlighting gaps in the literature and suggesting areas for future research. A framework for evaluating LLMs' code generation abilities [15] introduced the pass-ratio@n metric, effectively measuring test pass rates. Preliminary results indicated that the quality of generated code depends on prompt details and the recency of coding problems relative to the LLM's training date. While the framework showed potential for automation and efficiency, future studies with larger datasets are needed to generalise the findings. While another study [16] highlighted that ensuring the quality of training data and minimising errors are crucial for the responsible integration of LLMs. That it is essential to prevent biases, misinformation, and other risks. By prioritising these factors, we can achieve a more ethical and inclusive AI-driven future.

These findings underscore a significant gap in research addressing the integration of LLMs with automated machine learning (AutoML). While most existing studies have examined LLM code generation across various programming languages such as Java,

C++, and Python, there is a noticeable lack of research focusing exclusively on the generation of machine learning algorithms by LLMs or generative AI. The challenges identified, such as managing complex, repository-level code generation, developing novel model architectures to better capture code structures, and curating high-quality, diverse code datasets, are equally pertinent to the AutoML context. However, utilising AutoML to generate code could theoretically mitigate some of these issues. By automating the optimization and generation processes, AutoML could handle the intricacies of complex ML tasks more effectively than current LLM implementations. This approach could reduce the effort and knowledge required for writing ML programs in everyday scenarios, thereby addressing the inefficiencies observed in automated solutions compared to collaborative problem-solving on platforms like Kaggle. Moreover, AutoML's ability to continuously learn and adapt could enhance the practical applicability and performance of LLMs in generating machine learning models, thus filling a crucial research gap and advancing the field of AutoML.

3 Proposed Dataset

To the best of our knowledge there aren't any datasets available, and we utilised an original binary representation multi-classification dataset tailored specifically for our research objectives in this paper. Our focus necessitated a dataset[1] capable of classifying chatbot prompts exclusively within the domain of machine learning algorithms. At the time of conducting our study, no existing dataset met this precise criterion, compelling us to develop one from scratch. To ensure simplicity and scalability in dataset creation, we aligned our approach with the following initiatives:

Initially, the dataset will focus on describing 7 algorithms to facilitate experimentation while these algorithms are commonly recognized in both beginner and expert circles, particularly within the realm of classification problems. Starting with 7 algorithms is primarily a precautionary measure, considering the uncertain computational costs before conducting experiments. This approach allows for starting small while keeping future scalability in mind. Based on the outlined points, we proceeded to create the dataset by defining a subset Z from the universal set of all machine learning algorithms M, as described in Eq. 1.

$$M = \{All\ machine\ learning\ algorithms\} \tag{1}$$

$$Z = \{KNN, Decision\ Tree, Random\ Forest, MLP, SVM, Naive\ Bayes, AutoML\}$$

Using Eq. 2, we proceeded to create a dedicated set for each element within Z by replicating each element x two hundred times per set. This method resulted in 6 distinct sets, each containing 200 questions pertaining to a specific element from Z.

$$A\ set = \{x\ is\ a\ machine\ learning\ question\ |\ x\ relates\ to\ an\ element\ e\ |\ e \in Z\} \tag{2}$$

To help visualise this, we can express the contents of our sets as shown in Eq. 3.

$$Set_1 = \{x_1, x_2, x_3x_4x_5, ..., x_{200}\}, \ \forall x \ Set_1, x_n \rightarrow e_1 \mid e_1 \in Z,$$
$$Set_2 = \{x_1, x_2, x_3x_4x_5, ..., x_{200}\}, \ \forall x \ Set_2, x_n \rightarrow e_2 \mid e_2 \in Z,$$
$$...$$
$$Set_7 = \{x_1, x_2, x_3x_4x_5, ..., x_{200}\}, \ \forall x \ Set_7, x_n \rightarrow e_7 \mid e_7 \in Z$$

(3)

After this, we repeated the process once more. Using prompts that described none of the algorithms previously discussed. Following this methodology, we implemented a Bag of Words (BoW) approach by consolidating all sets into a unified dictionary. This dictionary encompassed the frequencies of each word across the entire dataset after eliminating stop words. This process gave us 1,600 records with 1,744 attributes, plus an additional column for the label appended at the dataset's end. Each element from a set was then vectorized, allowing us to encode occurrences that aligned with the dictionary's column headers—akin to an intersection—with a binary representation of 1 indicating a match and 0 otherwise. Subsequently, we assigned distinct labels to denote each algorithm type represented in the dataset[1].

4 Proposed Methodology

As outlined in Sect. 1, we focused on utilising a Convolutional Neural Network (CNN) for the NLP task. The CNN model was trained on the dataset described in Sect. 3, evaluated during training, and further tested to assess its performance on unseen data. Detailed metrics used for evaluation are provided in Sect. 5.

The primary objective of the CNN is to accurately interpret user prompts directed at an LLM/Generative AI system, identifying the specific machine learning model the user is requesting. This capability is essential for the model to generalise effectively to new data not included in the training set. After constructing and training the CNN model, we needed to evaluate its performance using new input data. Subsequently, to do this objectively, we generated 10 diverse prompts per model type and an additional 10 for non-model queries, asking for the creation of various models. These prompts encompassed a wide range of topics, including fraud detection, handwriting classification, business statistics, image classification etc. This variety ensured that the prompts were representative of the types of inquiries a user may present such a system. Furthermore, some prompts were intentionally alike to ensure the CNN was accurately trained to distinguish between different models or no model, even when the questions were closely related. This training was crucial for the CNN to correctly identify the intent from user prompts. Finally, Once the CNN identifies the desired intent from a user's prompt, the LLM can then engage the user with more detailed and specific questions about that model. Or in the case of no model, respond to their question normally. This would allow for the creation of a custom ML model tailored to the user's specific needs, rather than relying on a generic model generation process. The rationale for this approach stems from our preliminary observations during the creation of this paper. We found that

---

[1] Dataset available at: https://www.kaggle.com/datasets/jordanln/llm-prompts-in-the-context-of-machine-learning.

popular LLMs often provide code examples as a practical implementation, even when the user does not explicitly request it, and a copy of the evaluation questions is included with the dataset. The methodology is further detailed in the following Sect. 4.1 CNN, 4.2 Chatbot.

4.1 The Proposed Convolutional Neural Network

The first model we will examine is our Convolutional Neural Network (CNN). The architecture of the CNN is depicted in Fig. 1. CNNs share many similarities with standard neural networks, such as neurons with learnable weights and biases. Each neuron in the network receives inputs, performs a dot product, and may apply a non-linearity function. Additionally, CNNs include elements like a score function and a loss function, which are common in neural networks. The key difference lies in the CNN architecture's assumption that the inputs are images. This assumption allows CNNs to leverage specific properties that can be encoded into the architecture, enhancing their efficiency. As a result, CNNs require fewer parameters and achieve greater efficiency in the forward function compared to standard neural networks.

To construct our CNN model, we employed the AutoModel function available in the AutoKeras API [17]. This API streamlines hyperparameter tuning using its integrated Keras tuner, requiring minimal setup to begin building. Initially, we set the max_trials parameter to its default value, triggering the execution of 100 models with automated hyperparameter optimization. The objective was defined as maximising validation accuracy (val_accuracy), aligning with the objectives outlined in Sect. 1.

For efficiency, we opted for a Bayesian tuner instead of grid-search. The AutoModel initially ran with a simple configuration comprising one convolutional layer and one classification head, achieving a low accuracy of approximately 15%. To enhance performance, we suggested an additional convolutional layer to the AutoModel architecture. However, this led to increased computational demands, resulting in memory issues that necessitated the inclusion of SpatialReduction layers (dropout) to mitigate them. Subsequent iterations included further enhancements such as additional convolutional and dense layers to improve accuracy. Each addition required corresponding SpatialReduction layers to manage escalating memory requirements effectively. At the next step by adding more layers to the AutoModel function allows the API to explore a greater depth of model complexity within its predefined search space. This can potentially enhance the model's capability to capture and learn from more intricate patterns in data. Then we experimented with N-fold cross-validation to combat overfitting but discontinued it due to impractically long training times without recording conclusive results. Consequently, we settled on the CNN model as depicted in the preceding figure, which represents our optimised configuration.

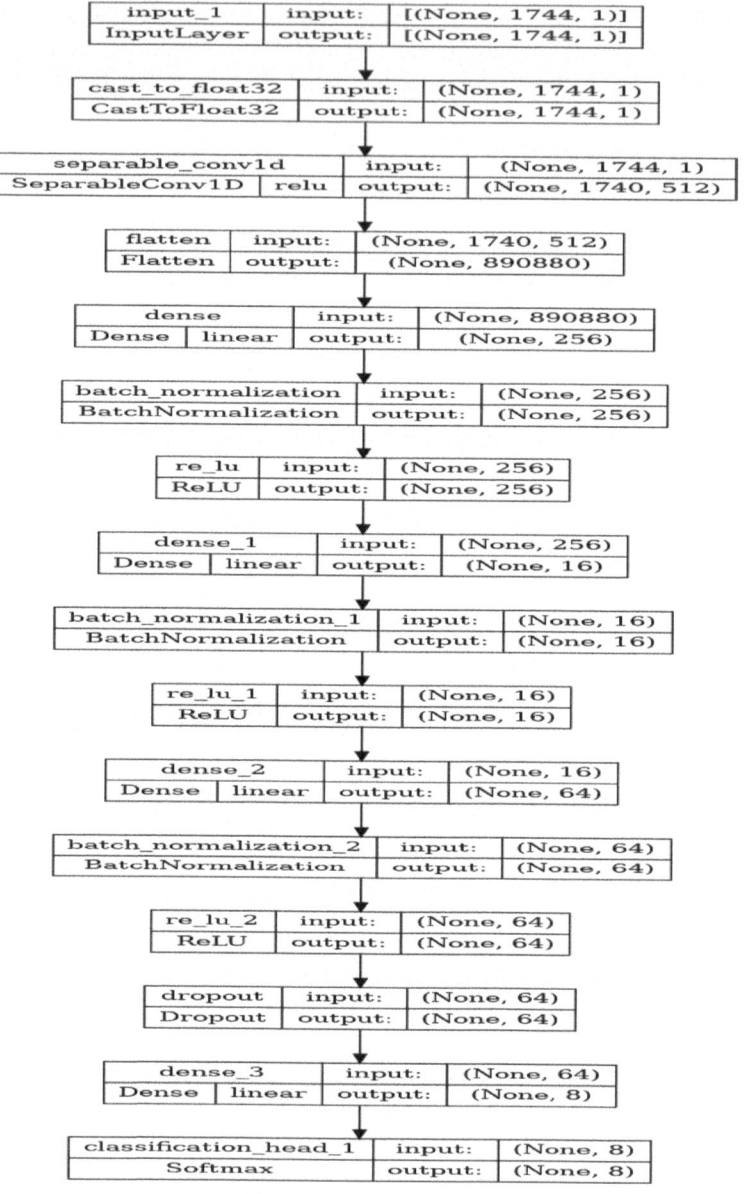

Fig. 1. The CNN model.

4.2 Chatbot

Following our experimentation with the CNN model, we integrated it with the OpenAI GPT API [18], utilising the 3.5-turbo version of GPT. This integration enabled us to initiate the incorporation of dialogue system elements. Initially, we implemented a basic preprocessing stage for user inputs, involving tasks such as removing punctuation, numbers, and converting text to lowercase to match the headings in our dataset. This method facilitated the evaluation of our CNN-based approach within the chatbot model, as illustrated in Fig. 2. Moving forward, this setup can be extended to integrate an AutoML system, leveraging the CNN to tailor custom questions that the language model (LLM) can pose to users. These questions aim to solicit input for creating personalised AutoML solutions or custom machine learning models, enhancing the machine learning models returned to the user.

```
client = OpenAI(
    # defaults to os.environ.get("OPENAI_API_KEY")
    api_key="hidden",
)
MODEL = "gpt-3.5-turbo" # GPT Model version
message = text
if message:
    messages = [
        {"role": "system", "content": 'You answer question
about machine learning'},
        {"role": "user", "content": message},
    ]
    response = client.chat.completions.create(
        model=MODEL,
        messages=messages,
        temperature=1
    )
response_message = response.choices[0].message.content
print(response_message)
```

Fig. 2. The Chatbot Code

To achieve this, we interrupt the API function to determine if the input text (variable: text) corresponds to any of our predefined classes, as detailed in Sect. 4 and Subsect. 4.1. Upon classification, we ascertain the specific model the user aims to create. Subsequently, we engage the user with tailored questions to validate parameters or settings, which can then be translated into parameters and model architectures for either an AutoML solution or pre-written templates of machine learning models, ensuring they are free of coding errors. If the input fits into the final non-generative category, we seamlessly return the question to the LLM for its normal response handling.

5 Experimental Evaluation

In our experimentation we used the python programming language exclusively, along with the deep learning API's AutoKeras and Keras. All experiments were executed on an Intel® Core™ i7-9750H CPU @ 2.60 GHz × 12 using 16 GB of DDR4 memory on the Linux Ubuntu 22.04.3 LTS system.

5.1 Evaluation Metrics

Predictions

When assessing our models for predictive performance, we employed three key metrics. The first metric measures accuracy, which evaluates how well the algorithm categorises new inputs into specific classes. The second metric evaluates the correctness of predictions themselves. Based on these metrics, our prediction results are formatted as demonstrated below:

- P@K of 2 & Probability (%) of the prediction

   To expand on this, when making a prediction, we set up our program to return the top two predictions and the corresponding probability converted into a percentage in descending order.

1. Precision @ K
2. This refers to the output probabilities of an input belonging to one of our classes, such that the P@K value is greater than zero, otherwise it is recorded as zero to indicate a failure.

   This scoring provides an initial indication of the model's performance by showing relevant items in the top 2 recommendations. If the correct model is not returned first, the system can suggest the second most probable model. Limiting this to the top 2 prevents scaling issues with incorrect recommendations. If the wrong answer is given twice, the system asks for a correction, justifying our choice of K = 2. The percentages reflect the algorithms' generalisability and predictive accuracy on new, unlabelled data. The Mean Reciprocal Rank (MRR) metric evaluates the quality of recommendations by determining the average position of the first relevant item across all recommendations. The equation for MRR can be seen in Eq. 4, where Q is the sample size and $rank_i$ is the position of the first relevant rank.

$$MRR = \frac{1}{Q} \sum_{i=1}^{Q} \frac{1}{rank_i} \tag{4}$$

5.2 Experimental Results

Training

To find the optimal number of epochs and batch size a grid-search like method was implemented, Fig. 3 presents a snapshot of our testing in the ranges of 10–50 and 2–24 respectively. As seen in Fig. 3, the optimal number of epochs was 40 & 50, with a batch size of 10 & 14 respectively as these displayed a notable higher degree of validation accuracy when compared to other values.

As we can see from Table 1, each trial had varying degrees of accuracy and loss depending on the combination of hyperparameter settings being used per trial. One trial of note in this table is trial 2 with an accuracy of 99.82%, and low loss values. This snapshot of the first ten trials is representative of the other 90 but they are not shown here to save space. The final iteration of the program is the evaluation of the test data using the best performing model, which ended with an accuracy of 98.93%.

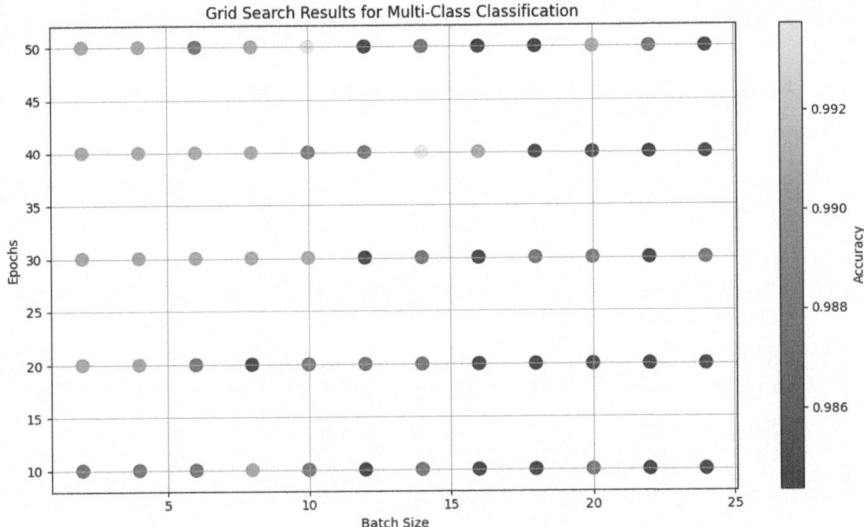

Fig. 3. Epoch and Batch size tuning results

Table 1. Results for loss, accuracy, validation loss and validation accuracy for the first 10 trials of the CNN model

Trial #	Loss	Accuracy	Validation Loss	Validation Accuracy
Trial 1	0.8663	0.1568	0.6587	0.1458
Trial 2	0.0191	0.9982	0.0211	0.9896
Trial 3	0.5483	0.2174	0.5785	0.2131
Trial 4	0.4516	0.2383	0.6144	0.2372
Trial 5	0.2206	0.3153	0.5926	0.3029
Trial 6	0.3778	0.2707	0.5031	0.2699
Trial 7	0.2328	0.4071	0.4317	0.3989
Trial 8	0.2937	0.3138	0.5430	0.3114
Trial 9	0.1247	0.5347	0.4191	0.5295
Trial 10	0.4883	0.2963	0.5047	0.295
Testing:	**Accuracy: 0.9893**			

Predictions

As shown in Table 2 and 3, the CNN model performed at a high degree of accuracy. It correctly predicted the correct model in the top two probabilities with its P@2 score of 0.5 across the board. As well as this the probability percentage remained high throughout all tests, showing a high degree of reliability.

Table 2. Prediction results of the CNN per algorithm showing the P@K and probability (%)

%	KNN	Decision Tree	Random Forest	MLP	SVM	Naive Bayes	AutoML	None
Test 1	0.5 & 98.57	0.5 & 98.68	0.5 & 98.82	0.5 & 98.17	0.5 & 98.88	0.5 & 98.34	0.5 & 98.79	0.5 & 98.45
Test 2	0.5 & 98.23	0.5 & 99.23	0.5 & 99.09	0.5 & 98.92	0.5 & 98.21	0.5 & 99.08	0.5 & 99.11	0.5 & 99.07
Test 3	0.5 & 98.89	0.5 & 98.87	0.5 & 98.36	0.5 & 98.43	0.5 & 99.03	0.5 & 98.72	0.5 & 98.92	0.5 & 98.62
Test 4	0.5 & 99.12	0.5 & 99.45	0.5 & 99.33	0.5 & 99.05	0.5 & 98.56	0.5 & 98.95	0.5 & 98.86	0.5 & 99.13
Test 5	0.5 & 98.45	0.5 & 98.09	0.5 & 98.71	0.5 & 98.61	0.5 & 98.47	0.5 & 98.13	0.5 & 99.02	0.5 & 98.77
Test 6	0.5 & 99.01	0.5 & 99.78	0.5 & 98.57	0.5 & 98.78	0.5 & 99.18	0.5 & 99.21	0.5 & 98.96	0.5 & 98.81
Test 7	0.5 & 98.76	0.5 & 98.32	0.5 & 99.17	0.5 & 99.37	0.5 & 98.65	0.5 & 98.88	0.5 & 99.24	0.5 & 99.07
Test 8	0.5 & 98.34	0.5 & 98.94	0.5 & 98.15	0.5 & 98.29	0.5 & 99.27	0.5 & 98.59	0.5 & 98.82	0.5 & 98.54
Test 9	0.5 & 99.54	0.5 & 99.67	0.5 & 99.27	0.5 & 98.74	0.5 & 98.79	0.5 & 99.34	0.5 & 98.69	0.5 & 98.86
Test 10	0.5 & 98.91	0.5 & 98.76	0.5 & 98.64	0.5 & 99.16	0.5 & 98.92	0.5 & 98.66	0.5 & 99.10	0.5 & 99.25
AVG	**0.5 & 98.78**	**0.5 & 98.98**	**0.5 & 98.81**	**0.5 & 98.75**	**0.5 & 98.80**	**0.5 & 98.79**	**0.5 & 98.95**	**0.5 & 98.86**

Table 3. Overall average of the CNN's performance across all classes

Model	Average Score
KNN	0.5 & 98.78
Decision Tree	0.5 & 98.98
Random Forest	0.5 & 98.81
MLP	0.5 & 98.75
SVM	0.5 & 98.80
Naive Bayes	0.5 & 98.79
AutoML	0.5 & 98.95
None	0.5 & 98.86

(continued)

Table 3. (*continued*)

Model	Average Score
Average	**0.5 & 98.84**

Finally, looking at Table 4, we have our results for MRR per model for the CNN. We can see from this that the CNN is an excellent performing model, with a guarantee to present a relevant response first and foremost every time with a score of 100% using the MRR metric.

Table 4. MRR for the CNN overall (%)

Model	MRR Score
KNN	100
Decision Tree	100
Random Forest	100
MLP	100
SVM	100
Naïve Bayes	100
AutoML	100
None	100
Average	**100**

6 Conclusions and Future Work

The CNN model demonstrated exceptional performance with consistently high accuracy and MRR scores during training and testing phases, establishing its reliability for production environments. Integrating this model into the chatbot architecture serves as an initial step towards enhancing the generation process of machine learning models through LLMs. This integration enables the chatbot to engage users with specific model-related inquiries before generating or returning machine learning models. For instance, it can prompt for parameters like the k value in KNN or preferred layers in an AutoML Keras solution. By incorporating these functionalities, the chatbot not only addresses potential code issues and errors highlighted in earlier sections but also leverages the reliability and success of AutoML frameworks. This approach underscores the system's capability to facilitate the creation of linguistically engineered neural networks (LENN) and custom machine learning models tailored to user specifications free of the prevalent issues currently in the generational processes of LLM and generative AI.

Some limitations of this approach include users potentially struggling to articulate their requirements precisely, especially if they lack a deep understanding of machine learning concepts. This can lead to ambiguous or incomplete inputs, making it difficult for the chatbot to generate appropriate models. Additionally, regular updates and maintenance of both the LLM and AutoML components are necessary to keep up with advancements and address emerging issues, which can be resource intensive.

Looking forward, future work may include optimising the architecture and model generation pipeline to enhance scalability and efficiency, enabling robust handling of larger datasets and more complex model configurations. Implementing stringent security measures aligned with AutoML security research to safeguard user data and ensure compliance. Introducing automated hyperparameter optimization techniques within the generational process for non-AutoML solutions to streamline model tuning and enhance performance. Additionally, developing capabilities for the chatbot to understand context-specific nuances in user queries, thereby offering more precise and relevant model generations and customization options based on contextual understanding.

References

1. Intellectual property in chatgpt. IP Helpdesk (2023). https://intellectual-property-helpdesk.ec.europa.eu/news-events/news/intellectual-property-chatgpt-2023-02-20_en
2. Nah, F.-H., et al.: Generative AI and ChatGPT: applications, challenges, and AI-human collaboration. J. Inf. Technol. Case Appl. Res. **25**(3), 277–304 (2023)
3. Liu, J., Xia, C.S., Wang, Y., Zhang, L.: Is your code generated by CHATGPT really correct? Rigorous evaluation of large language models for code generation. In: Proceedings of the 37th International Conference on Neural Information Processing Systems (NIPS), New Orleans, LA, pp. 21558–21572 (2023)
4. Chen, M., et al.: Evaluating large language models trained on code. arXiv:2107.03374 (2021)
5. LeDell, E., Poirier, S.: H2o AutoML: scalable automatic machine learning. In: Proceedings of the 7th ICML Workshop on Automated Machine Learning, San Diego, CA (2020)
6. Yin, W., Kann, K., Yu, M., Schütze, H.: Comparative study of CNN and RNN for natural language processing. arXiv:1702.01923 (2017)
7. Jin, H., Song, Q., Hu, X.: Auto-Keras: an efficient neural architecture search system. In: Proceedings of the 25th ACM SIGKDD International Conference on Knowledge Discovery & Data Mining (KDD), Anchorage, AK, pp. 1946–1956 (2019)
8. Khoury, R., Avila, A., Brunelle, J., Camara, B.M.: How secure is code generated by CHATGPT? arXiv:2304.09655 (2023)
9. Kocoń, J., et al.: ChatGPT: jack of all trades, master of none. Inf. Fusion **99**, 101861 (2023)
10. Liu, Z., et al.: No need to lift a finger anymore? Assessing the quality of code generation by ChatGPT. IEEE Trans. Softw. Eng.Softw. Eng. **50**(6), 1548–1584 (2024)
11. Yetiştiren, B., Özsoy, I., Ayerdem, M., Tüzün, E.: Evaluating the code quality of AI-assisted code generation tools: an empirical study on GitHub Copilot, Amazon CodeWhisperer, and ChatGPT. arXiv:2304.10778 (2023)
12. Jiang, J., et al.: A survey on large language models for code generation. arXiv:2406.00515 (2024)
13. Xu, J., Liu, Z., Suryanarayanan, N.A.V., Iba, H.: Large language models synergize with automated machine learning. arXiv:2405.03727 (2024)
14. Cruz-Benito, J., Vishwakarma, S., Martin-Fernandez, F., Faro, I.: Automated source code generation and auto-completion using deep learning: comparing and discussing current language model-related approaches. AI **2**(1), 1–16 (2021)

15. Yeo, S., et al.: Framework for evaluating code generation ability of large language models. ETRI J. **46**(1), 106–117 (2024)
16. Junior, S.B., et al.: Are large language models the new interface for data pipelines? In: Proceedings of the International Workshop on Big Data in Emergent Distributed Environments (BiDEDE), Santiago, Chile, pp. 1–6 (2024)
17. Keras. AutoModel Example, AutoKeras (2024). https://autokeras.com/auto_model
18. OpenAI. OpenAI – GPT (2024). https://chat.openai.com/

LaSSI: Logical, Structural, and Semantic Text Interpretation

Oliver Robert Fox[ID], Giacomo Bergami[✉][ID], and Graham Morgan[ID]

Newcastle University, Newcastle Upon Tyne, UK
{o.fox3,giacomo.bergami,graham.morgan}@newcastle.ac.uk

Abstract. This paper proposes LaSSI, a Natural Language Processing (NLP) pipeline contextualising verified Artificial Intelligence (AI) by transforming text via Montague Grammars (MGs). We are approaching from the point-of-view of graphs and logic, in which we achieve explainable sentence similarity in terms of Knowledge Base (KB) expansion and possible worlds semantics. Experiments in the present paper excel current state-of-the-art, Graph Retrieval-Augmented Generation (RAG)-based technologies, through a novel method surpassing vector-based and graph-based sentence similarity metrics.

Keywords: Natural Language Processing · Montague Grammar · Sentence Similarity · Verified Artificial Intelligence · eXplainable AI

1 Introduction

Sentence similarity is the cornerstone of full-text Question Answering (QA), as it first determines whether the question matches the original by using a scoring function prior to resolving the variable [18]. When representing sentences as graphs, this is achieved through a graph matching problem over a Knowledge Base (KB) [8] where we can express the amount a sentence was matched by how many nodes and edges syntactically correspond [14]. Recent Retrieval Augmented Generation (RAG)-based approaches try to combine the two [9], by associating vectors to both nodes and relationship labels, which are used to enrich the syntactic match with more semantic information.

These techniques are not completely fool proof (Sect. 2.1): first, neither vectors nor graphs alone can represent semantics and structure, as vectors lose structural information through averaging [18], while *graph representations* of sentences do not usually convey similar structure for semantically-similar sentences. Second, metrics conceived for determining similarity are usually symmetric by definition, and cannot convey asymmetric notions of entailment: e.g., if there is traffic in the city, then there is traffic also in the city centre, while the latter implication is in doubt. While vector-based systems use cosine similarity, *graph-edit distances* are also designed to be symmetric [21]. Third, neither vectors nor graphs faithfully express logical connectives: for the latter, none of the current *graph-based representations* use nested nodes to represent a group of entities under a specific logical connector [9,23]. Our ideal representation for sentence similarity is related

R. Chbeir et al. (Eds.): IDEAS 2024, LNCS 15511, pp. 106–121, 2025.
https://doi.org/10.1007/978-3-031-83472-1_8

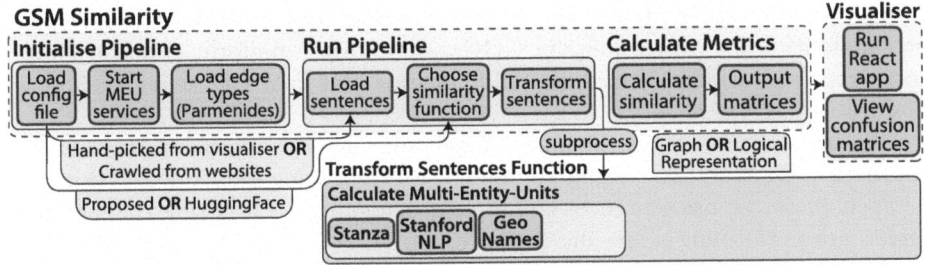

Fig. 1. Characterising LaSSI as a sentence processing pipeline.

to a human-explainable notion of confidence, which already provides a numerical characterization for entailment: given $\mathcal{W}$ the set of the *possible worlds* and $\mathcal{W}(s)$ a subset of $\mathcal{W}$ where the sentence s is true, we re-use the notion of confidence to express the similarity of two sentences s and t as:

$$\mathsf{confidence}(s, t) = \frac{|\mathcal{W}(s) \cap \mathcal{W}(t)|}{|\mathcal{W}(s)|} \tag{1}$$

This motivates the provision of a Logical Representation (LR) $\varphi(s)$ of s, through which we explain and interpret the natural language sentence, decompose each formula into its atomic constituents, and then derive the possible worlds where such sentence should be true.

Despite the theoretical possibility of automating rewriting natural language sentences in First Order Logic (FOL), through Natural Language Processing (NLP) techniques [12], this is currently unprecedented in practical domains, and is considered a hard-to-formalise task within the realm of verified Artificial Intelligence (AI) [15] (Sect. 2.2). We propose approaching from a graph and logic point-of-view, thereby creating LaSSI[1] (Fig. 1), a novel NLP pipeline implementing both vector-based, graph-based, and Montague Grammar (MG)-based [12] (Sect. 2.2) sentence rewriting through which we can test sentence similarities via different sentence rewriting strategies. In our paper, we restrict our analysis to factoid sentences as per [23] that can be represented in *propositional logic*, in which no universal/existential quantifiers are given. If we cannot execute propositional logic, then we cannot carry out FOL as the latter is a more expressive logical fragment of the former. LaSSI works as follows (Sect. 3.1): after determining the entities occurring within the text via dictionary-based entity-matching, extract a Dependency Graph (DG) representation from the full-text. By expressing simple grammatical functions [11], these can be rewritten using Generalised Graph Grammars (GGGs) [3] into an intermediate representation ensuring a uniform representation of direct and indirect objects. Independent from the active/passive form of the sentences, we also capture logical connectives from propositional calculus. After reconciling the entities occurring within the

---

[1] https://github.com/datagram-db/LaSSI-pipeline/releases/tag/v1.0.

graph with the ones matched via the dictionaries, then applying graph semantic capabilities not expressible via GGGs, we obtain a uniform *graph* or *logical graph* representation. Then, optionally generate the *final logical representation* from the latter via MGs.

To test our research hypothesis postulating the need of logic-driven approaches for truly expressing common-sense reasoning embedded within factoid sentences, we use sentences conjoining, disjoining, or negating the fact that agents are performing a specific action (Sect. 4). By doing so, we highlight the limitations of state-of-the-art sentence similarity mechanisms, being vectorial embeddings and graph-based QA, where current work does not exploit semantics or acknowledge grammatical functions (Sect. 2.1). We also found limitations on Graph Query Languages (GQLs), thus including the recently established GQL standard, which is not general enough to express unique rewriting queries that can be applied to all possible sentences expressed as graphs [3]. For this, we implemented a new GQL and graph database which, due to space, will not be described in detail in this paper.

The paper is structured as followed: Sect. 2 discusses the related technologies for the system. Section 3 explains our proposed, novel pipeline, with discussions on the similarity metrics adopted throughout for each representation returned by LaSSI as per Fig. 3. Section 4 discusses the results from our work so far, remarking the superiority of the logical-based approach. Section 5 draws conclusions on the work carried out and what is planned for the future.

2   Related Works

We discuss the main technologies used in this paper, relating to QA, sentence similarity and NLP; limitations of these technologies are presented to address the difference between our methodology and existing techniques in these fields.

2.1   Question Answering and Sentence Similarity

Graph Representation. Structure-based graph metrics [21], mainly rely on *graph edit-distance*; the authors refine this with the notions of *quality*, determining how much the paths matched in the KB align with the ones in the query, and *conformity*, determining how much the combination of paths retrieved from the answers are similar to ones in the query. This approach does not encompass any semantic information within the nodes or edges in the graph, as it is not considering node similarity in terms of entity matching. More semantic-driven approaches such as HypoGator [23] extend the former by introducing the notion of *node similarity*. It can also evaluate three types of queries: class-based, zero-hop, and graph-queries. The algorithm uses zero-hop queries to target one single entry-point matching an entity within the query semantically, for then extending this to the entire graph with graph-queries. As the KB might contain multiple documents representing identical information, HypoGator uses graph clustering to eliminate duplicate data, by computing a similarity score between subgraphs.

To mitigate any possible effects for improper QA not acknowledging the logical structure represented within the graph-based representation, the system performs a post-hoc QA cleaning through inconsistency detection. This identifies conflicting information, such as negated tuples from the KB, or events associated with the wrong field in the project's ontology (Sect. 2.1).

Vector Representation. Milvus[2] enables QA by converting full-text datasets into vectors using indexed BERT embeddings, as well as the users' question. The system uses vector cosine similarity $\cos(\theta) = A \cdot B / \|A\| \|B\|$, searches for the most similar question vector and returns the corresponding answer to such similarity. Nowadays, transformer-based QA technologies struggle with *hallucinations*, in which a response generated by an AI contains false or misleading information presented as fact, such as not being able to consistently calculate elementary mathematical problems [6]. This is mainly due to their nature of being "stochastic parrots" providing responses to stimuli independent from the presence of the answer within the training data [2], rather than effectively working on an open world assumption with rules as classical Machine Learning (ML) methods. Empirical evidence also suggests that this is mainly related to the impossibility of correctly representing in-depth semantic representations of full sentences within a real-valued vectorial representation [7]. As these representations are averaging the ones obtained from the last layer, we are losing possible core semantic information for the input sentence. Regardless of the latest attempts to also consider semantic considerations within the embedding [20], these methodologies are not dealing with inconsistent and contradictory data, as they are assuming to always deal with *clean data* containing rightful information. If, on the other hand, we want to apply these techniques on real opinion data being intrinsically contradictory, then we still lose precious information while requiring logical-driven metrics to amend for the propagated issues [23]. These training-based algorithms might also rely on datasets containing misclassified data mischievously increasing the perceived accuracy of the proposed methodology[3], further contributing to hallucination problems. As all these techniques are not necessarily capturing profound semantic understanding of the sentence deriving from its grammatical structure, this potentially invalidates approaches for document clustering relying on vectorial representations [10,19].

2.2 Natural Language Processing

Dependency Parsing. *Dependency parsing* refers to the extraction of language-independent grammatical functions expressed through a minimal set of binary relationships connecting part-of-speech components within the text, thus allowing a semi-structured, graph-based representation of the text. Generally speaking, Neural Network (NN)-based approaches are not proficient in precisely capturing relationships within the text, as they fall down the same limitations in

---

[2] https://milvus.io/.

[3] https://www.technologyreview.com/2021/04/01/1021619/ai-data-errors-warp-machine-learning-progress/.

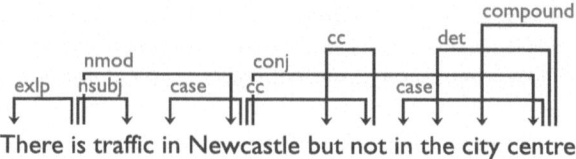

Fig. 2. Dependency structure for a sentence.

vector-based representation of sentences. Figure 2 shows how AI struggles with understanding negation from full-text, this sentence was fed into a natural language parser[4] and the result shows no sign of a **neg** (negated) dependency, despite "*but not*" being contained within the sentence. Still, we can easily determine these issues and fix them before returning the DG to our LaSSI pipeline.

Montague Grammar. MG works under the grammatical assumption of the language, through which we can freely assume that (both natural and programming) languages can be rewritten when their grammar is understood [12]. By leveraging this, we can preserve any occurring logical connective plus express verbs as logical predicates across subjects and (in)direct objects. As none of the available GQLs are able to express such rewriting mechanism through their query language, we used a prototypical graph database supporting an advanced graph rewriting mechanism [3]. Regardless, these rewritings acknowledge neither a correct representation of multi-named entities, nor the possibility of reasoning, on top of such sentences using common-sense knowledge [16]. While the former requires the adoption of specific dictionaries and entities for recognising such entities, the latter requires the adoption of a suitable KB containing the relevant information. Yet, these addressed limitations require the adoption of semantic and reasoning mechanisms that cannot be encompassed by sentence-rewriting.

3   Methodology

We now describe the key steps from Fig. 1 deriving the machine readable sentence representation from Fig. 3 via entity matching and rewriting steps. Due to space limitations, we describe the similarity metrics for each resulting graph representation, while the transformation steps are narrated.

LaSSI uses our *Parmenides ontology* for deriving: *(e1)* a list of transitive verbs, *(e2)* a list of "rejected edges", and *(e3)* a list of non-verbs. While *(e2)* identifies the multi-named entities from the DGs to be merged, where matches extracted from the raw sentences above a given threshold value are stored per sentence in an object-oriented database (MEUdb). *(e3)* identifies Universal Dependencies (UDs)[5] and other prepositions to be discarded from the edges occurring within the graphs. This is used to derive our LR.

---

[4] https://nlp.stanford.edu/software/lex-parser.shtml.
[5] UDs are a framework for consistent annotation of grammar in human language [11].

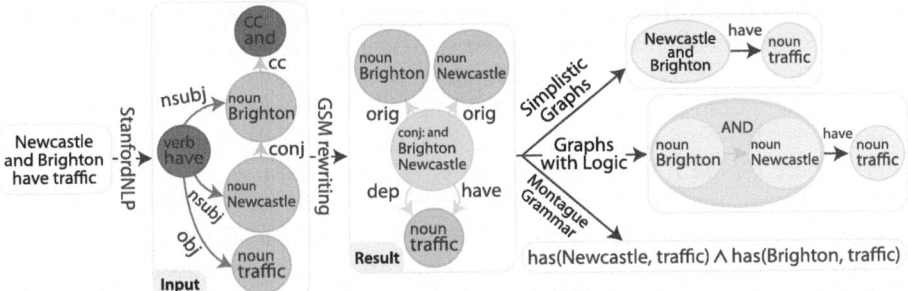

Fig. 3. Sentence transformation in LaSSI. "Input" shows red nodes - to be removed, "Result" shows a purple node in the centre- an added node. (Color figure online)

3.1 Run Pipeline

Calculate Multi-word Entity Units. At warm-up, we start services used to calculate the Multi-Word Entity Units (MEUs): Stanza [13] and SUTime [5] for deriving generic entity types and specific temporal annotations respectively, and PostgreSQL's fuzzy string matching for retrieving GeoPolitical Entities (GPE) defined in the GeoNames gazetteer [1] and for retrieving the common-sense entities within ConceptNet [16]. We associate each (*start, end*) span within the text to an entity recognised via the aforementioned services (MEUdb); this is used for associating a type to MEUs identified by StanfordNLP when generating Graphs with Logic (GwL), where logical connectives are implemented (Sect. 3.1).

GSM Representation (Applying Graph Grammars). We apply the preliminary syntactic Generalised Semistructured Model (GSM) (sentence) rewriting (Result from Fig. 3) by exploiting the grammatical structure of the sentence as derived by the UDs returned from StanfordNLP (Input). This representation pertains, for each node, the (*start, end*) character offsets reflecting its location within the original sentence. By leveraging the grammatical structure of the sentence, we also mark *kernel* as the edge reflecting the main subject-verb-(object) component. This marking will be relevant to determine which edge shall be represented as a final logical predicate, and which other edges shall be represented as its data properties. E.g., The DG associated to "The cat eats the mouse" and "The mouse is eaten by the cat" are syntactically different albeit in active/passive form. We aim to rewrite these into the same graph by exploiting the grammatical information from the DGs, so they can be rewritten into the same *kernel*: the edge then defines a binary relationship between the actor (cat), and the recipient of such action (mouse), as determined by the action (eat). The deficiency of current graph models and query languages requires using DatagramDB [3] to automatically transform such sentences into an intermediate representation according to the grammatical structure.

Create Internal Representation. As the graph matching and rewriting phase cannot account for the MEUs and other semantic features through the grammatical structure of the sentence, we now need to rewrite such graphs while taking into account semantic information. LaSSI identifies the following:

Intermediate Graph Representation. As GSM representations never contain isolated nodes, we represent it as a list of edges $\mathcal{E}$ [21]; an edge $(s,t) \in \mathcal{E}$ maps entity sources s to targets t; we associate each edge (Relationship) to another Entity via a labelling function $\lambda(s,t)$ holding the edge label information as well as the information of whether the associated verb is negated. Entities can either be *Singletons* or *SetOfSingletons*. Each Singleton contains a name, properties containing key-value association, $(start, end)$ character offsets defining the character position within the original sentence, and a type and confidence value of the MEU match (Sect. 3.1). A SetOfSingletons is composed of entities as a list of Singletons and is defined by a grouping type: *and, or, neither, not, none* or *grouping*. We associate each Singleton node s and a BERT embedding $\phi(s)$ generated by a transformer specified in the configuration.

Simplistic Graphs. Simplistic Graphs (SGs) flatten out SetOfSingletons as simple Singletons by a character offset, ordered by its Singleton constituents, while also pertaining the logical function associated to the node as text. This reflects the representation from both HypoGator [23] and traditional RDF [21].

Graphs with Logic. Otherwise, we retain SetOfSingletons to capture simple logical connectives as retrieved from StanfordNLP and our rewriting phase. We now discuss the identification of MEUs associated to *grouping* nodes: we match each SetOfSingletons to a proposed entity within MEUdb via interval intersection; then pick the entity with the highest string similarity. The type is then ranked based on a curated entity-type hierarchy as follows: GPE <: LOC(ation) and Noun <: Entity, while all the remaining types (including Person, DATE) are deemed as mutually exclusive except from Entity or None, capturing the most general representation. Given this, we also differentiate multi-entity types between the entity information, associated to the most specific type, and its specification, associated to a more general type. E.g., If we have: "Newcastle city centre", we can recognise "city centre" is a LOC while "Newcastle" is a GPE, thus the former is a specification of the latter Singleton. We also recognise "Newcastle and Brighton" as two distinct entities with an *and* grouping.

Montague Grammar. Finally, we rewrite GwL via MG to obtain a LR $\varphi(s)$ for each sentence s. We first isolate the kernel and associate the properties to it, then rewrite this into its final logical form by moving the logical connectives outside each formula while finally generating binary predicates for transitive verbs, and unary predicates for intransitive ones. Existential kernels as detected by StanfordNLP are associated to a "be" verb expressing the existence of a specific condition. After identifying the kernel, we discard all the *(e2)* rejected edges and, for all the remaining edges, we distinguish between kernel properties

---

Algorithm 1. Montague Grammar (Step 2)

1: **function** STEPTWO(k:*kernel*,p:*properties*)
2:    **if** $k=v(\bigvee\!\!\bigwedge_i x_i, b)$ **then return** $\bigvee\!\!\bigwedge_i$STEPTWO($v(x_i, b), p$);
3:    **if** $k=v(\bigvee\!\!\bigwedge_i x_i)$ **then return** $\bigvee\!\!\bigwedge_i$STEPTWO($v(x_i), p$);
4:    **if** $k=v(a, \bigvee\!\!\bigwedge_i y_i)$ **then return** $\bigvee\!\!\bigwedge_i$STEPTWO($v(a, y_i), p$);
5:    **if** $k = \neg v$ **then return** $\neg$STEPTWO(v, p);
6:    **if** $k = be(noun_{adj})$ **return** TWOSTEP($have(noun, adj), p$);
7:    **if**
8:        **return** TWOSTEP($be(t), p \circ [s: T]$);
9:    **return** k_p;

---

and adjectives referring to specific entities: for the former, we consider only the nodes not appearing within the kernel and we associate them as the kernel's properties. While doing so, we also pertain the information regarding the negation of specific properties, adjectives, or entire kernels. E.g., Given the sentence "In Newcastle, traffic is slow", we express an existential condition represented as an unary predicate be(traffic). We identify slow as an adjective for traffic and being "in Newcastle" as a spatial property associated to the unary predicate. We obtain $be_{\mathsf{GPE:\ Newcastle}}(\mathtt{traffic_{slow}})$. As *have* is a transitive verb, the sentences from Fig. 3 will then be associated to binary predicates.

Algorithm 1 describes the last rewriting step prior to obtaining the final Montague Grammar Representation (MGR), which does not approximate over the logical operator and on the node similarity. After normalising the representation of *be* and *have* predicates, so that the former are always preferred to represent the latter when the subject appears to be a spatio-temporal information, we return the final LR of our sentence by associating the same property to all the predicates that occur.

3.2   Calculate Metrics

The similarity score is now determined from our pre-chosen similarity function, and a resulting matrix is written to a JSON file. We exploit the metrics from [14] to guarantee the soundness of their definition: we denote $d|_N = \frac{d}{d+1}$ the normalisation of a distance value between 0 and 1, and $d|_N^s = 1 - d|_N$ its straightforward conversion to a similarity score. Given that our graphs of interest can be expressed as a collection of labelled edges, we reduce our argument to edge matching. Given an edge distance function ϵ, an edge e, and a set of edges $\mathcal{E}$, the best match for e is an edge $e' \in \mathcal{E}$ minimising the distance ϵ, i.e. $m_\epsilon(e, \mathcal{E}) = \arg\min_{e' \in \mathcal{E}} \epsilon(e, e')$. We can then express the best matches of edges in $\mathcal{E}$ over another set $\mathcal{E}'$ as a set of matched edge pairs $M_\epsilon(\mathcal{E}, \mathcal{E}') = \{(e, m(e, \mathcal{E}')) | e \in \mathcal{E}\}$. Then, denote $D_\epsilon(\mathcal{E}, \mathcal{E}')$ as the set of edges not participating in any match. The matching distance between two edge sets shall then consider both the sum of the distances of the matching edges as well as the number of the unmatched edges [14]. Given an edge-based representation $\mathcal{E}_s$ and $\mathcal{E}_t$ for two sentences s and t (Sect. 3.1), we derive the following edge similarity metric being the basis of any subsequent graph-based matching metric:

$$E_\epsilon(\mathcal{E}_s, \mathcal{E}_t) = \left(1 - \frac{\epsilon(M_\epsilon(\mathcal{E}_s, \mathcal{E}_t))}{|\mathcal{E}_s|}\right) \cdot D_\epsilon(\mathcal{E}_s, \mathcal{E}_t)|_N^s \tag{2}$$

Simplistic Graph. Given a node ν, an edge label ε, and normalised similarity metric ignoring the negation information, we refine ϵ from Eq. 2 by conjoining the similarity among the edges' sources and targets, while considering the edge label information. We annihilate such similarity if the negations associated to the edges do not match by multiplying such similarity by 0; then, we negate the result for transforming such similarity into a distance:

$$\epsilon_{\nu,\varepsilon}((s,t),(s',t')) = \begin{cases} \nu(s,s')\nu(t,t')\varepsilon(\lambda(s,t),\lambda(s,'t')) & \mathsf{neg}(\lambda(s,t)) = \mathsf{neg}(\lambda(s',t')) \\ 0 & \text{otherwise} \end{cases}$$

$$\tag{3}$$

For SGs, we use ν and ϵ as the ReLU-normalised cosine similarity $\nu(s,s') = \mathrm{ReLU}(\simeq_E (\phi(s), \phi(s')))$ over a specific vector embedding ϕ for a textual representation s of ν and ϵ. At this stage, we still have a symmetric measure.

Graphs with Logic. For these graphs, we extend the definition of ν from Eq. 3 as a similarity $\nu'(u,v) := \delta_\nu(u,v)|_N^s$ where δ_ν is the associated distance function defined in Eq. 4, where we leverage the logical structure of SetOfSingletons; we approximate the confidence metric via an asymmetric node-based distance deriving from fuzzy-logic metrics in combination with matching metrics for score maximisation. We return the maximum distance 1 for all the cases when one logical operator cannot necessarily entail the other.

$$\delta_\nu(u,v) = \begin{cases} 1 - \nu(u,v) & \mathsf{singleton}(u), \mathsf{singleton}(v) \\ \delta_\nu(u, m_{\delta_\nu}(u,v)) & u \equiv \wedge_i x_i, \mathsf{singleton}(v) \\ 1 - \delta_\nu(x,v) & u \equiv \neg x, \mathsf{singleton}(v) \\ 1 - \delta_\nu(u,y) & \mathsf{singleton}(u), v \equiv \neg y \\ \delta_\nu(u, m_{\delta_\nu}(u,v)) & \mathsf{singleton}(u), v \equiv \vee_i y_i \\ \delta_\nu(x,y) & u = \neg x, v = \neg y \\ \mathrm{avg}\, \delta_\nu(M_{\delta_\nu}(u,v)) & u = \wedge_i x_i, v = \wedge_i y_i, |M_{\delta_\nu}(u,v)| = |u| \\ \mathrm{avg}\, \delta_\nu(M_{\delta_\nu}(u,v)) & u = \wedge_i x_i, v = \vee_i y_i \\ 1 - E_{\delta_\nu}(u,v) & u = \vee_i x_i, v = \vee_i y_i \\ 1 & \text{oth.} \end{cases} \tag{4}$$

$R_{a_i \Rightarrow b_j}$	
a_i	b_j
0	0
0	1
1	1

$R_{a_i ? b_j}$	
a_i	b_j
0	0
0	1
1	0
1	1

$R_{(a_i \wedge b_j) \Rightarrow \perp}$	
a_i	b_j
0	1
1	0

Fig. 4. Truth tables over admissible and compatible worlds for implying, indifferent, and inconsistent atoms.

Montague Grammar. For relational algebra, we denote $\times$ as the cross product, $\bowtie$ as the natural equi-join, σ_P as the select/filter operator over a binary predicate P, and π_L as the projection operator selecting only the attributes appearing in L. We define $\mathsf{Calc}_{f \text{ as } A}(T)$ as the non-classical relational algebra operator extending the relational table T with a new attribute A while extending each tuple $(v_1, \dots, v_n) \in T$ via f as $(v_1, \dots, v_n, f(v_1, \dots, v_n))$. Given a logical formula $\varphi(s)$ and some truth assignments to logical predicates through a function Γ, we denote $[\![\varphi(s)]\!](\Gamma)$ as the *valuation function* computed over of $\varphi(s)$ via the truth assignments in Γ.

After determining all the binary or unary predicates $a_1, \dots, a_n$ associated to a LR $\varphi(s)$ of a factoid sentence s, we derive a truth table $T_s(a_1, \dots, a_n, s)$ for $\varphi(s)$ represented as a relational table $T_s = \mathsf{Calc}_{[\![\varphi(s)]\!] \text{ as } s}(\times_{a_i \in \{a_1, \dots, a_n\}} \{0, 1\})$ by assuming each atom to be completely independent from the other without any further background knowledge.

Given tables $T_s(a_1, \dots, a_n, s)$ and $T_t(b_1, \dots, b_m, t)$ representing the truth tables for s and t respectively, we derive whether each predicate a_i within $\varphi(s)$ entails, contradicts, or is independent from each predicate b_j occurring in $\varphi(t)$ by using semantic reasoning via our prototypical common-sense and linguistic ontology, Parmenides, from which we then associate a corresponding table $R_{a_i b_j}$ as in Fig. 4. By differentiating expansion TBox rules F leading to equivalent formulae and others Γ only generating weakly implying relationships, we say that a_i and b_j are inconsistent if $\exists x \in F(a_i). \exists y \in \Gamma(b_j). x \wedge y \Rightarrow \perp$ where $\perp$ is the universal falsehood, and we say that a_i implies b_j if $\exists x \in F(a_i). \exists y \in \Gamma(b_j). x = y$. Otherwise, we state that the two predicates are indifferent. From this rule derivation, we can then extract the explainable answer related to the sentence's unrelatedness. By natural-joining all the derived tables together into $\mathcal{T}$, we trivially reason paraconsistently by only considering the worlds that are not containing contradicting assumptions [4]. We express confidence from Eq. 1 as follows:

$$\mathsf{confidence}(s, t) = \mathrm{avg}\, \pi_t(\sigma_{s=1}(\mathcal{T})) \tag{5}$$

While the metric summarises the logic-based sentence relatedness, $\sigma_{s=1}(\mathcal{T})$ provides the full possible-world explanation for the number being derived.

Lemma 1. *Equation 5 provides the notion of confidence in Eq. 1 for LRs.*

Proof. Let $\mathcal{T}$ be $T_s \bowtie T_t \bowtie R_{a_1 b_1} \ldots \bowtie R_{a_n b_m}$. By selecting the rows where s is true with $\sigma_{s=1}(\mathcal{T})$, such selection represents the set of all possible worlds where s holds $\mathcal{W}(s)$. Then, summing up the truth values associated to t as $\sum \pi_t(\mathcal{W}(s))$, we obtain the number of worlds where both sentences hold $|\mathcal{W}(s) \cap \mathcal{W}(t)|$. We prove the correctness of Eq. 5 expressing 1 by definition expansion:

$$\operatorname{avg} \pi_t(\sigma_{s=1}(\mathcal{T})) = \frac{\sum \pi_t(\sigma_{s=1}(\mathcal{T}))}{|\sigma_{s=1}(\mathcal{T})|} = \frac{|\mathcal{W}(s) \cap \mathcal{W}(t)|}{|\mathcal{W}(s)|} = \operatorname{confidence}(s,t)$$

$\square$

4   Experimental Analysis

The following controlled experiments compare different sentence representations for the common-sense understanding of simplistic factoid sentence similarities. The rationale for this relies on the following observation: if those fail to capture subtleties within the sentence, those will fail to capture articulated and ambiguous ones in FOL. For addressing the possibility of using transformer-induced similarity metrics, we consider for our benchmarks two from HuggingFace: *all-MiniLM-L6-v2* and *all-mpnet-base-v2*. These transformers were chosen due to being two of the most popular sentence similarity transformers on the site at the time of writing and therefore most likely, the best performing. According to preliminary experiments, which were removed to due space, we also tested *all-roberta-large-v1* and *all-MiniLM-L12-v2* which performed worse than the ones here. To acknowledge the possibility of embedding LRs in graphs to consider more logical information, we considered in our experiments both SGs reflecting the graphs as expressed in [23] and their extension considering logical connectives (GWL). In both these scenarios, we expressed node similarity via *all-MiniLM-L6-v2*, thus mimicking RAG-based approaches. We compare all of these settings with the similarity values derived from the confidence values obtained via the MGR of the logical sentences. To better control the limitations of the competing approaches, we resort to a controlled experiment using three different sets of factoid sentences with no dependent sub-sentences, each with their own semantic qualities to test the limitations of the transformers and our approach. Set 1 tests using similar entities but with different roles to ensure understanding of actions on subjects with negation, Set 2 tests logical structure and implication thus how one sentence might imply the other but not the other way round, and Set 3 tests spatio-temporal reasoning by including locations and time. The full dataset for Fig. 5 and more detailed results can be found at OSF.io[6].

Set 1: (1) The cat eats the mouse (2) The mouse is eaten by the cat (3) The mouse eats the cat (4) The cat is eaten by the mouse (5) The cat doesn't eat the mouse (6) The mouse doesn't eat the cat. **Set 2:** (1) Alice plays football (2) Bob plays football (3) Alice and Bob play football (4) Alice or Bob play football (5) Alice doesn't play football (6) Bob doesn't play football (7) Dan

---

[6] https://osf.io/czehx/.

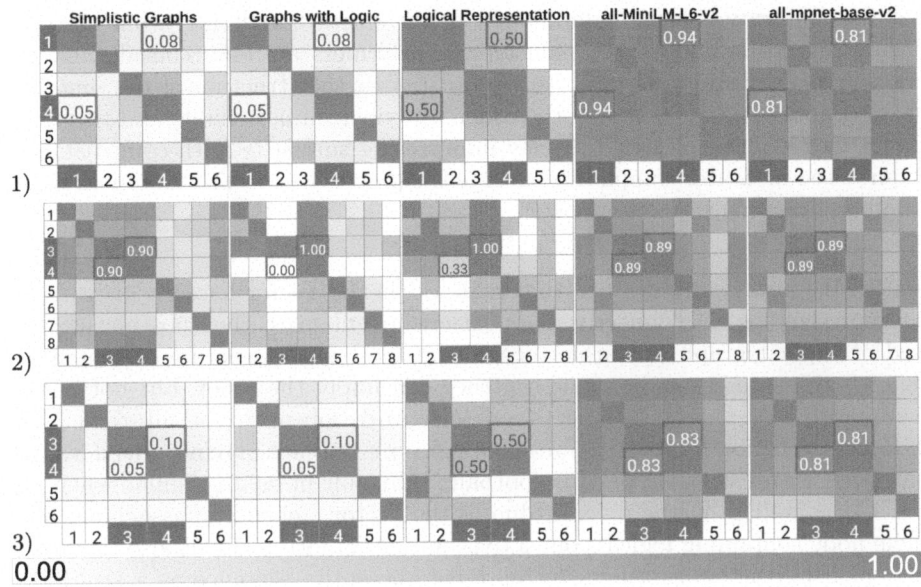

Fig. 5. Similarity matrices across different representations of Set 1–3.

plays football (8) Neither Alice nor Bob play football. **Set 3:** (1) There is traffic in the Newcastle city centre (2) Newcastle city centre is trafficked (3) There is traffic but not in the Newcastle city centre (4) In Newcastle city center on Saturdays, traffic is flowing (5) It is busy in Newcastle (6) Saturdays have usually busy city centers

4.1 Results

Set 1. The sentences in Set 1 are all variations of the same subjects (a cat and mouse), with different actions and active/passive relationships, where the cat either eats the mouse (1) or is eaten by it (4). We would expect these sentences to be opposite, and not similar. Figure 5-1 shows the similarity for graph approaches is low, between ≈5–8%, suggesting these sentences are not at all similar by a matching similarity between cat and mouse; SGs and GWL exhibit the same scores, as no logical connectors appear in sentences. Our LR shows that these sentences are 50% similar, as the cat eating the mouse does not completely exclude the possibility of the converse happening; furthermore, sentences associated to the same meaning and only differing by the active/passive form are deemed as identical, thus remarking the benefit. The transformers score ≈94% and ≈81% respectively, suggesting these two sentences are almost identical: this is completely incorrect, as both sentences do contain the same subjects, with the same action reversed on each subject. The pipeline for these given transformers will ignore stopwords which may also impact its resulting scores; we recognise that the similarity is heavily dominated by the entities occurring, while the sentence structure is almost not reflected: transformers are only considering one sentence as a collection of entities without taking the structure of the sentence into account, whereby changing the order will yield similar results and therefore cannot derive sentence structure. By interpreting similarity with compatibility, graph-based measures completely exclude the possibility of this to happen, while logic-based approaches are agnostic.

Set 2. We now want to prove that state-of-the-art vector-based semantics do not represent logical connectives appearing in sentences, thus postulating the need of explicit graph operators recognising the logical structure of sentences. Figure 5-2 demonstrates how one sentence can imply the other, but not necessarily vice versa. Given: (3) and (4), if Alice **and** Bob play football, then Alice **or** Bob are playing football. However, if Alice **or** Bob are playing, this does not imply Alice **and** Bob are playing football. SG result in ≈90% similar when (3) ⟹ (4) as well as in the reverse, due to the vector embedding associated to a single node containing logical connectives in textual form. This is also remarked by exploiting the transformers to represent the entire sentence as a vector, as these sentences are perceived as very similar with an high similarity score, thus remarking the limitations of symmetric-based similarity scores. This effect is mitigated by GWL, by both taking into consideration the logical structure of the sentence via SetOfSingletons, as well as through the design of an asymmetric node distance function (Fig. 4): we get a 100% similarity when (3) ⟹ (4), which is correct, but 0% for the vice versa, thus avoiding to acknowledge that, without any further background knowledge, the actions of Alice and Bob are completely independent, and so when Alice plays football, Bob might also be playing as well. Our LR result reflects this by exploiting the possible world semantics with 100% and ≈33% respectively.

Set 3. We have multiple scenarios involving traffic in Newcastle and whether it may be on a Saturday. From our graph-based similarity, we see concepts that should have been considered as mutually exclusive ((1) vs. (3)) are ranked higher if compared to indifferent concepts (1) vs. (2), thus remarking their inappropriateness in complex reasoning; (3) and (4) are also very dissimilar, while transformer-based are completely opposite, and still wrong. These are incorrect as (3) remarks that there is *no* traffic in Newcastle city centre, while (4) mentions that traffic is flowing in the city centre, thus suggesting little traffic. Our LR shows that (3) ⟹ (4) prompts a 50% similarity, as having traffic doesn't entail that the traffic is fast and (4) is not expanded. The vice versa is similar, as having traffic on Saturdays does not entail having traffic in general. Still, having traffic but in the city centre (3) does not necessarily entail having slow traffic throughout the city (5), while the vice versa is inconsistent.

5   Conclusions and Future Work

Preliminary results suggest the practicality of MG rewritings by combining dependency parsing, graph rewriting, MEU recognition, as well as semantic post-processing operations. We show that, by exploiting a curated KB having both common-sense knowledge (ABox) and rewriting rules (TBox), we can easily capture the essence of similar sentences. Future works will test this pipeline on real-world crawled data. Preliminary experiments suggests the inability of transformers to solve problems as closed boolean expressions expressed in natural language and to reason paraconsistently when using cosine similarity metrics [17].

Our future works will also require to enrich our Parmenides ontology to further acknowledge more fragments of language and common-sense knowledge, while ensuring its correctness through a human-in-the-loop approach. In fact, we identified several errors in GeoNames [1] that needed to be manually corrected to avoid problems (e.g., *center* and *city* alone were associated to proper location names, or GPE, which had to be fixed). Furthermore, despite ConceptNet containing most of the language information being expanded with common-sense notions extracted from texts, its automated extraction did not encompass the entirety of the information being available from Wiktionary: we then integrated a subset of ConceptNet with the linguistic information extracted from Wiktionary extracted automatically through Wiktextract [22]. We plan to also automate this integration process with human curation. Despite the slight glitches from StanfordNLP dependency parsing, our project amended some of the issues by patching their outputs in our implementation; future works will need to provide a more accurate dependency parser to ameliorate the currently-available state-of-the-art, thus improving the overall rewriting via MG. State-of-the-art NLP methods cannot capture propositional logic, while our approach can. As propositional logic is a fragment of FOL, then the former cannot express FOL entirely. FOL semantics requires going beyond a truth table approach, therefore future works will require to change the tabular semantics to also encompass variable interpretation and knowledge base grounding. Future works will further investigate on the logical expressiveness of transformers [17] and on their ability to reason by contradiction paraconsistently as in [23]. We also plan to further extend this publication to provide more details regarding our implementation from an algorithmic standpoint.

References

1. Ahlers, D.: Assessment of the accuracy of GeoNames gazetteer data. In: Proceedings 7th Workshop on Geographic Information Retrieval (GIR), pp. 74–81 (2013)
2. Bender, E.M., Gebru, T., McMillan-Major, A., Shmitchell, S.: On the dangers of stochastic parrots: can language models be too big? In: Proceedings ACM Conference on Fairness, Accountability, & Transparency (FACCT), pp. 610–623 (2021)
3. Bergami, G., Fox, O.R., Morgan, G.: Matching and rewriting rules in object-oriented databases. Mathematics **12**(17) (2024)

4. Carnielli, W., Esteban Coniglio, M.: Paraconsistent Logic: Consistency, Contradiction and Negation. Springer, Cham (2016)
5. Chang, A.X., Manning, C.D.: SUTime: a library for recognizing and normalizing time expressions. In: Proceedings 8th International Conference on Language Resources & Evaluation (LREC), pp. 3735–3740 (2012)
6. Davis, E.: An ill-designed study of math word problems in large language models: review of (ye, xu, li, and allen-zhu). Technical report, New York University (2024). https://cs.nyu.edu/~davise/papers/PhysicsOfLLMs.pdf
7. Devlin, J., Chang, M., Lee, K., Toutanova, K.: BERT: pre-training of deep bidirectional transformers for language understanding. In: Proceedings Conference of the N.American Chapter, Association for Computational Linguistics: Human Language Technologies (NAACL-HLT), vol. 1, pp. 4171–4186 (2019)
8. Fan, C., Chen, W., Wu, Y.: Knowledge base question answering via path matching. Knowl.-Based Syst. **256**, 109857 (2022)
9. He, X., et al.: G-retriever: retrieval-augmented generation for textual graph understanding and question answering (2024). https://arxiv.org/abs/2402.07630
10. Jovita, Linda, Hartawan, A., Suhartono, D.: Using vector space model in question answering system. In: Proceedings International Conference on Computer Science & Computational Intelligence (ICCSCI), pp. 305–311 (2015)
11. de Marneffe, M.C., Manning, C.D., Nivre, J., Zeman, D.: Universal dependencies. Comput. Linguist. **47**(2) (2021)
12. Montague, R.: English as a formal language. In: Linguaggi nella Societa e nella Tecnica, pp. 189–224. Edizioni di Communità, Milan, Italy (1970)
13. Qi, P., et al.: Stanza: a Python natural language processing toolkit for many human languages. In: Proceedings 58th Annual Meeting of the Association for Computational Linguistics: System Demonstrations (2020)
14. Raedt, L.D.: Logical and Relational Learning. Springer, Cham (2008)
15. Seshia, S.A., Sadigh, D., Sastry, S.S.: Toward verified artificial intelligence. Commun. ACM **65**(7), 46–55 (2022)
16. Speer, R., Chin, J., Havasi, C.: Conceptnet 5.5: an open multilingual graph of general knowledge. In: Proceedings 31st AAAI Conference on Artificial Intelligence (AAAI), pp. 4444–4451 (2017)
17. Strobl, L., et al.: What formal languages can transformers express? A survey. Trans. Assoc. Comput. Linguist. **12**, 543–561 (2024)
18. Tenghao, J.: FAQ question answering method based on semantic similarity matching. In: Proceedings 6th International Symposium on Computer Science & Intelligent Control (ISCSIC), pp. 93–100 (2022)
19. Teofili, T.: par2hier: towards vector representations for hierarchical content. In: Proceedings International Conference on Computational Science (ICCS), pp. 2343–2347 (2017)
20. Vaswani, A., et al.: Attention is all you need. In: Proceedings 31st Conference on Neural Information Processing Systems (NIPS), pp. 5998–6008 (2017)
21. Virgilio, R.D., Maccioni, A., Torlone, R.: Approximate querying of RDF graphs via path alignment. Distrib. Parallel Databases **33**(4), 555–581 (2015)

22. Ylonen, T.: Wiktextract: Wiktionary as machine-readable structured data. In: Proceedings 13th Language Resources and Evaluation Conference (LREC), pp. 1317–1325 (2022)
23. Zhang, T., et al.: GAIA - a multi-media multi-lingual knowledge extraction and hypothesis generation system. In: Proceedings Text Analysis Conference (TAC) (2018)

ClustEm4Ano: Clustering Text Embeddings of Nominal Textual Attributes for Microdata Anonymization

Robert Aufschläger(✉), Sebastian Wilhelm, Michael Heigl, and Martin Schramm

Deggendorf Institute of Technology, 94469 Deggendorf, Germany
`robert.aufschlaeger@th-deg.de`

Abstract. This work introduces `ClustEm4Ano`, an anonymization pipeline that can be used for generalization and suppression-based anonymization of nominal textual tabular data. It automatically generates value generalization hierarchies (VGHs) that, in turn, can be used to generalize attributes in quasi-identifiers. The pipeline leverages embeddings to generate semantically close value generalizations through iterative clustering. We applied KMeans and Hierarchical Agglomerative Clustering on 13 different predefined text embeddings (both open and closed-source (via APIs)). Our approach is experimentally tested on a well-known benchmark dataset for anonymization: The UCI Machine Learning Repository's Adult dataset. `ClustEm4Ano` supports anonymization procedures by offering more possibilities compared to using arbitrarily chosen VGHs. Experiments demonstrate that these VGHs can outperform manually constructed ones in terms of downstream efficacy (especially for small k-anonymity) and therefore can foster the quality of anonymized datasets. Our implementation is made public.

Keywords: anonymization · privacy · nlp · text embedding · clustering

1 Introduction

Background. Anonymizing tabular datasets with textual attributes typically requires generalizing textual attributes that qualify as quasi-identifier (QI), i.e., as set of attributes that in combination allow to reidentify individuals [39]. For this purpose, VGHs are commonly created to group similar values into categories to prevent privacy attacks, such as record linkage in microdata [39], and to maintain usability. However, creating suitable VGHs is usually a manual process that demands domain knowledge, presenting a challenge for human data processors.

Approach. We introduce `ClustEm4Ano`, a novel approach that employs embeddings to construct VGHs for anonymization. In particular, when considering the uncertainty penalty-based methods to generalize with less information loss, hierarchies of nominal attributes have to be known in advance. This is where text

R. Chbeir et al. (Eds.): IDEAS 2024, LNCS 15511, pp. 122–137, 2025.
https://doi.org/10.1007/978-3-031-83472-1_9

embeddings come in handy due to their intrinsic semantics. In fact, embeddings can be used to generate VGHs for nominal data that otherwise need to be defined manually or rely on domain orderings like in [6]. We experiment with averaging word embeddings and contextualized text embeddings. We demonstrate that clustering can automate the anonymization of nominal textual data in *microdata*, i.e., tabular data, where each entry is assigned to a single individual. Figure 1 visualizes the idea of how embeddings relate to clustering and VGH.

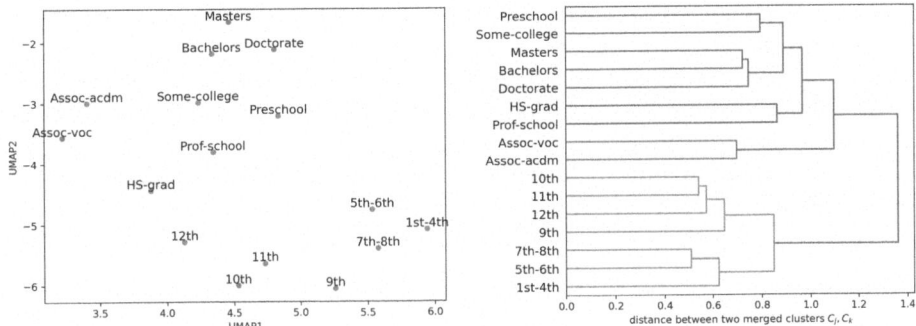

Fig. 1. Visualizing `Mistral AI` embeddings with Uniform Manifold Approximation and Projection (UMAP) (left) and the corresponding VGH (right) obtained by `Agglomerative Hierarchical Clustering` of embeddings of the values from the Adult's [7] attribute `education`.

A further motivation for the idea to cluster embeddings for anonymization of nominal data is presented in Fig. 2. In brief, the idea is to use open information about value similarities to maintain data utility while aiming for distribution-invariant anonymization using embeddings. In both plots, the x-axis represents the similarity between embeddings of nominal attribute values. The plots would

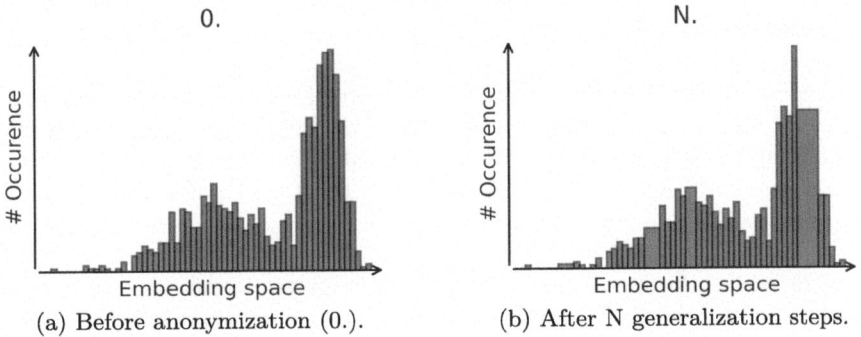

(a) Before anonymization (0.).   (b) After N generalization steps.

Fig. 2. Visualization of the embedding similarity and distribution-based motivation for `ClustEm4Ano`.

vary for different attributes, and even for the same attribute across different datasets.

We conduct experiments to evaluate the effectiveness of different anonymization strategies on datasets used for training machine learning (ML) models. The experiments involve 13 different text embeddings and various ML models to assess the classification performance using anonymized data.

Research Questions

- How can semantically accurate VGHs be created using clustering of text embeddings?
- How can text embeddings be integrated into existing value generalization and suppression-based anonymization algorithms to meet specific privacy thresholds, such as k-anonymity?
- How does using different text embeddings affect the VGH-based anonymization process?

Contributions

- We introduce an anonymization pipeline `ClustEm4Ano`, designed to anonymize microdata that have nominal textual quasi-identifier attributes.
- As part of that, we propose the use of automatically generated VGHs derived from clustering text embeddings from nominal attribute values.
- We experimentally compare the use of different text embeddings to create VGHs for anonymization and compare them to manually defined VGHs (baseline) with promising results in retained percentage (after suppression), normalized average group size metric, and ML efficacy (test accuracy and $F1$ Score after training ML classifiers with anonymized Adult datasets).

Availability. Our code is made available as open-source on GitHub at https://github.com/EAsyAnon/ClustEm4Ano.

2   Related Works

Text Anonymization. Due to the intricate nature of natural language, anonymizing textual data presents a complex challenge [21,29]. While numerous methods for anonymizing structured data with well-established privacy models are available, they are not directly transferable to text data, so particular approaches are necessary [25]. One of the earliest contributions to text data anonymization dates back to [42], which details the identification and replacement of personally identifying information in medical records using various pattern-matching algorithms. Over the years, various strategies such as removal, tagging, generalization, and substitution have been developed. The choice of strategy depends on the context of the application and varies significantly between anonymizing free-form text, like letters, and structured categorical data, like microdata.

VGHs for Anonymization of Microdata. The generalization process of textual attributes for microdata anonymization relies primarily on a specific VGH [39], typically obtained from a knowledge base [4,28], and the algorithms to obtain k-anonymity are based on a subsequent generalization lattice for the quasi-identifier attributes [16]. The creation and influence of such VGHs have been extensively discussed, but existing research on creating VGHs either uses specific knowledge bases, as in [5], focuses on numerical data [13], or uses semantic rules obtained from human expert knowledge [32].

NER-Based Anonymization for Textual Data. Named Entity Recognition (NER) is widely used for identifying sensitive entities, including categories such as individuals, organizations, and geographic locations in free-form text. NER employs both rule-based systems and ML techniques, such as Hidden Markov Models and Conditional Random Fields [45]. Post-processing often includes contextual analysis to identify additional sensitive data [28]. However, these methods are less effective for structured data and often lead to significant information loss or distortion [8,21,25]. The methods also depend on extensive volumes of annotated datasets, particularly data that has already been correctly anonymized.

Using Text Embeddings for Anonymization. Word embeddings represent words in a high-dimensional space, capturing semantic relationships, which are useful in NER and potentially for anonymization [2,8,21]. Text embedding models are typically learned on extensive domain-specific text corpora and primarily rely on neural networks. The vectorization process results in a token representation that is subsequently mapped to probabilities for specific entities – for example, to rate sensitivity [8]. Nevertheless, this approach also relies on the definition of subjects to be protected or on expert-annotated data, but in much smaller amounts than for NER-based technologies [21]. In [21], entities in text documents are represented as word vectors that capture semantic relationship. Then sensitive entities can be protected by removing other entities co-occuring, where the word vectors are similar. In [2], tokens are replaced with tokens that have neighboring word vectors in the embedding space, and in [8], token embeddings are classified and sensitive tokens are replaced by class tokens.

Using Clustering for Anonymization. Clustering methods have been applied to various data types, including text. Referring to free-text data, in [30], the authors use word embeddings from *BERT* and propose a clustering method for large text documents. Another example of clustering-based anonymization of text data is provided in [18], where the authors fine-tuned a NER tool and solved co-references between named entities by clustering. Further, a relevant work is [11], where the authors propose a greedy k-member clustering algorithm that adds records to a randomly chosen record individually while always adding the candidate record with the smallest information loss in the cluster after adding. This process is repeated while always using a record whose distance is most far away from the previous starting record. In [24], the authors use KMeans clustering to simultaneously partition records into clusters with similar records and then adjust the group members to obtain k-anonymity.

Placement of This Work. Our analysis identifies a gap in the literature concerning the automatic generation of VGHs using word embeddings and clustering. Current methods mainly focus on NER systems and do not fully utilize the potential of VGHs.

3 Basic Definitions

Anonymization. Let $D = \{R_1, R_2, \ldots, R_n\}$ be a multiset of row records with $n \in \mathbb{N}^+$ not necessarily distinct records composed of attribute values. We refer to an *attribute* as a peace of information about an individual. A QI is a set of attributes that allows re-identification of individuals in data records. *Sensitive Attributes* (SAs) are attributes in T, where it should be impossible to assign corresponding attribute values after potential attribute inference (attacks). In *domain* generalization [39], the domains of QI's attributes are mapped to a more general domain. Domain Generalization can be achieved via different VGHs, where each value in a non-generalized domain is mapped to a unique value in a more general domain. If all generalized values are on the same level in the value generalization hierarchy, it is called *full-domain* generalization. In general, VGHs are a subclass of ontologies where only the is-a (sub-type) relationships are considered [4]. In particular, in our experiments, we use the generalization algorithm FLASH [22], which, among others, is implemented by the ARX Software [36]. This algorithm offers fast execution times and seeks minimal information loss while traversing the generalization lattice in a bottom-up breadth-first search. Also, the anonymization algorithm allows to predefine certain required thresholds for *privacy models* such as k-*anonymity* [39,40], l-*diversity* [27], and t-*closeness* [23]. Thereby, records with identical QI attributes are summarized into a single equivalence class *group*. We denote the set of all equivalence classes as *groups*. Then, the privacy models are defined as follows: k-anonymity (k): $k := \min_{group \in groups} |group|$, l-diversity (l): $l := \min_{group \in groups} |\{\text{values for SA in } group\}|$, t-closeness (t): checks granularity of SA values in single groups in comparison to the distribution in the dataset. Each of the privacy models has its advantages and limitations [3]: k-anonymity effectively prevents singling out individuals but depends on k. l-diversity improves on k-anonymity by adding protection against inference. t-closeness provides additional defense against inference by ensuring that the distribution of sensitive attributes in each group closely matches the overall distribution.

Word Embeddings. Dealing with nominal textual attributes we transform text into numerical representations. A distributed word representation that is dense, low dimensional, and real-valued, is called *word embedding* [43].

Clustering. Clustering is an unsupervised ML method that takes as input n-dimensional real-valued feature vectors, where $n \in \mathbb{N}$, $K \in \mathbb{N}$ the number of desired cluster centers, and outputs n-dimensional real-valued cluster centers $c_1, c_2, \ldots, c_K$ and assigns each feature vector exactly one cluster center. We utilize two bottom-up clustering techniques. First, a clustering sequence that incorporates the non-hierarchical clustering method `KMeans` [26] and, second, Ward's `Hierarchical Agglomerative Clustering` [44], where two clusters are successively merged and the variance of the clusters being merged is minimized. While `KMeans` is not designed for hierarchical clustering, Ward's `Hierarchical Agglomerative Clustering` has a hierarchical structure, and among the Agglomerative Clustering approaches, Ward provides the most regular sizes[1], i.e., most similar cluster sizes. It is expected that `KMeans`' properties are beneficial: Separates samples in K groups (allows to specify number of groups, which is the only input we need), creates disjoint clusters, clusters are chosen to have similar variance, scales well to large number of samples, is widely used in many different fields. However, on the other hand, `KMeans` minimizes so-called inertia, which assumes clusters in the embedding space to be isotropic and convex. When it comes to isotropy in the encoding clusters, the authors in [12] found that *BERT* has a smaller average Local Intrinsic Dimension (LID) than larger language models such as *GPT* and *GPT2* in almost all hidden layers (resp. contextual embeddings). Thereby, smaller LIDs indicate local anisotropy that potentially could lead to a less qualitative clustering. We do not pursue further research in this direction.

4 Proposed Pipeline

Figure 3 shows the dataflow diagram of `ClustEm4Ano`. First, a data processor (denoted by the human symbol) needs to define a set of QIs with at least one entry and optionally a sensitive attribute. Then, the attribute values from the QI(s) column(s) are extracted. Given the set of extracted values, a predefined clustering method (`Hierarchical Agglomerative Clustering`/iterative applying `KMeans`), and an embedding model (for example, one of the 13 tested experimentally in this pipeline), can be used to generate VGH(s). Using the generated VGHs, predefined privacy models like k-anonymity, and a suppression limit (which defines the maximum percentage of records where all QI attributes are replaced with '*' to handle rare values or outliers), anonymized tabular data can be produced.

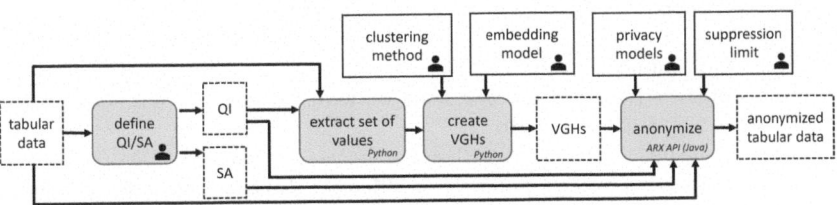

Fig. 3. `ClustEm4Ano`: Dataflow diagram. The human piktogram denotes variables and processes that need to be given by a user.

---

[1] https://scikit-learn.org/stable/modules/clustering.html#hierarchical-clustering, accessed on 13/02/2024.

We propose the anonymization pipeline in Algorithm 2. One novelty is introduced in Algorithm 1, which utilizes clustering of text embeddings to obtain multi-level (i.e., with multiple hierarchies), full-domain generalization (i.e., mapping a chosen value to the same generalized value or value set across all records). Our implementation uses KMeans and Agglomerative Hierarchical Clustering to group similar values together based on their distance in the embedding space. KMeans works by partitioning the whole dataset into clusters such that each embedding belongs to the cluster with the nearest center. The proposed algorithm first transforms each unique value in the column to an initial set of raw values and a parallel set of preprocessed values. Initially, each unique value is its own cluster and is associated with a unique numeric label. Over time, in each iteration of the while loop in Algorithm 1, the algorithm merges clusters into fewer and fewer clusters. The process is repeated until achieving a fully generalized state, where all values belong to a single cluster – represented by '*'. When using KMeans, we noted that, when fitting center vectors in early stages in the generalization algorithm, the KMeans implementation in sklearn sporadically did not converge, leading to inconsistent labels and cluster centers. For example, in the experiments, when clustering the word2vec embeddings of the attribute native-country with 42 values, in the first generalization step 7 values were generalized in one cluster, and in the 4 subsequent steps there was no generalization (only from the 5th step the KMeans algorithm input n_clusters was small enough that the algorithm could converge). This, however, did not result in worse performance in the experimental results. Besides KMeans, we applied Agglomerative Hierarchical Clustering, where pairs of clusters are recursively merged based on linkage distance between clusters, not only the distance between the cluster centers. In the anonymization (Algorithm 2), in line 2, we use the algorithm FLASH [22] – a globally-optimal anonymization algorithm with minimal information loss.

Algorithm 1: create_vgh_for_attribute

Input : T: Table of records

q: Attribute column in T

Output: vgh: Value generalization hierarchy

1 values = [preprocess($value$) for value in values($T[q]$)];

2 vgh = [values];

3 labels = [0, ..., num_of_values - 1] ;

4 cluster_centers = [vectorize($value$) for value in values];

5 **while** #$cluster\_centers > 1$ **do**

6 $\quad$ labels(, cluster_centers) = getCluster($labels$, n_clusters = $len(values)$ - i, $vectors = cluster\_centers$); // KMeans: labels, cluster_centers

7 $\quad$ categories = getCategories($values$, $labels$);

8 $\quad$ vgh.append(categories);

9 **return** vgh;

Algorithm 2: `ClustEm4Ano`

Input : $T_{original}$: Original table of records
$QI = \{q_1, q_2, \ldots, q_m\}$: QI attributes in T
k: Desired k-anonymity
l: Desired l-diversity
sup_lim: Suppression limit
Output: T^*: Table of anonymized records
1 $vghs = \{\texttt{create\_vgh\_for\_attribute}\ (T,q_j) \ : \ j \in 1..m\}$;
2 $T^* = \texttt{FLASH}(T, QI, SA, vghs, k, l, sup\_lim)$; // anonymization
3 **return** T^*;

---

5 Experimental Evaluation

5.1 Methodology

In the context of anonymization, evaluation can be applied in several steps. Before applying anonymization, a domain expert must choose IDs, QIs and SA(s). To the best of our knowledge, there is no automatic evaluation assigning these, and they must be manually defined. After applying the anonymization algorithm, we evaluate the data by measuring its efficacy in a downstream classification task. Additionally, we check the number of suppressed records and how close the data matches the desired k-anonymity. Also, from a privacy perspective, privacy risks are assessed at this step. We only perform a quantitative research.

5.2 Implementation Details

Embeddings. It can be assumed that there is no "one [embedding] fits all" because no particular text embedding method dominates across all downstream tasks, as shown in the MTEB: Massive Text Embedding Benchmark [33]. Besides using open-source embeddings, we also test the possibility of retrieving embeddings by accessing proprietary APIs. We experiment with OpenAI's text embedding models `text-embedding-3-small` ($0.02 / 1M tokens) and `text-embedding-3-large` ($0.13 / 1M tokens) - released on January 25, 2024 and Mistral AI's text embedding model `mistral-embed` ($0.1 / 1M tokens). Overall, due to the small number of tokens needed, the price of using OpenAI and Mistral AI embeddings in our experiments was negligible (<0.1$ in total). We use pre-trained word embedding based on *word2vec* [31] that was trained on the text corpus obtained from `Google News`. Thereby, we use the implementation from the `Python` library `Gensim` [37]. Further, we test our approach on other pre-trained word embeddings based on *BERT* [14,20,38], *GloVe* [35], *MPNet* [41], OpenAI's closed-source models `text-embedding-3-small` and `text-embedding-3-large`, and *fastText* [19]. We accessed most of the embeddings through Hugging Face's `Sentence-Transformers Python` framework². The complete list of embeddings is given in

---

² https://huggingface.co/sentence-transformers, accessed on 27/01/2024.

Sect. 5. With the embedding models `word2vec-google-news-300`, `bert-base-uncased`, and `fastText/cc.en.300.bin`, we use averaged word embeddings. Otherwise, we use embeddings that transform the whole textual attribute value at once. Whereas in *word2vec* and *GloVe*, a word is mapped to exactly one embedding vector, making them non-contextual word embeddings, in the embeddings obtained from the transformer-based language model *BERT*, or from OpenAI's embeddings, the context of a word can be used to find a corresponding embedding. Therefore, the same word can lead to different embedding vectors. However, in our experiments when we use pre-trained contextual embeddings, we simplify by considering the single words without contextual information and average their embeddings. Only when we use Mistral AI and OpenAI's embedding models we use text embeddings that take into account the contextual information of the whole textual attribute value.

Clustering. The clustering is implemented in `scikit-learn` [34]. When using the `scikit-learn`'s clustering algorithms, we use the default parameters in the constructor except for the number of initial clusters. For all implementations except for the anonymization algorithm we used `Python` (3.11.8).

Anonymization. Once given the VGHs we can conduct anonymization through generalization and suppression automatically by using the ARX Software [36]. We use the anonymization algorithm implemented in `Java` (java.version=17.0.10) in the ARX Software[3]. The tool conducts an efficient globally-optimal search algorithm for transforming data with full-domain generalization and record suppression. For practical reasons and plausibility, we fix one QI set that implies higher or equal k-anonymity for any QI that is a subset of the larger QI. The approach to choose only one (large) quasi-identifier is intrinsically most easy and most secure when it comes to k-anonymity. From a technical perspective, we only experiment using textual and numerical attribute values with the data types string and float, respectively.

5.3   Setup

Dataset. In this study, we utilize the Adult dataset [7] and its official train-test split. In our experiments, we use nominal attributes with expected high feature importance, define QI:={workclass, education, occupation, native-country}, and the attribute salary-class is the sensitive attribute SA.

Configuration. We use several values $k = 2, 5, 10, 15, 20, 25, 30, 50, 100, 150, 200$ for certain k-anonymity as input for the anonymization algorithm in our evaluation. We use a suppression limit of 50% which increases speed in anonymization and masks outlier values. Further, we enforce l-diversity ≥ 2. Because there are only two possible values for SA, we obtain $l = 2$ in all anonymizations. We do not enforce t-closeness because, given only two possible values, the distributions of groups tend to be more concentrated. In such cases, if a group ends up with an overrepresentation of one value, it can significantly skew the distribution away

---

[3] https://arx.deidentifier.org/development/algorithms/, accessed on 30/01/2024.

from the dataset's overall distribution. On the contrary, if a sensitive attribute has many possible values, the distribution in a group can be quite spread out and diverse even if some values are overrepresented. As a baseline, we use the same VGHs as applied in [15], where the hierarchies are extracted from the official ARX Software GitHub repository. The baseline VGHs have significantly smaller number of levels in the hierarchies $(3-4-3-3)$ compared to our approach where the number of levels equal the number of attribute values (see Algorithm 1). We only find hierarchies for textual attributes. After anonymization, the order of the rows is not altered.

Evaluation criteria. In our study, we evaluate various measures to assess the trade-off between privacy preservation and utility in the anonymization process. Additionally, we employ several ML models to gauge the impact of anonymization on predictive performance in binary classification. We calculate the measures given the anonymized data D', where suppressed records were removed. For ML efficacy, we include all data, including the suppressed records, where the QI attributes have value '*'. D denotes the un-anonymized dataset. Besides rating privacy preservation in terms of *privacy models* as discussed in Sect. 3, we evaluate the anonymized dataset in terms of utility and ML efficacy. *Utility criteria*: Percentage of records retained (perc_recs): Ratio of anonymized records to original records, indicating the data suppression: $perc\_recs := |D'|/|D|$, Normalized average group size metric citelefevre2006mondrian: $c_{avg} := |D'|/(|groups| \cdot k)$.

Machine Learning Efficacy. In advance, before using the data as input for ML models, we perform (multi-label) one-hot encoding on nominal textual data. We do not create VGHs for numerical data as we focus on textual attribute values in our approach. We evaluate by training ensemble ML models with anonymized data (features: QI $\cup$ {`capital-gain`, `capital-loss`, `hours-per-week`}, label: SA). We perform simple hyperparameter tuning in our experiments with `sklearn's GridSearchCV` with 5-Fold Cross Validation. We use the classification accuracy (ACC) for scoring, and after finding the best hyperparameters, we calculate ACC and $F1$ Score applying the trained binary-classification ML model on the unseen test dataset. For the Adult dataset, we use the sensitive attribute salary-class as target variable for classification as a downstream ML task. We use $F1$ Score, because it helps with high class-imbalance in the salary-class as it combines both precision and recall. We include the Random Forest classifier (RF) [10], AdaBoost [17] with base classifier decision tree, and Bagged Decision Trees (BAG) [9] in our experiments. We tuned the RF model using the following hyperparameter grid: [n_estimators: [10, 50, 100] (number of trees), max_depth: [None, 10, 20, 30], min_samples_split: [2, 5, 10] (min. samples required to split an internal node), min_samples_leaf: [1, 2, 4] (min. samples required to be at a leaf node)]. We configured the AdaBoost/BAG models with a `DecisionTreeClassifier` as `base_estimator` and used the following hyperparameter grid for tuning:[`base_estimator__criterion`: ['gini', 'entropy'], `base_estimator__max_depth`: [4, 5, 6, 7, 8, 9, 10, 11, 12, 15, 20, 30, 40, 50, 70, 90, 120, 150], `n_estimators`: [10, 50]].

5.4 Experimental Results

Figure 4 compares ML efficacy in terms of ACC using different clustering methods and different ML models in our approach. To speed up training and to increase the importance of generalized values in the ML efficacy evaluation, we only use the QI columns and the SA in training and testing. For $k \leq 30$ the `KMeans` based approach (right) shows higher mean ACC over all embeddings. Generally, `KMeans` plots show higher variance in the ACC results. The high variance at the AdaBoost model (yellow) for small k (< 25) is caused by some outliers in the ACC results. When comparing the different ML methods, AdaBoost generally performs best (yellow). Comparing the embedding based results with the baseline generalizations, for $k < 100$, the embedding labeled plots generally have higher ACC. When comparing to un-anonymized ACC obtained by the different ML methods, only few embedding-labeled ACC scores obtained by AdaBoost have higher scores for results at $k \leq 50$, meaning the ACC for anonymized data is generally lower than the non-anonymized baseline but remains competitive and can even outperform. Anonymized data show a noticeable decline as k increases.

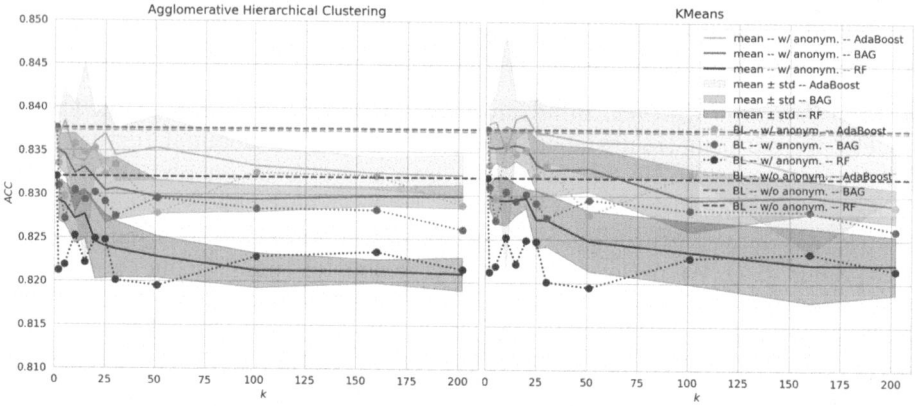

Fig. 4. Accuracy comparison between models trained on anonymized data obtained using VGHs obtained by `Agglomerative Hierarchical Clustering` (left) and `KMeans` clustering (right). (Color figure online)

Figure 5 shows privacy, utility, and ML efficacy when using iterative `KMeans` clustering to obtain the VGHs. Lower k values (2, 5, 10) tend to show better performance, with some variations depending on the specific embedding. In comparison to [15], ACC behaves similar with growing k. We only plot $k \leq 50$ due to its higher practical relevance compared to greater k values. Note, that obtained k-anonymity sometimes differs to input k in the anonymization algorithm. In Subfig. 5(c), the plots show a declining trend in t-closeness as k increases, which is expected since higher k typically means more generalization and potential

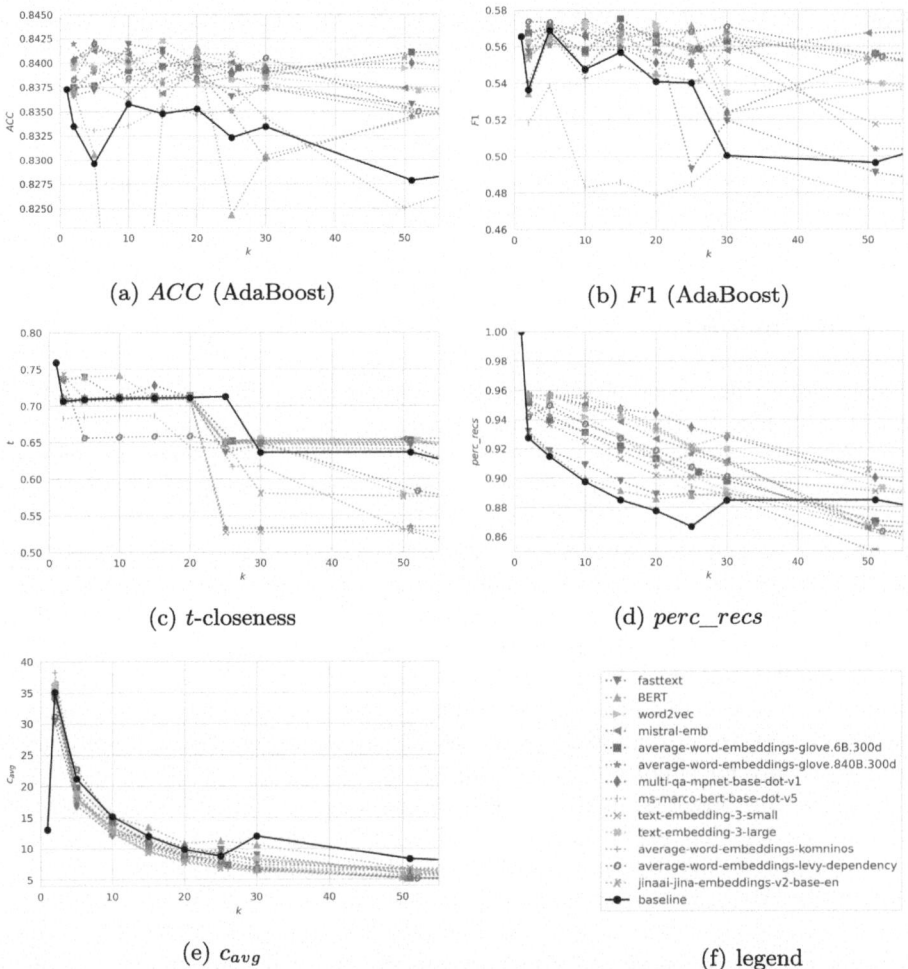

(a) *ACC* (AdaBoost)

(b) *F*1 (AdaBoost)

(c) *t*-closeness

(d) *perc_recs*

(e) c_{avg}

(f) legend

Fig. 5. Performance measures. The VGHs used for anonymization were generated using `KMeans` clustering.

information loss. Comparing embedding labels with the baseline label, no clear tendency can be seen. We assume that generalizing semantic similar values has stabilizing impact on sensitive attribute distributions in groups. Only the growing group sizes lead to similar group and global distributions. While increasing k impacts baseline *ACC* negatively, certain embeddings can maintain or even improve *ACC* (Subfig. 5(s)). k does not directly correlate with *ACC* – especially for small k. We assume, that generalization and suppression to obtain decent k might positively influence accuracy by removing outliers and promoting better model generalization capability when training. Except for outliers (*BERT*) at $k = 10$ and $k = 25$, the anonymized data labeled by embeddings mostly yields higher ACC up to $k = 30$. The baseline starts high (with no anonymization applied) and decreases steeply, then slightly recovers at $k = 30$. Other settings follow a similar trend but maintain higher $perc\_recs$ compared to the baseline as k increases (Subfig. 5(d)). From $k = 5$, other settings generally show better c_{avg} values compared to the baseline, especially for $2 \leq k \leq 30$ (Subfig. 5(e)). Smaller values indicate better performance in anonymization as group sizes are closer to k. If all groups in the anonymized data have group size k, $c_{avg} = 1$ is optimal.

6   Conclusions

In summary, this paper introduces an anonymization pipeline, `ClustEm4Ano` (dataflow in Fig. 3), that utilizes embeddings of nominal textual attribute values. The paper explores the use of generating VGHs from open and closed-source embeddings in the context of microdata anonymization. The VGHs are obtained through hierarchical clustering of the embeddings by iteratively applying `KMeans` or `Agglomerative Hierarchical Clustering`. In Sect. 5, it tests the effectiveness of clustering text embeddings to obtain VGHs for generalization and suppression-based anonymization. We show that various embeddings can be used to find VGHs that seem to be semantically correct (cf. Fig. 1). The experimental results obtained show that using automatically generated VGHs with many levels can yield anonymizations that outperform the use of manually constructed VGHs in anonymization and can even outperform un-anonymized data, especially for small k ($2 \leq k \leq 30$).

Despite the progress made, several limitations and areas for future work remain. Current clustering methods might not be effective at finding clusters in embeddings. While the number of clusters is predefined, the optimal number might differ. Also, our method primarily focuses on the specific privacy models k-anonymity and l-diversity and the anonymization algorithm `FLASH` [22], leaving room for exploration with other privacy models and other anonymization algorithms. Furthermore, in future research, domain-specific embeddings present an interesting area for exploration, as the choice of embedding models could be tailored to individual domains. Fine-tuning transformer-based word embeddings based on specific domains holds promise. However, in this regard, it is important to note that sharing word embeddings trained on sensitive data can compromise privacy, c.f. [1].

Acknowledgments. This research work results from the research projects EAsyAnon ("Empfehlungs- und Auditsystem zur Anonymisierung.") - funded by the German Federal Ministry of Education and Research (BMBF) and the European Union (grant number 16KISA128K) - and jbDATA - funded by the German Federal Ministry for Economic Affairs and Climate Action of Germany (BMWK) and the European Union (grant number 19A23003H).

References

1. Abdalla, M., Abdalla, M., Hirst, G., Rudzicz, F.: Exploring the privacy-preserving properties of word embeddings: algorithmic validation study. J. Med. Internet Res. **22**(7), e18055 (2020)
2. Abdalla, M., Abdalla, M., Rudzicz, F., Hirst, G.: Using word embeddings to improve the privacy of clinical notes. J. Am. Med. Inform. Assoc. **27**(6), 901–907 (2020)
3. Aufschläger, R., et al.: Anonymization procedures for tabular data: an explanatory technical and legal synthesis. Information **14**(9) (2023)
4. Ayala-Rivera, V., et al.: Enhancing the utility of anonymized data by improving the quality of generalization hierarchies. Trans. Data Priv. **10**(1) (2017)
5. Ayala-Rivera, V., Murphy, L., Thorpe, C.: Automatic construction of generalization hierarchies for publishing anonymized data. In: Proceedings of the 9th International Conference Knowledge Science, Engineering & Management (KSEM), pp. 262–274 (2016)
6. Bayardo, R., Agrawal, R.: Data privacy through optimal k-anonymization. In: Proceedings of the 21st International Conference on Data Engineering (ICDE), pp. 217–228 (2005)
7. Becker, B., Kohavi, R.: Adult. UCI Machine Learning Repository (1996)
8. Biesner, D., et al.: Anonymization of German financial documents using neural network-based language models with contextual word representations. Int. J. Data Sci. Anal. **13**(2), 151–161 (2022)
9. Breiman, L.: Bagging predictors. Mach. Learn. **24**, 123–140 (1996)
10. Breiman, L.: Random forests. Mach. Learn. **45**, 5–32 (2001)
11. Byun, J.W., Kamra, A., Bertino, E., Li, N.: Efficient k-anonymization using clustering techniques. In: Proceedings of the 12th International Conference on Database Systems for Advanced Applications (DASFAA), pp. 188–200 (2007)
12. Cai, X., Huang, J., Bian, Y., Church, K.: Isotropy in the contextual embedding space: clusters and manifolds. In: Proceedings of the 9th International Conference on Learning Representations (ICLR) (2021)
13. Campan, A., Cooper, N., Truta, T.M.: On-the-fly generalization hierarchies for numerical attributes revisited. In: Proceedings of the 8th VLDB Workshop on Secure Data Management (SDM), pp. 18–32 (2011)
14. Devlin, J., Chang, M.W., Lee, K., Toutanova, K.: BERT: pre-training of deep bidirectional transformers for language understanding. In: Proceedings of the Conference of the N.American Chapter of the Association for Computational Linguistics: Human Language Technologies, vol. 1 (Long and Short Papers), pp. 4171–4186 (2019)
15. Díaz, J.S.P., García, A.L.: Comparison of machine learning models applied on anonymized data with different techniques. In: Proceedings of the IEEE International Conference on Cyber Security & Resilience (CSR), pp. 618–623 (2023)
16. Emam, K., et al.: A globally optimal k-anonymity method for the de-identification of health data. J. Am. Med. Inform. Assoc. **16**(5), 670–82 (2009)
17. Freund, Y., Schapire, R.E.: A decision-theoretic generalization of on-line learning and an application to boosting. J. Comput. Syst. Sci. **55**(1), 119–139 (1997)
18. Garat, D., Wonsever, D.: Automatic curation of court documents: anonymizing personal data. Information **13**(1) (2022)
19. Grave, E., et al.: Learning word vectors for 157 languages. In: Proceedings of the 11th International Conference on Language Resources & Evaluation (LREC) (2018)

20. Günther, M., et al.: Jina embeddings 2: 8192-token general-purpose text embeddings for long documents. arxiv:2310.19923 (2023)
21. Hassan, F., Sánchez, D., Soria-Comas, J., Domingo-Ferrer, J.: Automatic anonymization of textual documents: detecting sensitive information via word embeddings. In: Proceedings of the 18th IEEE International Conference on Trust, Security and Privacy in Computing (TrustCom) & Communications/13th IEEE International Conference on Big Data Science & Engineering (BigDataSE), pp. 358–365 (2019)
22. Kohlmayer, F., et al.: Flash: efficient, stable and optimal k-anonymity. In: Proceedings of the International Conference on Privacy, Security, Risk (PASSAT) and Trust and International Conference on Social Computing (SocialCom), pp. 708–717 (2012)
23. Li, N., Li, T., Venkatasubramanian, S.: t-closeness: privacy beyond k-anonymity and l-diversity. In: Proceedings of the 23rd IEEE International Conference on Data Engineering (ICDE), pp. 106–115 (2006)
24. Lin, J.L., Wei, M.C.: An efficient clustering method for k-anonymization. In: Proceedings of the International Workshop on Privacy and Anonymity in Information Society (PAIS), pp. 46–50 (2008)
25. Lison, P., et al.: Anonymisation models for text data: state of the art, challenges and future directions. In: Proceedings of the 59th Annual Meeting of the Association for Computational Linguistics and the 11th International Joint Conference on Natural Language Processing, vol. 1: Long Papers, pp. 4188–4203 (2021)
26. Lloyd, S.: Least squares quantization in PCM. IEEE Trans. Inf. Theory **28**(2), 129–137 (1982)
27. Machanavajjhala, A., Gehrke, J., Kifer, D., et al.: l-diversity: privacy beyond k-anonymity. In: Proceedings of the 22nd IEEE International Conference on Data Engineering (ICDE) (2006)
28. Mamede, N., Baptista, J., Dias, F.: Automated anonymization of text documents. In: Proceedings of the IEEE Congress on Evolutionary Computation (CEC), pp. 1287–1294 (2016)
29. Medlock, B.: An introduction to NLP-based textual anonymisation. In: Proceedings of the 5th International Conference on Language Resources & Evaluation (LREC) (2006)
30. Mehta, V., Bawa, S., Singh, J.: WEClustering: word embeddings based text clustering technique for large datasets. Complex Intell. Syst. **7**, 1–14 (2021)
31. Mikolov, T., Chen, K., Corrado, G., Dean, J.: Efficient estimation of word representations in vector space. In: Proceedings of the 1st International Conference on Learning Representations (ICLR) (2013)
32. Mubark, A.A., Elabd, E., Abdulkader, H.: Semantic anonymization in publishing categorical sensitive attributes. In: Proceedings of the 8th International Conference on Knowledge & Smart Technology (KST), pp. 89–95 (2016)
33. Muennighoff, N., Tazi, N., Magne, L., Reimers, N.: MTEB: massive text embedding benchmark. In: Proceedings of the 17th Conference of the European Chapter of the Association for Computational Linguistics (EACL), pp. 2014–2037 (2023)
34. Pedregosa, F., Vet al.: Scikit-learn: machine learning in Python. J. Mach. Learn. Res. **12**, 2825–2830 (2011)
35. Pennington, J., Socher, R., Manning, C.: GloVe: global vectors for word representation. In: Proceedings of the Conference on Empirical Methods in Natural Language Processing (EMNLP), pp. 1532–1543 (2014)

36. Prasser, F., Kohlmayer, F., Lautenschlaeger, R., Kuhn, K.: ARX-A comprehensive tool for anonymizing biomedical data. In: Proceedings of the AMIA Annual Symposium, pp. 984–993 (2014)

37. Řehůřek, R., Sojka, P.: Software framework for topic modelling with large corpora. In: Proceedings of the LREC Workshop on New Challenges for NLP Frameworks, pp. 45–50 (2010)

38. Reimers, N., Gurevych, I.: Sentence-BERT: sentence embeddings using Siamese BERT-networks. In: Proceedings of the Conference on Empirical Methods in Natural Language Processing (EMNLP)and the 9th International Joint Conference on Natural Language Processing (IJCNLP) (2019)

39. Samarati, P.: Protecting respondents identities in microdata release. IEEE Trans. Knowl. Data Eng. **13**(6), 1010–1027 (2001)

40. Samarati, P., Sweeney, L.: Protecting privacy when disclosing information: k-anonymity and its enforcement through generalization and suppression. Technical report, SRI International (1998)

41. Song, K., et al.: MPNet: masked and permuted pre-training for language understanding. In: Proceedings of the 34th Annual Conference on Neural Information Processing Systems (NeurIPS), pp. 16857–16867 (2020)

42. Sweeney, L.: Replacing personally-identifying information in medical records, the scrub system. In: Proceedings of the AMIA Annual Symposium, pp. 333–337 (1996)

43. Turian, J., Ratinov, L., Bengio, Y.: Word representations: a simple and general method for semi-supervised learning. In: Proceedings of the 48th Annual Meeting of the Association for Computational Linguistics (ACL), pp. 384–394 (2010)

44. Ward, J.H., Jr.: Hierarchical grouping to optimize an objective function. J. Am. Stat. Assoc. **58**(301), 236–244 (1963)

45. Wellner, B., et al.: Rapidly retargetable approaches to de-identification in medical records. J. Am. Med. Inform. Assoc. **14**(5), 564–573 (2007)

Big Data

About Relationships in Data Lakes

Ahlame Diouan[1,2](✉)(iD), Eric Ferey[1], Jérôme Darmont[2](iD),
and Sabine Loudcher[2](iD)

[1] Université Lumière Lyon 2 UR ERIC, Lyon, France
a.diouan@univ-lyon2.fr, eric.ferey@bial-x.fr
[2] BIAL-X, Lyon, France
ahlame.diouan@bial-x.fr, {jerome.darmont,sabine.loudcher}@univ-lyon2.fr

Abstract. In the era of Big Data, managing voluminous and heterogeneous data presents significant challenges for organizations. To tackle
these challenges, the concept of a data lake has emerged as a promising
solution, allowing the storage of raw data from diverse sources in their
original format. An efficient metadata management system plays a crucial role in preventing data lake to turn into an unusable data swamp
by providing a structured framework for organizing, categorizing and
establishing relationships between data entities. In this paper, identify
the various relationships from diverse domains found in the literature.
Then, we categorize the types of relationships and propose a relationship
typology that classes relationships by similarity, containment, grouping
and provenance. Eventually, we also aim to check whether goldMEDAL,
a state-of-the-art generic metadata management model, adequately supports all such relationships. This evaluation is particularly relevant for
Bial-X, which seeks to implement a robust metadata management system
based on goldMEDAL's concepts.

Keywords: Data lakes · Data discovery · Semantic relationships · Big
data

1 Introduction

In recent years, there has been a huge increase in global data production and
organizations' decision-making processes have been revolutionized by the availability of large volumes of heterogeneous data, known as Big Data. This exponential growth not only presents real opportunities, but also challenges related
to data volume, velocity and variety that exceed the capabilities of traditional
data storage and management systems [18].

To address this issue, James Dixon proposes the concept of a data lake as a
practical solution [6]. A data lake allows storing raw data from heterogeneous
sources in their original format. In the absence of a data schema, the presence of
a robust metadata system is crucial for enabling data queries and thus preventing
the data lake from becoming a data swamp, i.e., an unusable data lake. Moreover,
an efficient metadata system provides users with a unified interface to search,

© The Author(s), under exclusive license to Springer Nature Switzerland AG 2025
R. Chbeir et al. (Eds.): IDEAS 2024, LNCS 15511, pp. 141–155, 2025.
https://doi.org/10.1007/978-3-031-83472-1_10

explore, and understand the available data entities and the relationships between them. This challenge is particularly relevant to Bial-X, a company specializing in business intelligence and data science. Bial-X's customers require a metadata management system to effectively manage a data lake and establish semantic relationships between data entities. Note that there are many terms similar to relationship, e.g., relation, link, linkage and connection. However, after reviewing the literature, relationship appears to be the most frequent term. Finding relationships provides users with a global view of metadata, through which they can interpret said relationships and gain valuable context into how various data entities are interconnected, facilitating a deeper understanding of their significance within the data lake. Since the data lake literature seems unanimous about the importance of a metadata system, we benchmarked state-of-the-art metadata management systems, i.e., DataGalaxy[1], Atlas[2], Open Data Discovery[3] and OpenMetadata[4]. These tools offer various forms of relationships, including operational and structural relationships, e.g., "entity In" and "aggregation"), but also lineage (provenance) relationships, which are important for understanding the origins and transformations of data entities.

Lineage relationships belong to so-called semantic relationships, but there are still other semantic relationships that metadata management systems do not support. Semantic relationships are defined as "any form of hierarchical, generic or predefined semantic relationships (semantic connections between data sets, e.g., for provenance or governance)" [14].

Eventually, our contribution is threefold.

1. We survey the various relationships between data entities found in the literature, notably aiming to pinpoint all semantic relationships that meet Bial-X's specific needs.
2. We categorize the types of relationships and propose a relationship typology that classes relationships by similarity, containment, grouping and provenance.
3. We hypothesize and check that goldMEDAL, a state-of-the-art generic metadata management model [24], can adequately support all the relationships we identify in our survey. This evaluation is crucial for Bial-X, as the company funded a PhD thesis that was part of goldMEDAL's design.

In the remainder of this paper, we first explicit our survey methodology and present the metadata metamodel goldMEDAL that we use throughout this paper (Sect. 2). Next, we present and discuss our relationship typology, i.e., similarity, containment, grouping and provenance relationships (Sect. 3). Finally, we conclude this paper and hint at future works (Sect. 4).

---

[1] https://www.datagalaxy.com.
[2] https://atlas.apache.org.
[3] https://opendatadiscovery.org.
[4] https://open-metadata.org.

2   Preliminaries

2.1   Survey Methodology

We conduct a systematic literature review to analyze relevant articles addressing specific questions related to relationships between data entities within data lakes. We adopt the Preferred Reporting Items for Systematic reviews and Meta-Analyses (PRISMA) protocol [19] to ensure rigor and transparency throughout the whole process.

Research Questions. First, we identify several key questions to guide our literature review.

- How are relationships between data entities defined in the literature?
- What types of relationships exist between data entities?
- Are there any semantic relationships between data entities?
- How relationships can enhance metadata management models or systems?
- Can goldMEDAL's concepts support all identified types of relationships?

Sources. To address our research questions, we conduct extensive searches across several academic databases, i.e., the ACM Digital Library, SpringerLink and IEEE Xplore. Moreover, Google Scholar is occasionally used to access papers not available in the above databases. Yet, Google Scholar is not a primary source, because the results it yields includes numerous non-peer-reviewed papers.

Search Strategy. Our search strategy involves selecting key terms designed to capture the full breadth of literature related to data lakes, relationships between data entities, and dataset discovery. Our search query below incorporates a range of terms and Boolean combinations to cover all relevant facets of the topic.

```
"Data lakes" AND
("relationships" OR "Semantic relationships" OR "Dataset
    discovery") AND
("data entities" OR "datasets" OR "tables") AND
(("Similarity" OR "Related" OR "Proximity") OR ("Containment"
    OR "Inclusion" OR "Encapsulation") OR ("Provenance"  OR "
    Lineage" OR "Tracking") OR ("Grouping" OR "Clustering" OR
    "Categorization"))
```

These results yield about 500 papers retrieved from ACM (118), IEEE (11), and Springer (371) databases.

Filtering. Applying a date filter onto articles published between 2016 and 2024 reduces the set to 315 papers. Next, we filter by publication type, narrowing down to 80 from ACM, 9 from IEEE and 199 from Springer. Then, we conduct an initial screening based on titles and abstracts, focusing on relevant keywords from our search query and assessing the alignment of abstracts with the scope of our study. This process leaves us with 50 articles from ACM, 6 from IEEE and 70 from Springer for further evaluation. Eventually, we apply inclusion and exclusion criteria (Table 1). The final result yields 10 papers from ACM, 2 from IEEE and 7 from Springer, for a total of 19 papers in the final review.

Table 1. Inclusion and exclusion criteria

Inclusion criteria	Exclusion criteria
Publication type: conference/journal	Brief papers or limited in scope
Publication date: from 2016 to 2024	Language: non-English
Article type: research/survey	Named relationship without definition
Relevant keywords	Accessibility: paper cannot downloaded
	Duplicates

2.2   GoldMEDAL and Relationships

goldMEDAL is a generic metadata model that bears a high level of abstraction to be very flexible for any data lake use case [24]. It encompasses three levels of modeling (conceptual, logical and physical) and is built upon four core concepts: data entity, grouping, link and process (Fig. 1).

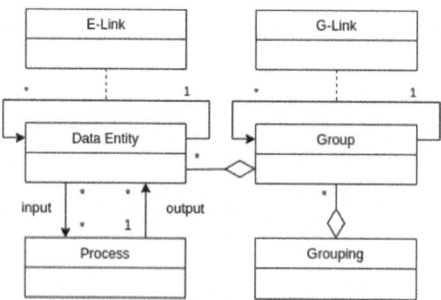

Fig. 1. goldMEDAL conceptual metadata model [24]

Data Entities are the basic units of the metadata model. For example, a data entity can represent both raw data and transformed data. It might be a spreadsheet file, a textual file, a semi-structured document, an image, a database table, a tuple or an entire database. This flexibility enables goldMEDAL to seamlessly handle data at various levels of granularity. Furthermore, the introduction of any new element such as file, document, image, etc., into the data lake triggers the creation of a new data entity.

A Grouping involves organizing data entities into sets denoted as groups based on common properties. For instance, within data lake architectures, the raw and preprocessed data zones constitute two groups within a zone grouping. Another example is a grouping of textual documents according to the language of writing.

Links associate either data entities with each other or groups of data entities with each other. Such links may be either directed or undirected.

A Process refers to any transformation or update applied to one or several data entities to produce a new data entity. It is used to track the relationships between data entities. Each process connects one or more "parent" data entities to "children" data entities. Yet, unlike links, processes represent the execution context of a transformation or modification operation (user, script, etc.).

By leveraging these core concepts, goldMEDAL facilitates the exploration of relationships between data entities within a data lake ecosystem. For example, processes can be employed to trace the lineage and parenthood relationships among data entities, while groupings aid in structuring related data entities, resulting in the establishment of containment relationships between different groups within the same grouping. Links can depict hierarchical or semantic connections between data entities or groups.

However, the practical application of these concepts needs to be assessed. While goldMEDAL offers a high level of abstraction and flexibility, it is essential to determine whether its core concepts can effectively support the implementation of all relationship types, including similarity, containment, grouping and provenance.

Ensuring that goldMEDAL's theoretical framework can guide the development of a metadata management system is vital for our future works. Bial-X especially needs a tailored and robust metadata management system.

3   Relationship Typology

We architecture our typology in four types of relationships: similarity (Sect. 3.1), containment (Sect. 3.2), grouping (Sect. 3.3) and provenance (Sect. 3.4) relationships. Each type of relationships is synthesized in a table with four columns:

- *Relationship name* as per the source article;
- bibliographical *Ref.*;
- a quote by the authors characterizing the relationship (*Authors' quote*);
- the goldMEDAL concept associated with the relationship (*gM concept*).

The relationships identified in the literature are categorized by the authors' definitions and descriptions within each paper. Some papers address multiple types of relationships. However, we do not include in our study named relationships that are not sufficiently defined.

Each identified relationship is classified according to our typology. Finally, we thoroughly analyze relationship definitions to determine what goldMEDAL concept can (or not) implement a particular relationship. We now present the selected papers and discuss them.

3.1 Similarity Relationships

In Table 2, we observe that the similarity relationships found in the literature may be based on different aspects such as content, structure or both a combination of content and structure. By classifying relationships into these three types, we can discern both differences and commonalities between them.

Content Similarity. Content similarity refers to resemblance or overlap in the information contained within data entities, particularly in terms of attributes, values or content. It indicates how closely related or similar two data entities are based on the content they store. Such similarity can be measured through various metrics, such as shared attributes, common values or semantic overlaps.

For Halevy et al., content similarity focuses on checksums and Locality Sensitive Hashing (LSH) values as indicators of content similarity [13]. Alserafi et al. emphasize the overall similarity of real-world objects or concepts stored in datasets [2]. On the other hand, Ravat and Zhao introduce the concept of partial overlap, suggesting that content similarity can exist even when data entities overlap rather than bearing strictly identical attributes [21]. Additionally, Eltabakh et al. define content similarity as establishing connections between a text document and table based on various criteria such as overlapping values, semantic similarity or metadata similarity. Moreover, they assess relatedness by assigning a score to each relationship between the document and table columns [8].

Furthermore, Kaminsky et al. expand the concept of content similarity by introducing joinability, which refers to the ability to link two similar columns from the same domain [16].

Structural Similarity. Structural similarity refers to the resemblance between data entities based on their structural aspects. It evaluates how data entities are similar based on their structure, including factors such as the types of variables, attributes names and data constraints. For Hai et al., a structural similarity involves clustering similar schemas together and selecting the core of each cluster as its representation. This process is based on the similarity between schemas calculated over imported data entities [12]. Alserafi et al. emphasize on identifying overlaps between data entity schemas by detecting related objects or attributes and matching instances between different schemas [1]. Moreover, the use of proximity mining adds another criterion by employing proximity scores to identify similar data entities with respect to structural similarity [1]. Eltabakh et al. propose joinable tables based on syntactic or semantic overlaps between columns [8]. Finally, Fernandez et al. introduce a property constraint that focus on selecting data entities according to specific properties, such as unique values or schema names, which represent inherent structural features [11].

Table 2. Similarity relationships

Relationship name	Ref.	Authors' quote	gM concept
Content similarity	[13]	... find datasets with content that is similar or identical to the given dataset, or columns from other datasets that are similar or identical to columns in the current dataset.	Link
Schema grouping	[12]	... clusters the schemas and picks up the core of each cluster as its presentation. The necessity of grouping depends on the schema similarity calculated over the imported data sources.	Link
Related	[2]	... Related pairs of datasets describe similar real-world objects or concepts from the same domain of interest. These datasets store similar information in (some of) their attributes.	Link
Proximity	[1]	... We utilise a novel proximity mining approach to assess the similarity of datasets.	Link
Similarity relationship	[5]	It's used to present that an object is "similarTo" another object.	Link
Relationship constraint	[11]	... the analyst may be interested in finding similar datasets to the ones found so far to make sure no information is missing (content similarity). Or, having already found a handful of relevant datasets, the analyst may want to find a join path to join them together (a primary-key/foreign-key (PK/FK) candidate).	Link
Property constraints	[11]	... selecting columns with unique values, or columns with a string in the schema name, which are all properties of the data. For instance, the analyst who is building the stock change prediction model may start with a search for tables that include metrics of relevance... (schema similarity).	Link
Content similarity	[20]	... Which means that different datasets share the same attributes	Link
Partial overlap	[20]	... Partial overlap which means that some attributes with corresponding data in different datasets overlap.	Link
Similarity Link	[23]	... Reflect the strength of the similarity between two objects. Unlike object groupings, similarity relationships refer to the intrinsic properties of objects, such as their content or structure.	Link
Schema matching	[1]	... It seeks to identify schematic overlaps between datasets. This involves detecting related objects (instances or attributes) and matching instances between two different schemata.	Link
Union	[9]	... Table union search aims to find all tables that are unionable with the query table. To determine whether two tables are unionable, existing solutions first identify all pairs of unionable columns from the two tables based on column representations, such as bag of tokens or bag of word embeddings.	Link
Doc to Table (From Document to Tables)	[8]	... A Table T with column set A is related to a text document D if there exists $A_i \in A$ such that D and A_i are related via overlapping values, semantic similarity, or metadata similarity, each with a relatedness score.	Link
Table j Table (Joinable Tables)	[8]	... Table T with column set A is joinable to Table T' with column set A' if there exists $A_i \in A$ and some $A'_j \in A'$ such that: 1. A_i and A'_j have value overlap suggesting syntactic join, or 2. A_i and A'_j have semantic overlap suggesting semantic join.	Link
Table U Table (Unionable Tables)	[8]	... Table T with column set A is unionable to Table T' with column set A' if a one-to-one mapping $H : A \rightarrow A'$ exists wherein there exists $h \in H$ such that the column pair given by h exhibits name, value, or semantic similarity.	Link
Joinability relationship	[16] [26]	... Joinability means that two columns can be linked together because they contain similar data from the same domain	Link

Hybrid Similarity. Hybrid similarity combines elements from both content and structure, offering a more comprehensive perspective on finding relevant and similar data entities. For Diamantini et al. [5] and Sawadogo et al. [23] similarity relationship focus on establishing a relationship between data entities, considering both their content and structural characteristics. Moreover, Fernandez et al. introduce constraints based on specific properties. Such as columns with unique values or columns with a particular string in the schema name. These constraints aim to select data entities by integrating both content and structural aspects [11]. Eventually, Fan et al. [9] and Eltabakh et al. [8], focus on identifying unionable tables, considering both content and structural overlaps between their columns to determine similarity.

Despite the fact that different terms used to describe similarity relationships, they share the common understanding that content or structural similarity is determined by shared attributes, values, semantic overlaps or structural aspects, using different methodologies and metrics. The comparison highlights that goldMEDAL's concepts can handle diverse similarity relationships. For example, the notion of data entities corresponds well with content similarity, focusing on the similarity or overlap in the data they contain. Furthermore, goldMEDAL's link concept facilitates the implementation of similarity relationships between data entities. After analyzing the various similarity relationships outlined in Table 2, it becomes evident that goldMEDAL is a robust metamodel to implement these relationships within a data lake.

3.2   Containment Relationships

In the context of data lake management, containment relationship refers to the hierarchical structure of data entities (Table 3). This relationship indicates how a data entity can be encapsulated or nested with in another, such as sub-data entities or sub-tables within its structure. It illustrates how data entities are grouped or organized in a hierarchical manner, with some data entities being contained within others.

According to Halevy's et al., a containment relationship refers to how data entities may contain other data entities, such as bigtable column families [13]. Deng et al. focus on the subsumption relationship, providing a function to identify data entities or groups of data entities that are contained in or contain other data entities [4]. Additionally, Diamantini et al. present structural relationship, indicating how an object contains another object, like a relational database containing tables and the same way tables contain attributes [5].

Eichler et al. introduce the granularity indicator entity, enabling the collection of metadata on multiple granularity levels, closely tied to some kind of structure in the data, such as object instances or key-value pairs within a JSON document [7].

Huang et al. describes containment relationships as a parenthood relationship between two data entities, i.e., a parent entity contains a child entity. Shah et al. quantify how much a data entity is contained in another, either in terms of their structure or content [25]. Finally, Fernandes et al. focus on inclusion dependency, where data values from one data entity are contained within another, either fully or partially [10].

Table 3. Containment relationships

Relationship name	Ref.	Authors' quote	gM concept
Dataset containment	[13]	...Some datasets may contain other datasets.	Link
Subsumption relationship	[4]	...a list of tables or groups of tables that have some form of subsumption relationship (i.e., are contained in or contain) with respect to the reference table;	Link
Structural relationship	[5]	...Which is used to present that an object "contains" another object.	Link
Granularity Indicator	[7]	...collecting metadata on different granular levels. These levels are closely tied to some kind of structure in the data.	Link
Containment relationship	[15]	...containment relationship, i.e., a parent entity T may contain another child entity Ti.	Link
Containment fraction	[25]	...If A and B are schemas, $n(B)$ refers to the length of the flattened schema set in B, and $|A \cap B|$ refers to the length of the intersection between the flattened schema sets. If they are tables, $n(B)$ refers to the number of rows in B and $|A \cap B|$ refers to the number of rows common to both tables.	Link
Inclusion dependency	[10]	...$T_u.A_v \subseteq_{\text{level}} T_q.A_r$, where each T_i is a table, each A_j is an attribute, and level is the fraction of the values in $T_u.A_v$ that are contained in $T_q.A_r$. When the level $= 1$, there is a full inclusion dependency, and when the level < 1, there is a partial inclusion dependency.	Link

Overall, these authors agree in their interpretation, collectively defining containment relationship as a hiearchical link between data entities within data lake, which align closely with goldMEDAL's Link concept. Each relationship, focusing on different aspects of containment within data entities, finds resonance in goldMEDAL's approach of using links to represent hierarchical and structural connections between data entities. Consequently, goldMEDAL's Link concept effectively captures the essence of these relationships, facilitating their implementation within data lake.

3.3 Grouping Relationships

Grouping relationships signify how data entities are grouped and classified together, based on various criteria as outlined in Table 4, enabling a more effective data management, discovery and analysis within data lakes.

Table 4. Grouping relationships

Relationship name	Ref.	Authors' quote	gM concept
Schema grouping	[12]	...clusters the schemas and picks up the core of each cluster as its presentation. The necessity of grouping depends on the schema similarity calculated over the imported data sources.	Grouping
Logical cluster	[13]	...We identify datasets that belong to the same logical cluster.	Grouping
Logical cluster	[20]	...Which means that some datasets are from the same domain (different versions, duplication etc.).	Grouping
Objects groupings	[23]	...Organize objects into collections, each object being able to belong simultaneously to several collections.	Grouping
Categorization	[7]	...The categorization entity is a label assigned according to the metadata element's context.	Grouping
ZoneIndicator	[7]	...The zoneIndicator entity is a label on the data entity supplying information on the location of the data element in the data lake's zone architecture.	Grouping
Outlier datasets	[1]	...Which have no similarity with any other dataset in the DL (i.e., no similar attributes in the DL).	Grouping

For Hai et al., grouping focuses on clustering schemas based on common attributes and similarities [12]. Halevy et al. [13] and Ravat et al. [20] group data entities from the same domain or with similar attributes due to duplication.

Sawadogo et al. propose to organize data entities into collections, where each data entity can belong to several collections simultaneously. These groups are generated automatically based on semantic metadata, including tags and business categories [23]. Eichler et al. introduce two types of grouping relationships [7]. The first one aims to categorize data entities using their metadata elements with labels based on their context. For example, operational labels for metadata elements storing access information. The second one aims to assigning labels to data entities to indicate their location within the data lake's zone architecture. In both Categorization and ZoneIndicator, the grouping is only based on metadata.

Moreover, Al-serafi et al. propose another approach which involves identifying data entities exhibit no similarity with any other data entities. This lack of similarity can be used as a criteria to categorize data entities that belong to no grouping of shared attributes [1]. Despite the diversity in approaches, these authors converge to the same goal: categorizing data entities into groups or collections within data lake using different criteria. Furthermore, these different approaches are not mutually exclusive, rather, they can complement each other, contributing to an efficient data lake management.

As a conclusion, goldMEDAL's grouping concept effectively aligns with these various relationships by providing mechanisms to organize and categorize data entities into clusters or collections based on their similarities, context, or other criteria within the data lake.

3.4   Provenance Relationships

Provenance relationships refer to the lineage or origin of data entities, tracing their evolution within a data lake. Table 5 shows these relationships, which offer a comprehensive understanding of data origins and transformations, and provide insights into data entities' history.

In the context of identifying relationships between data entities, Halevy et al. suggest content similarity that aims in identifying data entities with similar content [13]. Content similarity indirectly contributes to data provenance by highlighting data entities that may have originated from the same source or undergone similar transformations. For Halevy et al. and for Ravat et al., logical cluster helps in organizing related data entities within the data lake, particularly those with shared attributes or versions [13,20]. This organization facilitates the tracing of data lineage by grouping together data entities that are likely to have similar origins or same transformations.

Alserafi et al. highlight the importance of recognizing duplicated data entities, as they may reveal common sources or transformations [2]. Which can helps trace back to the original sources and gain insights into the data entities's history and transformations.

Provenance, as emphasized by Halevy et al., involves tracking the production, consumption and dependencies of datasets, offering direct insights into their lineage and origins [13]. This explicit documentation provides information on how datasets are created and used, aiding in understanding their provenance within the data lake. Beheshti et al. advocate aggregating tracing metadata to build a thorough provenance graph, facilitating the reconstruction of data lineage within the data lake [3]. Sawadogo et al. describe the parenthood relationship, which reflects the connections between combined data entities and their resulting data entities, offering insights into their dependencies and lineage [23]. Additionally, Sawadogo et al. highlight the importance of tracking data entities versions and representations, which provide insights into data evolution and transformations over time [23]. Versioning and representation tracking contribute to data provenance and data lineage by documenting changes to data entities and their structures, allowing for a comprehensive understanding of their history and evolution.

Provenance of data entities holds significant importance in the context of managing data lakes. goldMEDAL's process, enables the tracking of data entities changes over time, their origin, usage, status in the life cycle, aligning well with the notion of documenting data origins advocated in the literature. Despite the existence of various approaches, goldMEDAL's process concept demonstrates flexibility in implementing different provenance relationships outlined from literature.

Table 5. Provenance relationships

Relationship name	Ref.	Authors' quote	gM concept
Content similarity	[13]	. . . Content similarity-both at the level of dataset as a whole and at the level of individual columns-is another graph relationship that we extract. . . . we rely on approximate techniques to determine which datasets are replicas of each other and which have different content.	Process
Logical cluster	[13]	. . . Datasets that are versions of the same logical dataset and that are being generated on a regular basis; datasets that are replicated across different data centers; or datasets that are sharded into smaller datasets for faster loading.	Process
Duplicated	[2]	. . . Duplicate pairs of datasets describe the same concepts. They convey the same information in most of their attributes, but such information can be stored using differences in data.	Process
Provenance	[13]	. . . For each dataset, we maintain the provenance of how the dataset is produced, how it is consumed, what datasets this dataset depends on, and what other datasets depend on this dataset.	Process
Tracing and Provenance	[3]	. . . Collect and aggregate tracing metadata (including descriptive, administrative and temporal metadata and build a provenance graph) for both data and the contextualized data.	Process
Logical cluster	[20]	. . . Which means that some datasets are from the same domain (different versions, duplication etc.).	Process
Parenthood relationship	[23]	. . . Reflect the fact that an object can be the result of joining several others. There is a "parenthood" relationship between the combined objects and the resulting object, and a "co-parenthood" relationship between the merged objects.	Process
Versions	[23]	. . . Raw data in the lake are often modified through updates that result in the creation of new versions of the initial data, which can be considered as metadata.	Process
Representations	[23]	. . . Raw data (especially unstructured data) can be reformatted for a specific use, inducing the creation of new representations of an object	Process

4   Conclusion and Perspectives

One key challenge in data lakes is to find and discover relationships between different data entities, which facilitate the process of data integration, discovery and analysis. While various metadata management systems exist, they often do not address relationships and particularly semantic relationships.

Our primary contribution is an extensive literature review and analysis, where we identify and categorize relationships based on their underlying characteristics and implications for data management. The outcome is a relationship typology that shed light on the diverse semantic relationships between data entities within data lakes.

Furthermore, we had hypothesized that goldMEDAL could support all the relationships found in our survey. Tables 2, 3, 4 and 5 show that goldMEDAL's concepts cover all the types of relationships identified in our survey. It is somehow a validation that goldMEDAL's conceptual model provides a flexible and comprehensive framework for metadata management and a promising solution for enhancing data discovery, exploration and analysis in data lake environments.

In future research, we plan to design a metadata management system that not only supports operational and structural relationships, but also semantic relationships. As of today, we have not ruled between:

1. contribute to the open source metadata management systems available, i.e., Open Data Discovery and OpenMetadata, and extend one of them to support semantic relationships;
2. build a metadata management system from scratch, based on the goldMEDAL metadata metamodel.

Furthermore, there are explicit relationships that are easy to spot, e.g., when designing a database schema. Yet, there are also implicit relationships that are hidden, especially in data lakes with highly heterogeneous data. Such high-potential relationships, e.g., similarity relationships, can be mined by machine learning or Large Language Models (LLMs). The ultimate goal is to interlink data entities so as to navigate and search data within a whole data lake.

Eventually, we lately identified additional relationships, i.e., causality [17] and correlation [22]. Of course, they are definitely different, so we need to investigate these relationships further.

Acknowledgments. Ahlame Diouan's PhD is funded by BIAL-X. The authors thank the anonymous reviewers for their useful comments.

Disclosure of Interests. The authors have no competing interests to declare that are relevant to the content of this article.

References

1. Al-Serafi, A.M.M.: Dataset proximity mining for supporting schema matching and data lake governance. Ph.D. thesis, Universitat Politècnica de Catalunya (2021)
2. Alserafi, A., Calders, T., Abelló, A., Romero, O.: DS-prox: dataset proximity mining for governing the data lake. In: Proceedings of the 10th International Conference on Similarity Search & Applications (SISAP), pp. 284–299 (2017)

3. Beheshti, A., Benatallah, B., Nouri, R., Tabebordbar, A.: CoreKG: a knowledge lake service. Proc. VLDB Endow. **11**(12), 1942–1945 (2018)
4. Deng, D., et al.: The data civilizer system. In: Proceedings of the 8th Biennial Conference on Innovative Data Systems Research (CIDR) (2017)
5. Diamantini, C., et al.: A new metadata model to uniformly handle heterogeneous data lake sources. In: Proceedings of the 22nd ADBIS Short Papers & Workshops, AI*QA, BIGPMED, CSACDB, M2U, BigDataMAPS, ISTREND, Doctoral Consortium, pp. 165–177 (2018)
6. Dixon, J.: Pentaho, Hadoop, and Data Lakes (2010). https://jamesdixon. wordpress.com/2010/10/14/pentaho-hadoop-and-data-lakes/
7. Eichler, R., et al.: HANDLE - a generic metadata model for data lakes. In: Proceedings of the 22nd International Conference on Big Data Analytics & Knowledge Discovery (DAWAK), pp. 73–88 (2020)
8. Eltabakh, M.Y., Kunjir, M., Elmagarmid, A., Ahmad, M.S.: Cross modal data discovery over structured and unstructured data lakes. Proc. VLDB Endow. **16**(11), 3377–3390 (2023)
9. Fan, G., et al.: Semantics-aware dataset discovery from data lakes with contextualized column-based representation learning. Proc. VLDB Endow. **16**(7), 1726–1739 (2022)
10. Fernandes, A.A., et al.: Data preparation: a technological perspective and review. SN Comput. Sci. **4**(4), 425 (2023)
11. Fernandez, R.C., et al.: Aurum: a data discovery system. In: Proceedings of the 34th IEEE International Conference on Data Engineering (ICDE), pp. 1001–1012 (2018)
12. Hai, R., Geisler, S., Quix, C.: Constance: an intelligent data lake system. In: Proceedings of the ACM International Conference on Management of Data (SIGMOD), pp. 2097–2100 (2016)
13. Halevy, A.Y., et al.: Managing Google's data lake: an overview of the Goods system. IEEE Data Eng. Bull. **39**(3), 5–14 (2016)
14. Hoseini, S., Theissen-Lipp, J., Quix, C.: A survey on semantic data management as intersection of ontology-based data access, semantic modeling and data lakes. J. Web Semant. 100819 (2024)
15. Huang, R., et al.: Effective and efficient retrieval of structured entities. Proc. VLDB Endow. **13**(6), 826–839 (2020)
16. Kaminsky, Y., Pena, E.H., Naumann, F.: Discovering similarity inclusion dependencies. Proc. ACM Manag. Data **1**(1), 1–24 (2023)
17. Liu, J., Sun, S., Nargesian, F.: Causal dataset discovery with large language models. In: Proceedings of the ACM SIGMOD Workshop on Human-In-the-Loop Data Analytics (HILDA), pp. 1–8 (2024)
18. Miloslavskaya, N., Tolstoy, A.: Big data, fast data and data lake concepts. In: Proceedings of the 7th Annual International Conference on Biologically Inspired Cognitive Architectures (BICA), pp. 300–305 (2016)
19. Moher, D., et al.: Preferred reporting items for systematic review and meta-analysis protocols (PRISMA-P) 2015 statement. Syst. Rev. **4**(1) (2015)
20. Ravat, F., Zhao, Y.: Data lakes: trends and perspectives. In: Proceedings of the 30th International Conference on Database & Expert Systems Applications (DEXA), vol. I, pp. 304–313 (2019)
21. Ravat, F., Zhao, Y.: Metadata management for data lakes. In: Proceedings of the 23rd ADBIS Short Papers & Workshops: BBIGAP, QAUCA, SemBDM, SIMPDA, M2P, MADEISD, and Doctoral Consortium, pp. 37–44 (2019)

22. Santos, A., et al.: Correlation sketches for approximate join-correlation queries. In: Proceedings of the ACM International Conference on Management of Data (SIGMOD), pp. 1531–1544 (2021)
23. Sawadogo, P.N., et al.: Metadata systems for data lakes: models and features. In: Proceedings of the 23rd ADBIS Short Papers & Workshops: BBIGAP, QAUCA, SemBDM, SIMPDA, M2P, MADEISD, and Doctoral Consortium, pp. 440–451 (2019)
24. Scholly, E., et al.: Coining goldMEDAL: a new contribution to data lake generic metadata modeling. In: Proceedings of the 23rd International Workshop on Design, Optimization, Languages and Analytical Processing of Big Data (DOLAP), vol. 2840, pp. 31–40 (2021)
25. Shah, R., et al.: R2D2: reducing redundancy and duplication in data lakes. Proc. ACM Manag. Data 1(4), 1–25 (2023)
26. Youngmann, B., Cafarella, M., Salimi, B., Zeng, A.: Causal data integration. Proc. VLDB Endow. 16(10), 2659–2665 (2023)

From Theory to Practice of Multidimensional Big Data Analytics over Big Healthcare Data: A Real-Life Case Study

Alfredo Cuzzocrea[1,2]($\boxtimes$), Abderraouf Hafsaoui[1], and Carmine Gallo[1]

[1] iDEA Lab, University of Calabria, Rende, Italy
alfredo.cuzzocrea@unical.it
[2] Department of Computer Science, University of Paris City, Paris, France

Abstract. This paper provides practical and experimental results coming from the real-life research project that focuses on Big Data Analytics over Big Healthcare Data about chronical pain of patients in real-life hospitals and medical centers of North Italy (Region Lombardy), named as Pain-RELife. In this context, we devise innovative Multidimensional Big Data Analytics tools and a reference Cloud-based architecture that implements our analytics paradigm. Finally, we provide our analytical results based on advanced tools such as Multidimensional Clustering and Multidimensional Regression.

Keywords: Multidimensional Big Data Analytics Tools · Healthcare Data Science · Advanced Big Data Systems

1 Introduction

Starting from the results of the real-life national project POR 2014–2020 Regione Lombardia "Pain-RELife - Ecosistema big data sostenibile e integrato per la continuità della cura e il supporto alla decisione dei pazienti con dolore" – "Pain-RELife - Sustainable and integrated big data ecosystem for continuity of care and decision support for patients with pain" (Project ID: 1173269), this paper reports on the practical and experimental results we obtained thanks to the application of the well-known Multidimensional Big Data Analytics paradigm (e.g., [1–4]), along with lessons learned.

More into details, in the *Pain-RELife* project, starting from a *kernel* real-life *chronical-pain-patient dataset*, with a special focus on important diseases such as fibromyalgia, ictus and cancer, we *artificially* build a *synthetic multidimensional database* named as *PainRELifeDB* via ad-hoc *random sampling techniques* (e.g., [12]) that strictly adhere to the multidimensional nature of the schema. One specific characteristic of the *PainRELifeDB* database is represented by the *pain-level questionnaire* where patients report on their levels of pain following the therapy. This questionnaire is

This research has been made in the context of the Excellence Chair in Big Data Management and Analytics at University of Paris City, Paris, France.

completely encoded in the database and, as a consequence, in the Multidimensional Big Data Analytics tools developed by our Cloud-based architecture, as a previous extension of classical big data analytics tools (e.g., [8]). As an immediate consequence, the knowledge derived from such a multidimensional analysis is of primary interest for the underlying healthcare analytics (e.g., [5]).

Inspired by this line of research, this paper proposes an overview of our Cloud-Based Multidimensional Big Data Analytics architecture, along with the tools we introduced and experimented based on this new big data analytics pattern, namely *Multidimensional Clustering* (e.g., [21]) and *Multidimensional Regression* (e.g., [22]), applied to the specific case of *big healthcare data analytics* (e.g., [9, 10]). The first one is a powerful extension of classical clustering tools directly applied to the *core* multidimensional modelling of the target dataset via meaningfully dimensions and measures. The second one is a nice generalization of the one-dimensional regression towards *simultaneous* dimensions of a given multidimensional space.

The remaining part of this paper is organized as follows. In Sect. 2, we provide an overview of our proposed architecture. Section 3 is devoted to the *PainRELifeDB* database, by describing its anatomy and generative process. In Sect. 4, we describe the multidimensional modelling embedded in our Multidimensional Big Data Analytics methodology. Section 5 moves the attention on the design and implementation of the Multidimensional Big Data Analytics tools embedded in our architecture, along with their experimental evaluation against real-life datasets coming from the *Pain-RELife* research project. In Sect. 6, we provide our remarks about the outcomes of our research. Finally, Sect. 7 reports conclusions and future work of our research.

2 A Cloud-Based Multidimensional Big Healthcare Data Analytics Architecture: Anatomy, and Main Layers

In this section, we present an extensive reference architecture for effective and efficient supporting Multidimensional Big Data Analytics; this framework clarifies the integration of specialized methods. We aim to provide a framework that can be used as a guide and help in extract valuable insights and performing multidimensional data analysis.

Figure 1 illustrates our framework, which consists of three essential hierarchical levels, representing different steps of our approach to extract valuable and actionable information, which also shows data structures, models, and data information like companies, drugs, hospital structures, medical personnel, etc. as well as different tools that we used in the process.

The proposed architecture demonstrates the usefulness of Multidimensional Big Data Analytics in specific use cases, such as healthcare. This real-life architecture consists of three fundamental layers that work together to enable an effective and efficient model for big healthcare data.

In greater depth, the reference architecture is composed of three hierarchical tiers that incorporate several large data repositories. Specifically, it is made of the following layers:

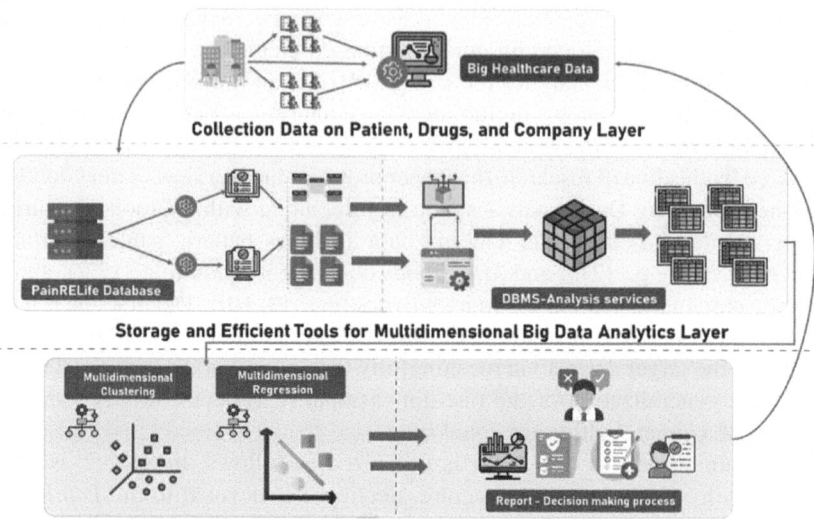

Fig. 1. Cloud-Based Multidimensional Big Healthcare Data Analytics Architecture.

- *Big Data Collection Layer.* In this layer, collecting data is an essential step, the healthcare data is generated from different data sources with various types and formats. The focus is on obtaining data about specific individuals and patients, including their demographics, medical history, and other potential ailments. Additionally, Data collecting involves gathering information on other actors in healthcare domains, such as pharmaceutical products (e.g., drug composition, dosage, side effects, manufacturers, suppliers, and so forth.). By capturing this data, users, providers, and companies involved in the healthcare system can gain valuable insights, which help improve the healthcare sector (e.g., provide patients with specific treatments, improve supply chain management, etc.).
- *Big Data Storage Layer.* In this layer, all data from the previous layer is stored in *PainRELifeDB* implemented on *Microsoft SQL Server*, which is used as a supporting DBMS for the analytical interpretation of multidimensional data, providing more realistic modeling for real-world databases aiming for high effectiveness and efficiency, notably scalability across big data. To support OLAP analysis, we create the so-called *Multidimensional Star Schema*, which enables us to perform OLAP complex analysis operations such as slice-and-dice, drill-down, and roll-up on multidimensional data cubes (e.g., [6, 7]). Therefore, these cubes store data in a structured format that is optimized for analytical queries, allowing for fast retrieval of aggregated data. Furthermore, *Microsoft SQL Server Analysis Services* (SSAS) is an essential component in the second layer, SSAS enables us to perform complex analysis on big data and gain valuable insights to support decision-making processes. Note that our architecture applies a variety of optimization approaches, advanced tools, and technologies, including data partitioning, *Machine Learning* (ML), and predictive

modeling. Additionally, pivot analysis is performed as a starting point, this analysis, provides an effective two-dimensional interface for exploring aggregate data in a multidimensional data model.

- *Multidimensional Reporting Layer.* The last layer is the *front- end* of our architecture; this layer includes different types of analysis. Mainly, OLAP analysis with the integration of proposed Multidimensional Data Mining techniques, such as (*i*) Multidimensional Clustering, which is a crucial tool in analyzing large volumes of multidimensional data efficiently; (*ii*) Multidimensional Regression, which is an ML structure used for predictive modeling, discover new insights and predict trends by analyzing current data distributions. Furthermore, various analyses are performed based on discovered patterns and analytical results to gain valuable and actionable information (e.g., best drug, common side effects, etc.). Different data visualization tools are used to well present the results graphically which helps users understand complex patterns, trends, and outliers.

In sensitive sectors like healthcare, the tools required to handle and interpret reporting data efficiently are mandatory for healthcare actors. This could represent a significant advancement for the industry. This could pave the way for significant innovations in medical practices, and provide new methods, technologies, and approaches that could revolutionize patient healthcare, and treatment protocols.

Apart analytics considerations, it should be noted that our proposed architecture must also take into account *performance* aspects (e.g., [23]). Indeed, it has been proved that, when big datasets are processed for analytics purposes, then relevant computational overheads are introduced, especially for the case of big multidimensional data. These drawbacks affect all the main big data management phases, including accessing, indexing and querying, so that effective optimization solutions must be devised.

3   The *PainRELifeDB* Database

In this section, we provide a detailed description about the design and the implementation of the database *PainRELifeDB*.

One part of the data is a patient survey, which is a questionnaire designed to evaluate patients' perceptions of pain, patients are affected by two specific disease conditions: *stroke* and *early breast cancer*. All patients live in the same Region Lombardy. The goal of this survey is to collect valuable information about patients that suffering from *chronic pain*. The pain can be divided into two categories: (*i*) *Neuropathic pain*: it is nerve pain that can happen if the nervous system malfunctions or is damaged; (*ii*) *Nociceptive pain*: it is a type of pain caused by damage to body tissue. Researchers can extract and analyze patients' experiences reported before and after various medical treatments by providing questionnaires. The questionnaire data is saved in a separate table called *Survey*, which acts as a reference point for measures.

The sections within the table *Survey* consist of the following fields:

- *Pain Catastrophizing Scales* (PCS). The PCS scale consists of 13 attributes (PCS1–PCS13) that represent patient responses to their thoughts and feelings that may be connected with pain. This part asks questioned patients to identify the frequency (on a scale of 0 to 4) with which they display these thoughts and feelings while they are in pain;

- *Brief Pain Inventory* (BPI). *BPI* scale includes 16 attributes, it requires to interview patients to indicate whether they have experience pain. The aim is to evaluate the impact of the pain in relation to the performance of daily activities and emotional states;
- *Patients' Global Impression of Change* (PGIC). *PGIC* scale allows patients to describe their opinion of a probable change in pain (yes or no), as well as a numerical value of this change from 1 to 10;
- *Health Assessment* (SF). The *SF* scale requests that patients provide their perceptions of their emotional condition, performance of work activities, and health status.

We use *Microsoft SQL Server* as a primary tool to create and manage the *PainRE-LifeDB* database. It allows users to perform various tasks such as creating, reading, and importing data, writing and executing queries, designing databases, managing security, and configuring server settings. Additionally, we use *SQL Server Management Studio* (SSMS) for the analysis tasks.

The *PainRELifeDB* database is stored on a virtual infrastructure that runs on the *VMware vSphere* platform with a *Foundation* license and a *3-physiscal-node cluster*. The *ESXi* hosts feature 256 GB RAM and two *Intel(R) Xeon(R) Gold 5120* 2.20 GHz CPUs with *Hyperthreading* (56 logical CPUs). The three Cluster nodes are designed for high reliability; hence, in case one of the three fails, all of the VMs in this node will be automatically restarted on one of the two remaining operational hosts.

Figure 2 shows a pie plot that highlights the percentage of disk space occupied by database *PainRELifeDB*, which corresponds to 2,72 GB, with the corresponding legend that shows how the space is occupied.

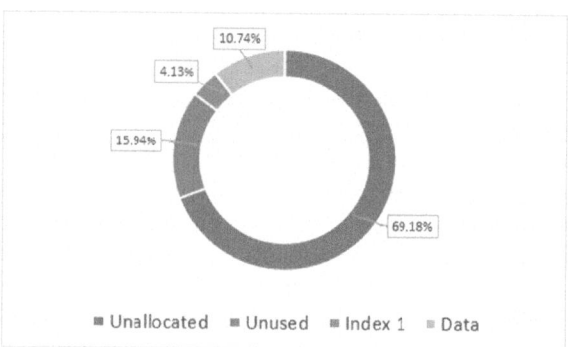

Fig. 2. Space Occupancy for the Database *PainRELifeDB*

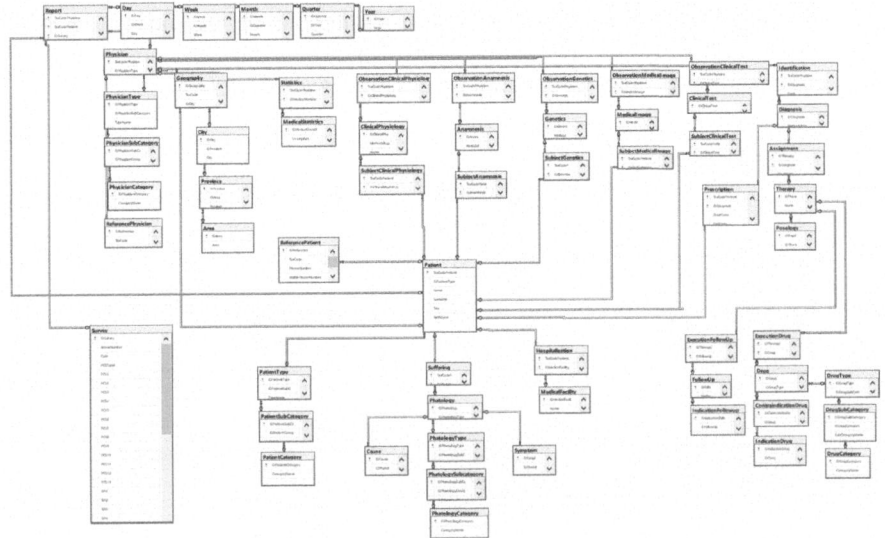

Fig. 3. *PainRELifeDB* Schema

Now, we offer a detailed illustration of the database *PainRELifeDB*, along with its dimensions and hierarchies. Subsequently, we highlight the effect of our Multidimensional Big Data Analytics tools.

As shown in the schema of *PainRELifeDB* in Fig. 3, several dimensions have been introduced in the *PainRELifeDB* design to allow for a complete analysis of complex data. The database includes additional tables such as *Report, Physician, Patient, Pathology, Geography,* and *Drug*, each of which serves as a different dimension that contributes to multidimensional data analysis, along with the previous fact table *Survey*. Note that we designed this database to facilitate the integration of multidimensional data.

Due to space limitation, here we just report an example of multidimensional dimension embedded in *PainRELifeDB*. The table *Report*, shown in Fig. 4, contains crucial information about patients; however, individual patients may require to complete multiple questionnaires, which may affect the data accessibility. Moreover, to make it easier to organize and access this data, *Report* includes a hierarchical structure: *Day, Week, Month, Quarter*, and *Year*. It is vital to record, in fact, the particular day on which each survey is done, along its OLAP dimensional hierarchy.

4 Populating *PainRELifeDB*

In our architecture, *PainRELifeDB* database is populated with *synthetic data*, such that generated data relies on the integration of real-world data obtained from patient questionnaires and medical professionals. Synthetic data are generated from real data using *algorithmic data generation techniques* via sampling, depending on the features of the characteristics of the original (real) dataset. Moreover, there are *random approaches* that create random samples from probability distributions or random numbers within a predefined range. Furthermore, some methods use interpolation techniques to generate synthetic data to fill gaps in a dataset that contains missing values. Figure 5 shows the process that describes the data generation process for *PainRELifeDB*.

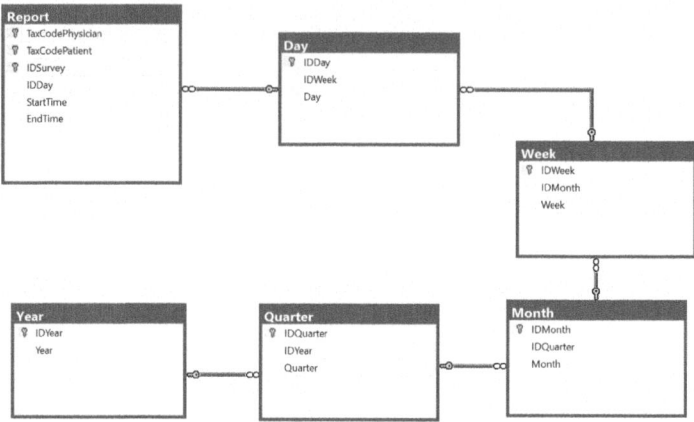

Fig. 4. Dimensional Table *Report*

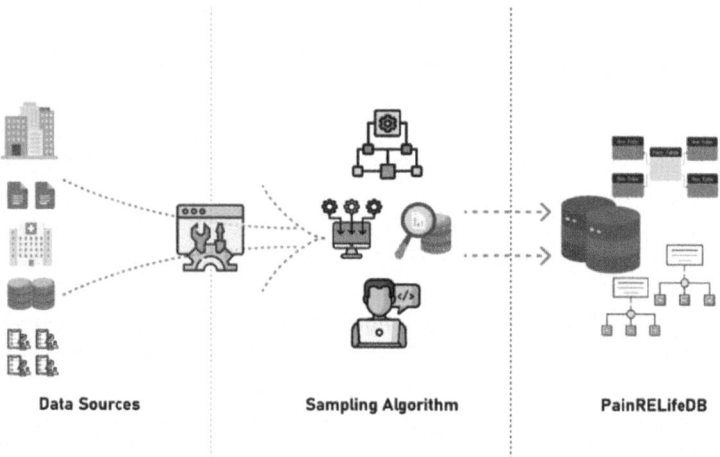

Fig. 5. Algorithm for Populating the Database *PainRELifeDB*

Methods such as averaging, linear regression, and so forth, are used to calculate and estimate missing or Incomplete values. Additionally, samples can be randomly generated from the real data and then manipulated to produce new data. ML models also can be used by taking the attributes of original data and producing new versions with the same style and distribution as the original data. Note that while producing synthetic data, it may not correctly reflect the complexity and diversity of the original data, therefore, we should be careful while dealing with the generated data, and it is critical to analyze synthetic data and properly evaluate its quality before using it for a given goal.

Synthetic data may be used for different purposes, including data analysis, training ML algorithms and testing new models. Generally, synthetic data can be used as a complementary or as a substitute for actual data. Specifically, when it is difficult or costly in time or money to gather real big data, generated data is beneficial in testing new models or algorithm capabilities, and exposes it to unique situations, scenarios, or events that do not regularly occur in reality or the data that rarely can appear in real life.

The first phase in our process is to collect patient surveys that include a range of information, such as patient demographics, medical history, and pain level. These questions are important in obtaining a thorough picture of the patient's pain experience. Thesurveys are thoroughly examined to ensure the data is accurate and consistent. This stage is critical because it guarantees that the data utilized to analyze the patient population is accurate and representative. The data is then organized and written into the *PainRELifeDB* database, and various data fields from the surveys must be allocated to the respective columns in the database. Note that the *PainRELifeDB* database was designed with scalability and data analysis flexibility in mind.

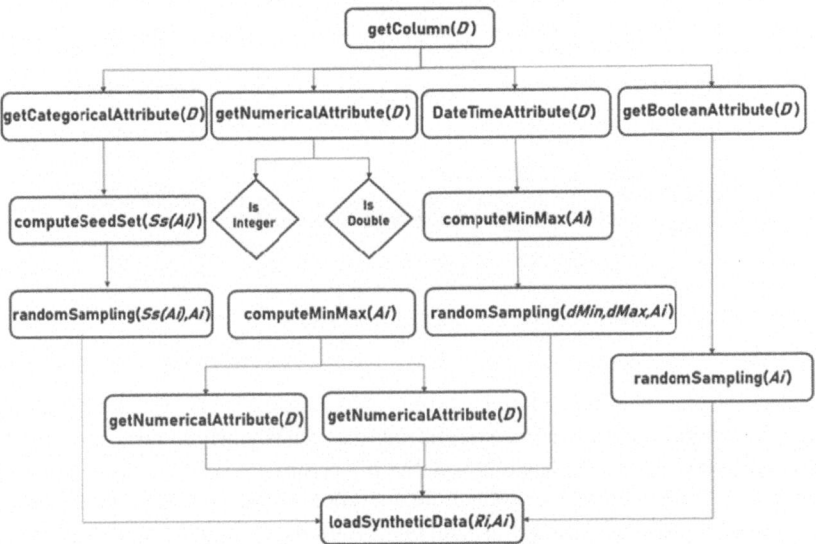

Fig. 6. Generation of Synthetic Data for the Database PainRELifeDB

Here, we present our methods for producing synthetic data using real data as input and using random sampling (e.g., [12]), which results in the synthetic database named Synthetic PainRELifeDB. Figure 6 shows through steps and functions that we utilize to create the data, we utilize specific configurations based on the characteristics of each type of data to be generated. Essentially, our functions generate the final synthetic data using the same database, PainRELifeDB that contains the real data. The functions developed depend on the type of data to be created, namely category data, numeric data (which is further divided into Integer data and Double data), Time data, and Boolean data. As shown in Fig. 6, depending on the type of data, we generate random data via sampling and load the result in the final synthetic multidimensional database.

5 Design and Implementation of Multidimensional Big Data Analytics Tools

In this section, we focus the attention on the design and implementation of Multidimensional Big Data Analytics tools embedded in our proposed architecture. Figure 7 shows the conceptual model for representing the *Dimensional Fact Model* (DFM) through which it is possible to represent data within the main OLAP cube of *PainRELifeDB*. This model captures the dimensions of analysis and the relevant facts pertaining to the modeled reality. The multidimensional model offers several key advantages: (*i*) provide support for conceptual design; (*ii*) create an environment for intuitive and guided querying.

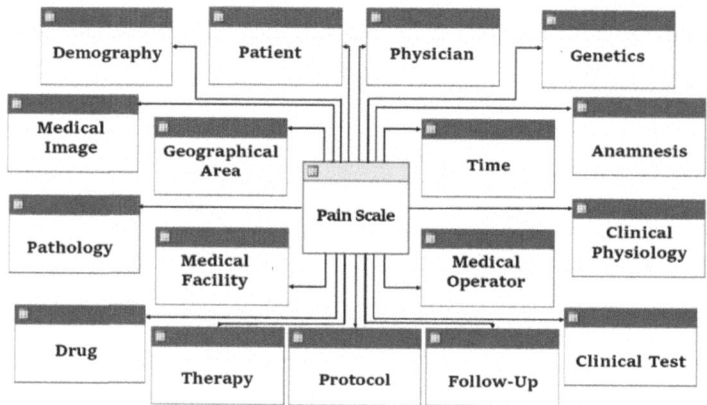

Fig. 7. DFM of the Database *PainRELifeDB*

In terms of graphical representation, the DFM is depicted as a star schema. At the center of the diagram, Pain Scale, also referred to as Survey, which holds prominence in the decision-making process. The measure Survey is intricately linked to 17 dimensions, by establishing relationships that reflect the analytical dimensions of the model.

The multidimensional data analysis involves utilizing prominent analysis and reporting tools, such as *Microsoft Excel*. These tools aim to simplify the analysis process by providing specific functions and formulas and enabling the creation of relevant visualizations. In addition, the analysis encompasses the employ of two specific Data Mining structures. The first structure, known as Multidimensional Clustering, clusters similar elements into distinct groups or clusters based on shared characteristics and attribute values in a multidimensional space. This clustering process facilitates the differentiation of clusters based on their proximity or distance from one another. The second structure, Multidimensional Regression is used in the application domain of Machine Learning, an important branch of artificial intelligence. ML focuses on a machine's ability to generalize experiences and perform inductive reasoning. Notably, Multidimensional Regression demonstrates its strength in autonomously generating examples, which it achieves by training on a set of learning data. By employing the DFM and utilizing the aforementioned structures, researchers and analysts benefit from a comprehensive and efficient approach to data analysis.

Indeed, the primary aim of multidimensional analysis is to extract pertinent information from a substantial volume of data, which, taken individually, could not be considered suitable for satisfying more in-depth analyses. However, analyzing the main cube in its entirety could cause anomalies that would disfavor or, in the worst-case scenario, prevent a very specific manipulation of the data itself. Consequently, the main cube is partitioned into three distinct target cubes, each representing specific dimensions. However, in our experiments, due to space limitation, we present the results on the *PatientCube* cube only.

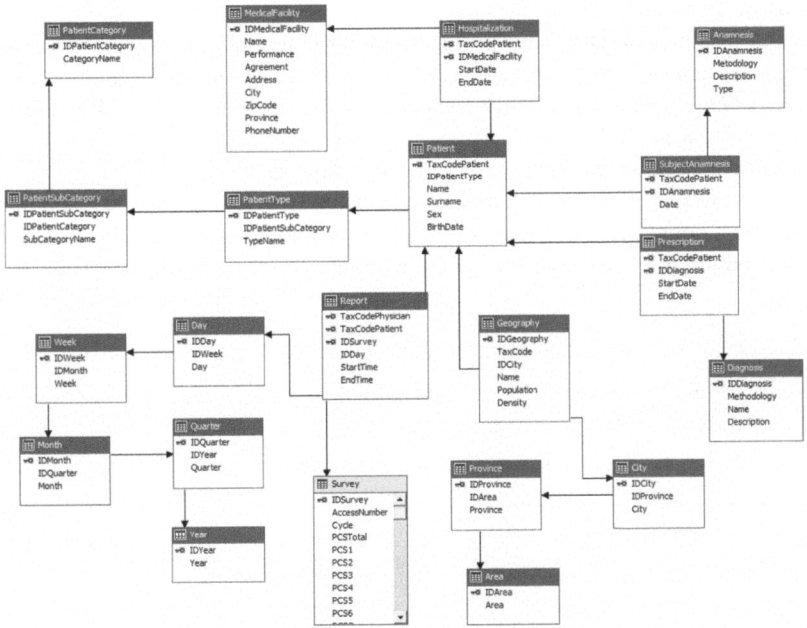

Fig. 8. Cube *PatientCube*

PatientCube, which is shown in Fig. 8, is designed to store and organize data specifically related to patients. The fact table *Survey* contains various key attributes and measures related to patient information. *Survey* is linked to important dimensions such as *Patient, Geography, Anamnesis, Diagnosis,* and *Report. Patient* is further divided into several hierarchies to provide more detailed analysis of patient data. The hierarchies include *PatientType, PatientSubCategory,* and *PatientCategory*, by allowing for different levels of classification and segmentation of patients. *Report* focuses on the temporal aspect of patient data. Like *Patient, Report* features a hierarchy that includes *Day, Week, Month, Quarter,* and *Year*. This enables analysis of patient information based on different time intervals. *Geography*, on the other hand, allows for analysis of patient data based on geographic locations. It includes a hierarchy that regards *City*, Province, and *Area*, by providing insight into the distribution of patients in different geographic regions. Through *PatientCube*, researchers and analysts can extract specific patient-related information from *PainRELifeDB*. This ensures that only data related to patients is included and are excluded information about doctors or other entities. Any analysis or relationships related to doctors will be addressed separately in another cube.

We aim to gather crucial information about patients through the cube *PatientCube*. A comprehensive evaluation of medical well-being cannot disregard the subject's medical history, the facility in which they were hospitalized, or the diagnoses provided by healthcare professionals. Additionally, for statistical and informative purposes, it is valuable to gather geographical details concerning patients, by including their city of origin, province, and reference area. The analysis will delve deep into extracting data regarding patients' specific concerns, as well as the impact that diagnosed conditions have had on various aspects of their lives.

Without loss of generality, our aim is to gather valuable insights about patients by analyzing the data present in the cube *PatientCube*. By examining the dimensions of *Patient* and *Geography*, specifically focusing on the attribute *City* within the hierarchy *Geography*, we can determine which cities have the highest and lowest number of patients. Figure 9 shows the histogram that reports the results.

According to Fig. 9, *Brescia* stands out as the city of origin for the largest number of patients, with a count of 117. Conversely, both *Lodi* and *Sondrio* have an equal number of patients, which happens to be the lowest in comparison, with 80 patients each.

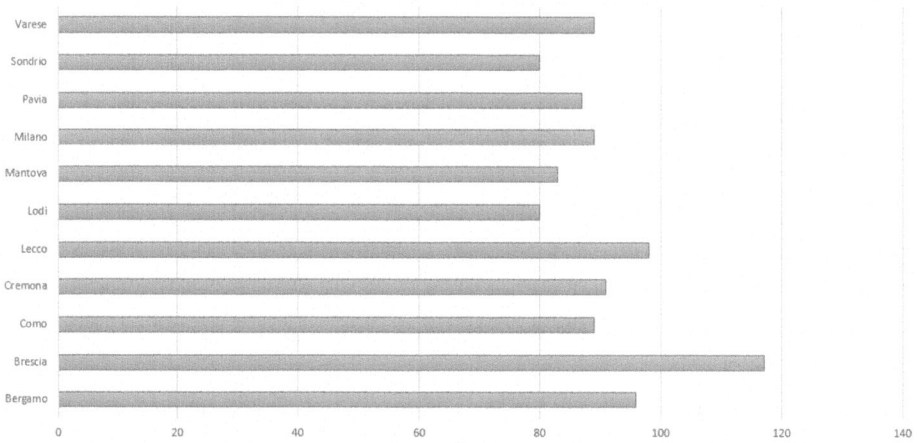

Fig. 9. *PatientCube_MDPivoting* on *PatientCube*

Multidimensional Clustering analysis is carried out as well. In Fig. 10, *Cluster* refers to *Population* of *PatientCube*. *Cluster 1* is the largest in terms of population, equal to 108. Followed, in descending order, by *Cluster 3* with 84, *Cluster 2* with 82, *Cluster 5* with 80, *Cluster 4* with 76, *Cluster 6* with 67, *Cluster 8* with 61, *Cluster 7* with 56, *Cluster 9* with 51, and *Cluster 10* with 35.

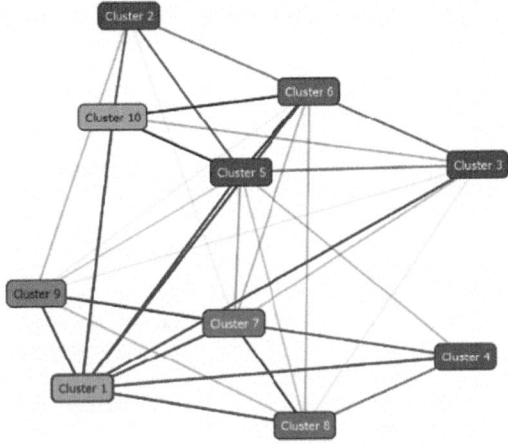

Fig. 10. *PatientCube_MDClustering* on *PatientCube* – Cluster Variable: *Population*

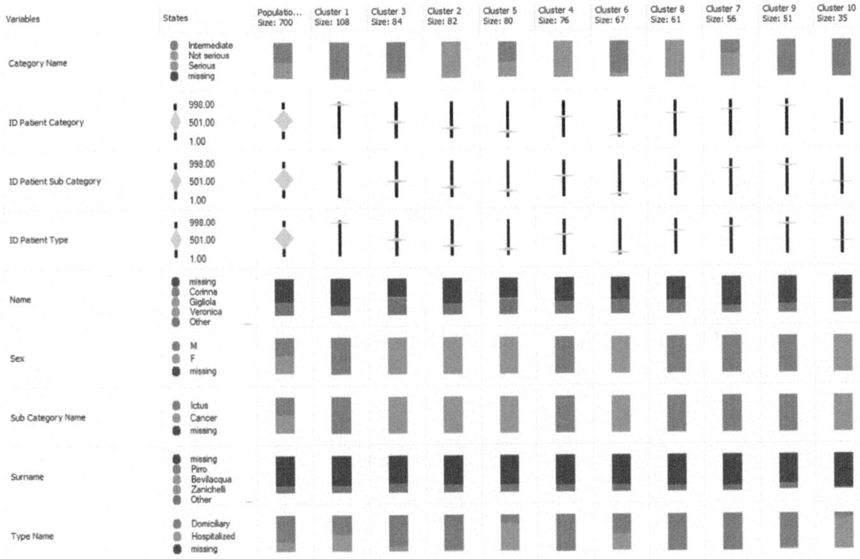

Fig. 11. *PatientCube_MDClustering* on *PatientCube* – Cluster Profiles

Figure 11 shows a screen captured from *Cluster Profiles* section, which summarizes in a simplistic and graphically intuitive way data relating to the various variables that previous cluster diagrams shown have indicated. Figure 12 shows *Cluster Discrimination* section in which values present in two different *Clusters* are compared. In particular, in the example in Fig. 12, *Cluster 1* and *Cluster 2* are compared, as the differences shown are common to all others. Relevant differences, therefore, concern female patients affected by *breast cancer* in *Cluster 2* and male patients affected by *stroke* in *Cluster 1*. The severity status is *intermediate* as regards *Cluster 1*, while not at all in *Cluster 2*. As regards *hospitalization* rate, it is present by about 15% in favor of *Cluster 1* compared to 15% of the *domiciliation* rate in *Cluster2*.

Variables	Values	Favors Cluster 1	Favors Cluster 2
ID Patient Type	1 - 552		
ID Patient Sub Category	553 - 998		
ID Patient Sub Category	1 - 552		
Sex	F		
Sex	M		
Sub Category Name	Cancer		
ID Patient Type	553 - 998		
Sub Category Name	Ictus		
ID Patient Category	553 - 998		
ID Patient Category	1 - 552		
Category Name	Intermediate		
Category Name	Not serious		
Type Name	Hospitalized		
Type Name	Domiciliary		
Birth Date	10/12/1971 3:39:38 PM - 5/21/1973...		
Birth Date	5/21/1973 6:35:30 PM - 12/26/1978...		

Fig. 12. *PatientCube_MDClustering* on *PatientCube* – Cluster Discrimination between *Cluster 1* and *Cluster 2*

It should be noted, here, that Clustering is a powerful methodology for analyzing real-life datasets. On the other hand, when combined with multidimensional paradigms (e.g., [24]), it further enhances its expressive power and its accuracy in advanced analytics methodologies.

Next, we perform the Multidimensional Regression analysis over *PatientCube*. *Mining Multidimensional Regression Framework* is able to predict what could be future values of the measures of pain through the trend of historical values present in *PainRELifeDB*. Figure 13 shows measures to which a different color is associated.

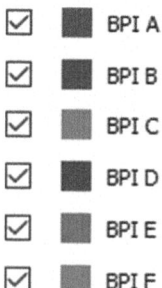

☑ ◼ BPI A

☑ ◼ BPI B

☑ ◻ BPI C

☑ ◼ BPI D

☑ ◼ BPI E

☑ ◻ BPI F

Fig. 13. *PatientCube_MDRegression* on *PatientCube* – Measures

Figure 14 shows predictive analysis based on the previously selected measures. Analysis reports percentages of last current values, represented by the last non-dashed values, and percentages of the last predicted values. The percentages are the following:

- *BPI A (blue)*: current value −34%, last predicted value 67%;
- *BPI B (red):* current value −80%, last predicted value 0%;
- *BPI C (green):* current value −30%, last predicted value −50%;
- *BPI D (purple):* current value −100%, last predicted value −28%;
- *BPI E (light blue)*: current value −75%, last predicted value −87%.

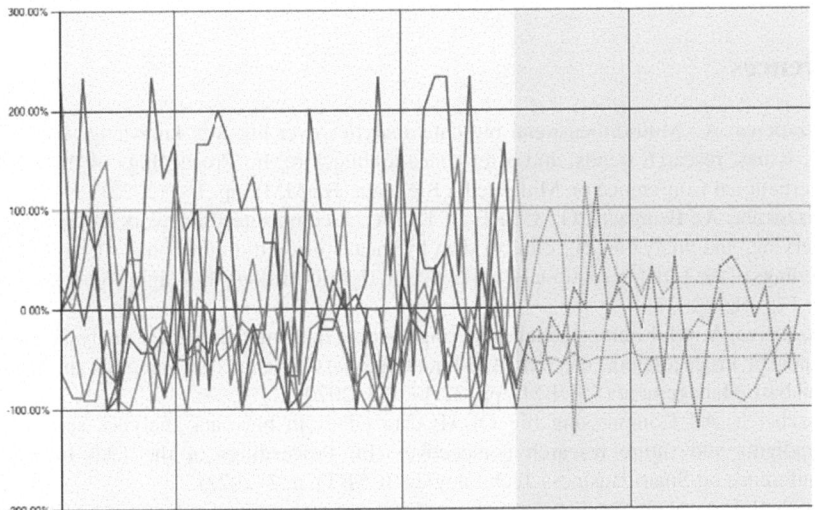

Fig. 14. *PatientCube_MDRegression* on *PatientCube*

6 Remarks

As shown by our experimental analysis provided in Sect. 5, our Cloud-Based Multidimensional Big Data Analytics methodology (and reference architecture) is really capable of providing powerful tools supporting healthcare analytics, according to several, rich perspectives of analysis, according to multiple and correlated dimensions. On the other hand, the proposed methodology exposes powerful predictive functionalities, which nicely complement the main characteristics of the framework.

7 Conclusions and Future Work

Focusing on the emerging healthcare analytics context, this paper has provided practical and experimental results coming from the real-life research project that focuses on Big Data Analytics over Big Healthcare Data about chronical pain of patients in real-life hospitals and medical centers of North Italy (Region Lombardy), *Pain-RELife*. In this context, we have provided e innovative Multidimensional Big Data Analytics tools and a reference Cloud-based architecture that implements our analytics paradigm, plus several experimental results that highlight the benefits in both big data analytics and big data predictive analytics.

Future work is mainly focused dealing with emerging open problems in big data analytics for healthcare (e.g., [16–20, 25]).

Acknowledgments. This research is supported by the ICSC National Research Centre for High Performance Computing, Big Data and Quantum Computing within the NextGenerationEU program (Project Code: PNRR CN00000013).

References

1. Cuzzocrea, A.: Multidimensional big data analytics over big web knowledge bases: models, issues, research trends, and a reference architecture. In: Proceedings of the 8th IEEE International Conference on Multimedia Big Data (BigMM), pp. 1–6 (2022)
2. Cuzzocrea, A., Bringas, P.G.: CORE-BCD-mAI: a composite framework for representing, querying, and analyzing big clinical data by means of multidimensional AI tools. In: Proceedings of the 17th International Conference on Hybrid Artificial Intelligent Systems (HAIS), pp. 175–185 (2022)
3. Cuzzocrea, A.: Multidimensional clustering over big data: models, issues, analysis, emerging trends. In: Proceedings of the 32nd ACM International Conference on Scientific and Statistical Database Management (SSDBM), pp. 32:1–32:6 (2020)
4. Cuzzocrea, A.: Compressing big OLAP data cubes in big data analytics systems: new paradigms and future research perspectives. In: Proceedings of the 19th International Conference on Smart Business Technologies (ICSBT), p. 7 (2022)
5. Marshall, E.A., et al.: Qualitative analysis of responses to a questionnaire via an EHR patient portal. In: Proceedings of the American Medical Informatics Association Annual Symposium (AMIA) (2015)
6. Cuzzocrea, A.: Aggregation and multidimensional analysis of big data for large-scale scientific applications: models, issues, analytics, and beyond. In: Proceedings of the 27th ACM International Conference on Scientific and Statistical Database Management (SSDBM), pp. 23:1–23:6 (2015)
7. Beuscart-Zéphir, M.C., et al.: User-centred, multidimensional assessment method of clinical information systems: a case-study in anaesthesiology. Int. J. Med. Inform. 74(2–4), 179–189 (2005)
8. Russom, P.: Big data analytics. TDWI Best Practices Report, Fourth Quarter 19(4), 1–34 (2011)
9. Ristevski, B., Chen, M.: Big data analytics in medicine and healthcare. J. Integr. Bioinform. 15(3) (2018)
10. Belle, A., et al.: Big data analytics in healthcare. BioMed Research International (2015)
11. Cuzzocrea, A., et al.: Approximate query answering on sensor network data streams. GeoSens. Netw. 93–108 (2003)
12. Olken, F., Rotem, D.: Random sampling from databases: a survey. Stat. Comput. 5, 25–42 (1995)
13. Cuzzocrea, A.: Scalable distributed pivot analysis over massive big data: models, paradigms, new advancements. In: Proceedings of the 20th IEEE International Conference on Data Mining Workshops (ICDMw), pp. 696–700 (2020)
14. Wan, S.J., Wong, S.K.M., Prusinkiewicz, P.: An algorithm for multidimensional data clustering. ACM Trans. Math. Softw. 14(2), 153–162 (1988)
15. Gabriel, K.R.: Least squares approximation of matrices by additive and multiplicative models. J. Roy. Stat. Soc.: Ser. B (Methodol.) 40(2), 186–196 (1978)

16. Cuzzocrea, A., Russo, V.: Privacy preserving OLAP and OLAP security. Encycl. Data Warehous. Min. (2009)
17. Cuzzocrea, A., Saccà, D.: Balancing accuracy and privacy of OLAP aggregations on data cubes. In: Proceedings of the 13th ACM International Workshop on Data Warehousing and OLAP (DOLAP), pp. 93–98 (2010)
18. Lee, Y.W., Choi, J.W., Shin, E.: Machine learning model for diagnostic method prediction in parasitic disease using clinical information. Expert Syst. Appl. **185**, 115658 (2021)
19. Rodriguez-Loya, S., Aziz, A., Chatwin, C.R.: Enabling healthcare IT governance: human task management service for administering emergency department's resources for efficient patient flow. Integr. Inf. Technol. Manage. Qual. Care **202**, 87–90 (2014)
20. Chen, S., et al.: Hypertension monitoring by a real time management system for patients in community and its data mining by vector autoregressive model. IEEE Access **11**, 12606–12621 (2023)
21. Poon, L., Zhang, N., Liu, T., Liu, A.H.: Model-based clustering of high-dimensional data: variable selection versus facet determination. Int. J. Approximate Reasoning **54**, 196–215 (2013)
22. Yixin, C., et al.: Multi-dimensional regression analysis of time-series data streams. In: Proceedings of the 28th International Conference on Very Large Data Bases (VLDB), pp. 323–334 (2002)
23. Li, J., Wang, S., Rudinac, S., Osseyran, A.: High-performance computing in healthcare: an automatic literature analysis perspective. J. Big Data **11**(1), 61 (2024)
24. Cuzzocrea, A.: Computing and mining clustcube cubes efficiently. In: Proceedings of the 19th Pacific-Asia Conference on Advances in Knowledge Discovery & Data Mining (PAKDD), pp. 146–161 (2015)
25. Sangaiah, A.K., Rezaei, S., Javadpour, A., Zhang, W.: Explainable AI in big data intelligence of community detection for digitalization e-healthcare services. Appl. Soft Comput. **136**, 110–119 (2023)

A Database Engineered System for Big Data Analytics on Tornado Climatology

Fengfan Bian, Carson K. Leung$^{(\boxtimes)}$ (ID), Piers Grenier, Harry Pu, and Samuel Ning

University of Manitoba, Winnipeg, MB, Canada
Carson.Leung@UManitoba.ca

Abstract. Facing the challenges with current tornado warning systems, we explore alternative approaches. Specifically, we present a database engineered system that integrates information from heterogeneous rich data sources, capturing both climatology data for tornadoes and those data just before a tornado warning. Such a system aids in predicting tornado occurrences by identifying the data points that form the basis of a tornado warning. Applications of this system to US data highlights the advantages of using a classification forecasting recurrent neural network (RNN) model in our system. The application results also highlight the effectiveness of our database engineered system for big data analytics on tornado climatology—especially, in accurately predicting tornado lead-time, magnitude, and location.

Keywords: Database Engineered Application · Data Science · Data Mining · Data Analytics · Climate Data · Tornado

1 Introduction

In the current era of big data, data science or database engineered systems [1, 2]—which make good use of machine learning (ML) [3, 4] (including deep learning [5, 6]), data mining [6–14] and analytics [15–17]—help reuse and/or integrate information in various real-life applications for public good. These include healthcare informatics [18–22] and social network analysis [23, 24].

In this paper, we focus on tornado climatology within the realm of meteorology analysis. Note that a tornado is a highly destructive, life-threatening natural disaster that can cause extensive damage to buildings, infrastructure, and agriculture resulting in significant property loss, economic hardship, as well as tragically, the loss of life. Tornadoes are most common in the USA than in any other countries. On average, there are about 1,200 tornadoes (including confirmed and unconfirmed) that touch down on the USA each year. Not only the USA has the highest quantity of tornado, it also has the highest quantity of intensive tornado. For instance, violent tornadoes—e.g., those rated EF4 or EF5 on the Enhanced Fujita scale (i.e., wind speed ≥ 166 mph) leading devastating or incredible damage—occur most often in the USA than in any other countries.

Current tornado warning systems provide only a limited lead-time for residents and authorities to prepare an effective response. Improving the accuracy and lead-time of

R. Chbeir et al. (Eds.): IDEAS 2024, LNCS 15511, pp. 172–185, 2025.
https://doi.org/10.1007/978-3-031-83472-1_12

tornado prediction is crucial for saving lives as well as minimizing damage. In the USA, tornadoes occur quite frequently and clearly, a specialized forecasting system is needed. These tornadoes often occur in regions with lower population density, making their impact on communities less predictable. As a result, the current warning system, while valuable, lacks the granularity required for precise preparedness and response.

ML within the realm of artificial intelligence (AI) is commonly employed to create this desired forecasting system, where the model typically incorporates a neural network. Here, the pre-storm environment is used to train the network and prediction involves classifying weather data points based on if they constitute a tornado event. This provides the motivation for this paper as it present a database engineered system—namely, a recurrent neural network (RNN) model for tornado forecasting.

When considering the benefits of a predictive forecasting system, it is important to acknowledge its potential real-life applications. Examples include:

- advancement in comprehensive information systems for tornado genesis,
- safe evacuation procedures to minimize casualties,
- standardization of prompt warning systems to increase tornado prediction lead-time,
- tornado protection for a broader range of regions, and
- real-time analyses on tornado speed and strength.

These are just some of the real-world database engineered applications of using a predictive forecasting model, and the advantages they offer are clear.

In this paper, we present a database engineered system to integrate information from multiple rich data sources such as climatology data for tornadoes and those immediately preceding a tornado warning. The resulting system helps predict the occurrence of tornadoes. It determines the data points that constitute a tornado warning. Our *key contributions* of this paper include:

- a new long short-term memory (LSTM) RNN predictive forecasting model with 86% cross-validation accuracy;
- active learning for labeling weather data points and dynamically updating when new data arrives, which is invaluable when considering the issuing of tornado warnings;
- classification of new data points to predict a tornado event or not;
- implementation of 10-fold cross validation for multiple training sets to reduce bias in results by testing on a more diverse range of data; and
- new insights for yearly model accuracy, the area under the curve (AUC) scores for the receiver operating characteristic (ROC) curve in each fold, a confusion matrix for evaluating our classification model, and model training accuracy vs. loss.

The remainder of this paper is organized as follows. The next section discusses some background and related works. Section 3 presents our database engineered system. Sections 4 and 5 shows our evaluation results and draws conclusions, respectively.

2 Background and Related Works

Existing works on tornado forecasting systems usually use statistical model [25], active learning with support vector machines (SVM) [26], ML classifiers [27], deep learning [28–31], fuzzy logic and classification [32], feature extraction classification [33], or spatiotemporal relational probability trees [34].

However, the statistical model can only make predictions in a short-term diagnostic mode because it does not incorporate active learning techniques, and instead uses passive learning [25]. Consequently, less significant data points for distinguishing a tornado event are labeled and placed into the training set for their model. In contrast, our database engineered system is naturally different because it does not incorporate passive learning, where predictions are made using old data points. In other words, the statistical model is static in that it labels data points randomly during training and it may not be able to update dynamically with new data.

Trafali et al. [26] applied active learning with SVM for tornado prediction. Their ML classifiers for imbalanced tornado data [27] deal with the minority class, in which tornadoes form from unusual pre-storm conditions as represented by the weather data. The objective being to maximize the accuracy of predicting tornadoes in the rare cases, which naturally makes the classifiers less applicable for standard tornado occurrences. In contrast, our database engineered system deals with the majority class for tornado prediction.

Many works involving deep learning for predicting tornado events use convolutional neural networks (CNNs) [28–30], which are multi-layered and designed to learn from spatial grids. These spatial grids must first be converted into scalar features, which can then be used to train the network. The model can then predict the outcome of unseen data using training knowledge. Due to the complexity of the data, CNNs are required to learn from a very large number of samples. They also need heavy tuning as well as hardening to prevent false-positives, and especially false-negatives. McGuire and Moore [30] tested a variety of CNNs based "on their ability to classify the strength of daily tornado outbreaks". The relevant CNNs include LeNet-5, VGG-16 and Resnet-50, which all use object recognition in image datasets. Soni et al. [31] implemented a counterfeit neural network where weather parameters can access future information on weather history. This means that the parameters of temperature, rain and air speed are expected without a lot of flexibility. In contrast, our database engineered system uses an RNN. The main difference being that an RNN has fewer layers and requires much less data for learning. In addition, our system does not compare multiple neural networks for tornado prediction, and is therefore different from related works that predict tornado days in the USA. Moreover, we do not allow parameter access to future information on weather history, and is therefore more flexible at the cost of accuracy.

Lakshmanan et al. [32] incorporated feedforward neural networks and SVM to classify radar signatures automatically for tornado prediction. Gradient estimation and fuzzy logic are integrated into the classifier. As such, a more probabilistic approach was taken as opposed to an ML one by viewing tornado prediction as a spatiotemporal problem "of estimating the probability of a tornado event at a particular spatial location within a given time window". Consequently, smaller timeframes was resulted for predictions on tornado events. In contrast, our database engineered system uses an RNN.

Coccomini and Zara [33] collected meteorological data from certain geological locations daily and placed them into climate metrics. Features were then extracted and stored as single-day features and multiple day features for classification. Their model had 84% accuracy for up to 5 days of prediction, which could be more accurate. Although our database engineered system is similar in the sense that it also aims to develop accurate early-detection tornado systems, it is different in its temporal analysis of the collected data and the features extracted.

McGovern et al. [34] used probability estimation trees, which learn using data temporally varies (also known as spatiotemporal relational probability trees, or SRPTs for short). Their probabilistic model uses SRPTs to automatically discover spatial temporal relationships in data. They also only discusses how to evaluate an ML model rather than incorporating one, and only covers prediction of tornadoes in state of Oklahoma, USA. In contrast, our database engineered system focuses on the USA as a whole and uses an ML approach.

3 Our Database Engineered System

3.1 Overview

Our database engineered system aims to address these challenges by designing and developing a tornado forecasting system that accounts for geographical and meteorological characteristics. By utilizing tornado data, meteorological parameters and advanced ML techniques, we aim to create a localized forecasting system capable of providing detailed forecasts—including specific timing, magnitude and geographical location of tornado events.

Our system also uses data mining to make tornado forecasting better. We use ML to look at old weather data and find patterns that might tell us when and where tornadoes will happen. Our method differs from related works as we implement a classification predictive model. First, we group the data based on the kind of weather that often precedes tornadoes and then use classification to figure out which of these weather groups usually lead to tornadoes. This helps users know information like how strong the tornado will be, where it will happen, and when it might hit. Using these data mining techniques enables residents to have more time to get ready for tornadoes, which can help keep them safe.

We focus on utilizing existing datasets sourced from the USA, where most tornadoes occur east of the Rocky Mountains, the Great Plains, the Midwest, the Mississippi Valley and the southern USA. Among the 48 continental US states, although Florida has the highest number of tornadoes per unit area, these are usually weak tornadoes. In contrast, Oklahoma has the highest number of *strong* tornadoes. With warm moist air from southeast (Gulf of Mexico) and warm dry air from southwest (Gulf of California and Equator in the Pacific Ocean) meeting cold dry air from the northwest (and Canadian Rockies), Tornado Alley—covering parts of Texas, Oklahoma, Kansas, Colorado, Nebraska and South Dakota—becomes an ideal environment for tornadoes to form within developed thunderstorms.

In our presented tornado forecasting system, we employ a binary classification approach integrated into an RNN architecture to overcome forecasting challenges. Unlike traditional feedforward neural networks that process input data in a single pass, RNNs handle sequential data by maintaining a hidden state that captures information about previous inputs. This sequential memory makes RNNs particularly suitable for time-series forecasting, such as tornado prediction, where the temporal order of meteorological events is crucial. The system leverages the capabilities of the RNN to discern patterns indicative of tornado occurrence based on meteorological data.

In the context of tornado classification prediction, RNNs offer advantages over other models due to their ability to capture temporal dependencies and patterns in the data. Our presented database engineered system for tornado forecast utilizes a binary classification approach integrated into an RNN architecture, allowing it to overcome forecasting challenges effectively. Tornado events unfold over time, with meteorological conditions evolving dynamically, and RNNs excel at capturing the temporal nuances in such sequences. By considering the sequential nature of meteorological parameters leading up to a tornado event, RNNs can potentially capture patterns that other models (e.g., those treating the data as independent samples) may fail to recognize. In addition, RNNs are well suited for handling varying lengths of input sequences, a critical feature when dealing with meteorological data with irregular time intervals or different sampling rates. This flexibility allows the model to adapt to the diverse temporal aspects of tornado formation, enhancing its ability to generalize across different scenarios. While other ML models—e.g., traditional feedforward neural networks, SVM—may perform well in certain applications, their limitations in handling sequential data make them less suitable for tornado prediction tasks. In contrast, the RNN's capacity to learn and remember temporal dependencies positions it as a promising choice for capturing the complex dynamics associated with tornado formation.

3.2 Design Details

We integrate information about USA weather and tornado. Within these data, we explore the following key features:

- Cloud and visibility: cloud coverage, and the distance of visibility.
- Moisture and dew: dew point, and the amount of moisture in the air.
- Moon phase: phase of the moon.
- Perceived temperature.
- Precipitation: amount of precipitation, its probability, and its coverage.
- Pressure and atmospheric conditions: atmospheric pressure at sea level.
- Temperature metrics: maximum, minimum, and average temperatures.
- Wind information: wind speed and direction.

As for the output of the model, it is designed for binary classification—specifically to predict whether a tornado will occur or not. The binary output variable is denoted as result, where a value of 1 indicates the occurrence of a tornado, and 0 signifies the absence of a tornado.

The RNN training process involves splitting the data into features and labels. We then standardize the numerical features (say, by using the Scikit-learn's StandardScaler) and reshape the data to suit the RNN format. The Synthetic Minority Over-sampling Technique (SMOTE) is applied to address potential imbalances in the target variable. Regarding the architecture, the RNN model consists of:

- an LSTM layer,
- a dropout layer for regularization, and
- a dense output layer with a sigmoid activation function.

Then, the model utilizes a 10-fold cross-validation approach for training and evaluation. This involves dividing the data into 10 subsets, training the model on 9 of them, and validating the remaining subset. This process repeats 10 times, and accuracy scores for each fold are collected. To prevent overfitting, 20% of the neurons in the dropout layer are randomly deactivated during each training iteration when creating a model. Lastly, the overall model performance is assessed by computing the mean and standard deviation of the accuracy scores across all folds.

3.3   Implementation Details

To prepare and clean data, we utilize the Python libraries Pandas (for data analysis) and NumPy (for supporting large multi-dimensional arrays). We meticulously process the weather data, discarding non-numeric and sparsely populated columns that could potentially skew our analyses. The "datetime" conversion is not just a procedural necessity but also a strategic move to ensure the sequential integrity of our dataset, which is vital for time-series analysis. By imputing the missing values with the mean for the numeric columns, we preserve every data point's potential to contribute to our findings.

Selecting the RNN as our predictive model is a deliberate choice, driven by the sequential and temporal nature of meteorological data. The RNN's innate ability to process sequences through its internal state and memory capabilities makes it an unparalleled tool for identifying the precursors to tornado events. The preprocessing phase involves not just standardizing the data but also transforming it into a format amenable to RNN processing. Addressing data imbalance with the SMOTE is also critical, as it ensures our model not being biased towards the majority class and can generalize well across all scenarios.

Our model's architecture is thoughtfully designed using Keras (an open-source library that provides a Python interface for ANNs) and incorporating LSTM layers (which are renowned for their sequence prediction capabilities). We introduce dropout layers as a precautionary measure against overfitting, and the binary nature of the output layer directly corresponded to the occurrence or non-occurrence of tornado events. The robustness of our model is further bolstered by a comprehensive 10-fold cross-validation, which ensures that every data point is utilized in both training and validation phases, thus minimizing any potential bias and enhancing the model's applicability across diverse data samples.

The coding implementation was a testament to the end-to-end process, from initial data cleaning to the final stages of model evaluation. By deploying Scikit-learn's KFold cross-validator, we not only divide our dataset for validation purposes but also set up a robust framework for tracking and analyzing the model's accuracy and AUC across different folds. This attention to detail allows our model's predictive performance to be fine-tuned and ensure its reliability. The resultant confusion matrices provide a cumulative view of the model's true positive and negative rates, presenting a clear picture of its predictive prowess. Moreover, the visual aids (e.g., ROC curves, plots) juxtaposing training accuracy with loss, served as intuitive indicators of the model's learning trajectory and performance benchmarks.

4 Evaluation

4.1 Setup

Our investigation into tornado prediction leverages a significant corpus of meteorological data amassed from Visual Crossing Weather Data Services.[1] This platform is a repository of historical weather data, meticulously curated to aid a wide array of climatological research. The service provides a wealth of information essential to the analysis—covering various weather-related parameters such as temperature, precipitation, and wind, which are critical factors in understanding and forecasting the conditions that lead to tornado occurrences. The granularity and precision of the data from Visual Crossing enabled our model to capture the nuanced environmental patterns that typically precede tornado events.

In conjunction with the Visual Crossing dataset, we incorporated the Storm Prediction Center (SPC) dataset[2] provided by the National Oceanic and Atmospheric Administration (NOAA). Renowned for its comprehensive documentation of severe weather events, the SPC dataset is particularly detailed in its accounts of tornado occurrences, including geographical and temporal markers. These data are invaluable for training a predictive model, as they allow for the calibration of outputs against well-documented historical instances of tornadoes, thus enhancing the model's predictive precision with respect to location and intensity.

The datasets collectively provide a robust analytical foundation, comprising over 178,997 records post-cleanup, covering a 10-year interval from 1998 to 2007. With 49 records per day, there are about $49 \times 365 = 17{,}885$ records per non-leap year and $49 \times 366 = 17{,}934$ records per leap year (e.g., 2000, 2004). Each record, regardless of leap year or not, encompassing 28 attributes reflecting various meteorological conditions. Figure 1 shows yearly distribution of tornado occurrences.The highest count is in 1998 with 491 tornadoes, while the lowest is in 2002 with 336 tornadoes. Other years show counts ranging between these values, indicating fluctuations in tornado occurrences over the decade.

Our system uses a RNN model, meticulously designed to process and learn from the intricate sequences inherent in weather data. The model was designed to classify data points for predicting the likelihood of tornado events. It uses a binary "result" output where a value of 1 signifies the prediction of a tornado. By using the RNN, the study harnessed the power of sequential data processing to uncover patterns across the temporal landscape of weather conditions that culminate in tornado genesis.

The datasets from Visual Crossing and the SPC are indispensable in the development of our system. The diverse and high-dimensional dataset enables the construction of a predictive model that not only aims to enhance the accuracy of tornado forecasts but also potentially extends the lead-time for warnings, thereby contributing to disaster preparedness and saving lives. The quality and depth of the datasets reflect the current state-of-the-art in meteorological data collection and severe weather forecasting, underscoring the value of reliable data sources in advancing the field of climatological research.

---

[1] https://www.visualcrossing.com/.

[2] https://www.spc.noaa.gov/.

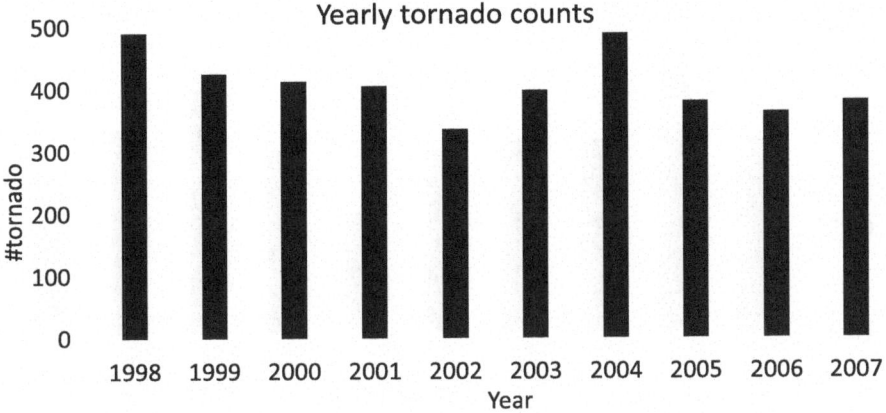

Fig. 1. Distribution of yearly tornado occurrences.

To evaluate and demonstrate the usefulness of our database engineered system, we test it with 10-fold evaluation. We randomize the data before splitting it into 10 folds, which is good for ensuring that each fold is representative of the whole. We set the random state of the 10-fold evaluation to 42 to ensure the reproducibility of our results. From the 10-fold validation results, we analyze the results using AUC score, confusion matrix and model training accuracy and loss.

4.2 Evaluation Results

Figure 2 shows the AUC scores for our ROC curve from the 10-fold cross-validation process. The AUC scores range from just under 0.856 to just over 0.864. This indicates a fairly consistent model performance across different folds as the variation is within a narrow band. There is a noticeable dip in AUC score in fold 6, which suggests that the data in fold 6 was different from the rest of the folds, possibly due to outliers in fold 6. Despite the variability, The AUC scores are quite high as they are all above 0.850, suggesting our model has a good ability to differentiate between the positive class (presence of tornado) and negative class (absence of tornado). Apart from the dip in fold 6, the model's performance is relatively consistent, which proves the robustness of our model. Overall, the AUC scores suggest our model has good discrimination to distinguish positive and negative outcomes of tornadoes.

Figure 3 shows the confusion matrix table of our 10-fold cross validation results. The row of the matrix represents the instances of an actual class from the results, while the column represents the instances in a predicted class. Here are the results analysis: Our model correctly predicted the negative class 14,412 times on average. The model also correctly predicted the positive class 15,644 times on average.

Based on the values in this confusion matrix, we compute the metrics shown in Table 1:

$$\text{Sensitivity} = \frac{\text{TP}}{\text{TP} + \text{FN}} \tag{1}$$

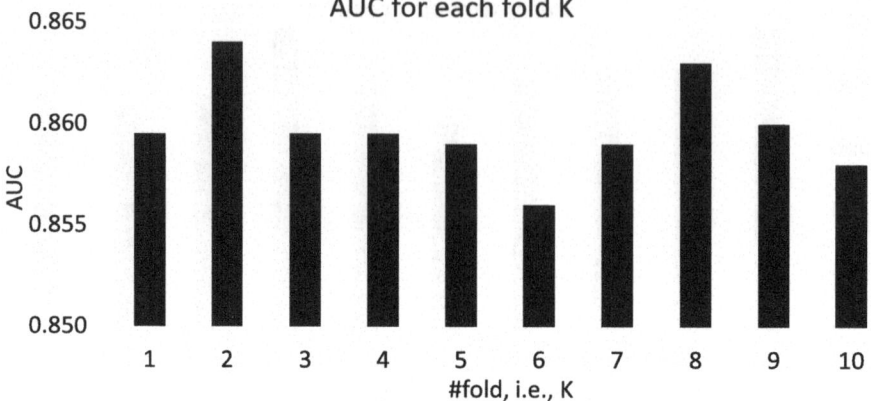

Fig. 2. AUC in each fold for 10-fold cross validation.

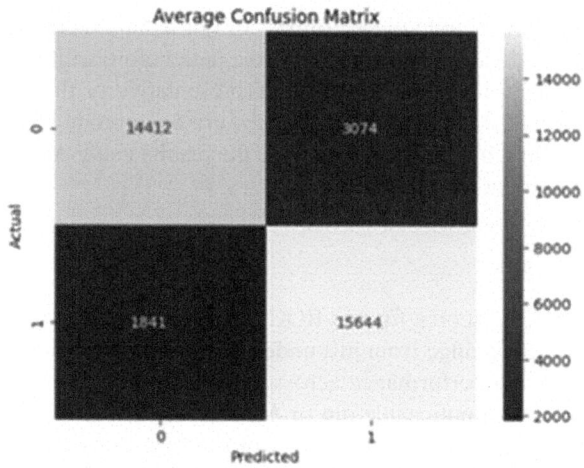

Fig. 3. Average confusion matrix for 10-fold cross validation results.

$$\text{False positive rate (FPR)} = \frac{FP}{FP + TN} \tag{2}$$

$$\text{Accuracy} = \frac{TP + TN}{TP + TN + FP + FN} \tag{3}$$

$$\text{Precision} = \frac{TP + TN}{TP + TN + FP + FN} \tag{4}$$

$$\text{Negative predictive value (NPV)} = \frac{TN}{TN + FN} \tag{5}$$

where TP is true positive, TN is true negative, FP is false positive, and FN is false negative. For instance:

- High sensitivity: The model demonstrates strong sensitivity, correctly identifying approximately 90% of the positive cases. This indicates a high true positive rate, suggesting that the model is effective in detecting the presence of the condition or characteristic being tested.
- Moderate false positive rate (FPR): With a FPR of 18%, there is a noticeable, but not overwhelming, tendency for the model to incorrectly classify negative cases as positive. This level suggests room for improvement in distinguishing non-cases more accurately.
- Good accuracy: An accuracy of 86% signifies that the model correctly classifies both positive and negative cases with high reliability in the majority of instances. This reflects its overall effectiveness in making correct predictions.
- Solid precision: The precision of 84% indicates that when the model predicts a positive result, there is an approximately 84% chance that it is correct. This points to the model's reliability in its positive predictions, though there is still some scope for reducing false positives.
- High negative predictive value (NPV): An NPV of 89% shows that the model is quite effective in predicting negative cases. When the model predicts a negative result, it is correct about 88.7% of the time, which is a strong indication of its ability to correctly identify true negatives.

Table 1. Metrics based on the confusion matrix

Metric	Equation	Value
Sensitivity	TP/(TP + FN)	90%
False positive rate (FPR)	FP/(FP + TN)	18%
Accuracy	(TP + TN)/(TP + TN + FP + FN)	86%
Precision	TP/(TP + FP)	84%
Negative predictive value (NPV)	TN/(TN + FN)	89%

To recap, the confusion matrix and the derived metrics suggest that the model is quite effective in identifying true positives and true negatives when it comes to predicting tornadoes, with a high degree of sensitivity and accuracy. The moderate false positive rate and good precision indicate that while the model is reliable in its predictions, there is potential for further refinement, particularly in reducing the number of false positives. Overall, the model demonstrates a robust performance in classification tasks, making it a valuable tool for tornado prediction.

Figure 4 shows the model training accuracy and loss. The graph illustrates the model's performance during the training phase, measured by accuracy and loss over a series of 10 epochs. Notably, the accuracy curve as shown in Fig. 4(a) increases from 0.80 to 0.86. More precisely, it ascends sharply at the onset and begins to level off as it approaches the final epochs, indicating that the model is making increasingly correct predictions as it processes more training data. The plateau towards the end of the training suggests that the model may be nearing its peak learning capacity given the current data and network architecture.

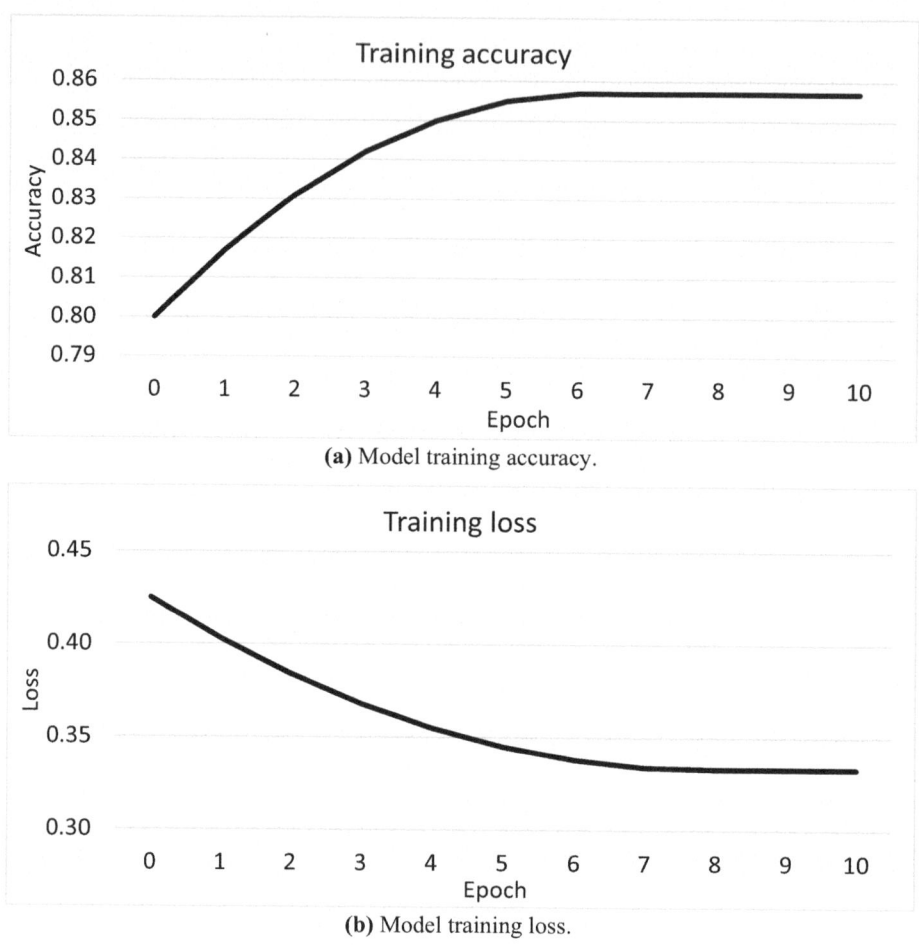

(a) Model training accuracy.

(b) Model training loss.

Fig. 4. Training accuracy and loss.

Conversely, the loss curve as shown in Fig. 4(b) decreases from 0.43 to 0.33. More precisely, it demonstrates a significant decrease, particularly evident in the initial epochs, reflecting a reduction in the error between the model's predictions and the actual targets. The loss curve's decline is steeper in the early stages of training, which is common as the model quickly corrects itself from initial random or uninformed predictions. As the epochs progress, the rate of loss reduction slows, suggesting that the model is starting to converge and is making fewer and smaller mistakes in its predictions.

The convergence of these two metrics—accuracy and loss—towards their respective high and low values indicates a successful training process. However, the graph also implies that further training beyond the 10th epoch may not yield substantial improvements and could lead to overfitting, where the model learns patterns specific to the training data that do not generalize well to unseen data.

5 Conclusions

In conclusion, this paper focused on the critical issue of tornado prediction, considering the high frequency of tornado occurrences in the USA and the limitations of existing warning systems. The motivation stems from the need to enhance the accuracy and lead-time of tornado prediction to minimize the devastating impact on lives and property. The presented database engineered system utilizes an LSTM-RNN predictive forecasting model, achieving an impressive 86% k-fold cross-validation accuracy. The real-life applications of such a predictive forecasting system are extensive, ranging from comprehensive information systems for tornado genesis to safe evacuation procedures, standardizing warning systems, and providing tornado protection for a broader range of regions. Our key contributions of this paper include active learning for labeling weather data points, dynamic updates with new data, classification of data points for tornado prediction, 10-fold cross-validation for diverse training sets, and insights into model accuracy metrics.

Comparisons with existing models underscore the uniqueness of the proposed LSTM RNN model in our database engineered system. For instance, our system incorporates active learning and dynamically updates with new data. It focuses on the majority class for tornado prediction, distinguishing it from models targeting imbalanced tornado data. While sharing similarities with deep learning CNN models, our RNN model offers advantages in terms of reduced complexity and data requirements. Moreover, the model prioritizes flexibility over accuracy. The existing spatiotemporal method for predicting tornado events (which views tornado prediction as a spatiotemporal problem) differs significantly from our RNN model, which employs ML for classification. Our system also contains a tailored tornado forecasting system for the USA, addressing challenges posed by unique geographical and meteorological characteristics. Leveraging US tornado data and advanced ML techniques, our system employs a binary classification approach integrated into an RNN architecture. This approach proves advantageous in capturing temporal dependencies and patterns crucial for tornado prediction, surpassing the limitations of traditional models in handling sequential meteorological data.

The model, trained on a dataset combining USA weather and tornado data, focuses on key meteorological parameters to predict tornado occurrence using a binary output variable. The methodology involves rigorous data cleaning, RNN model selection, and careful consideration of imbalances using SMOTE. The 10-fold cross-validation approach ensures robust training and evaluation, emphasizing the model's ability to generalize across diverse data samples. Datasets from Visual Crossing and the Storm Prediction Center (SPC) provide a comprehensive foundation for the model, capturing nuanced environmental patterns preceding tornado events. The model's evaluation involves AUC graph analysis, confusion matrix interpretation, and examination of model training accuracy and loss. The AUC scores consistently above 0.85 demonstrate the model's strong discrimination between positive and negative tornado outcomes. Confusion matrix analysis reveals the model's high sensitivity, moderate false positive rate, good accuracy, solid precision, and high negative predictive value, indicating effectiveness in identifying true positives and true negatives. The model's training accuracy and loss graphs suggest successful convergence, with potential caution against overfitting beyond the 10th epoch. Our database engineered system highlights a robust tornado forecasting system with promising predictive capabilities.

As *ongoing and future work*, we would integrate other relevant data with an aim to further enhance prediction accuracy and data analytics. This would further enhance our system in classifying tornado climatology data and predicting future disasters for sustainable cities. We would also like to transfer knowledge learned from designing and development the current database engineered system to analyze big tornado climatology data from other geographical locations (e.g., Canada).

Acknowledgments. This work is partially supported by NSERC (Canada) and University of Manitoba.

References

1. Özsu, M.T.: Data science - a systematic treatment. CACM **66**(7), 106–116 (2023)
2. Revesz, P.Z.: Data science applied to discover ancient Minoan-Indus Valley trade routes implied by common weight measures. In: IDEAS 2022, pp. 150–155 (2022)
3. Kim, J., et al.: Cluster-guided temporal modeling for action recognition. Int. J. Multim. Inf. Retr. **12**(2), 15 (2023)
4. Phiwhorm, K., et al.: Adaptive multiple imputations of missing values using the class center. J. Big Data **9**(1), 52 (2022)
5. Lin, E., et al.: ScGMM-VGAE: a Gaussian mixture model-based variational graph autoencoder algorithm for clustering single-cell RNA-seq data. Mach. Learn. Sci. Technol. **4**(3), 35013 (2023)
6. Zammit, J., et al.: Semi-supervised COVID-19 CT image segmentation using deep generative models. BMC Bioinform. **23**(7), 343 (2022)
7. Alam, M.T., et al.: Discovering interesting patterns from hypergraphs. ACM Trans. Knowl. Discov. Data **18**(1), 32:1–32:34 (2024)
8. Alam, M.T., et al.: Mining frequent patterns from hypergraph databases. In: PAKDD 2021, Part II. LNCS (LNAI), vol. 12713, pp. 3–15 (2021)
9. Bernhard, S.D., et al.: Clickstream prediction using sequential stream mining techniques with Markov chains. In: IDEAS 2016, pp. 24–33 (2016)
10. Chowdhury, M.E.S., et al.: A new approach for mining correlated frequent subgraphs. ACM ACM Trans. Manag. Inf. Syst. **13**(1), 9:1–9:28 (2022)
11. Czubryt, T.J., et al.: Q-Eclat: vertical mining of interesting quantitative patterns. In: IDEAS 2022, pp. 25–33 (2022)
12. Leung, C.K.: Pattern mining for knowledge discovery. In: IDEAS 2019, p. 34:1–34:5 (2019)
13. Roy, K.K., et al.: Mining sequential patterns in uncertain databases using hierarchical index structure. In: PAKDD 2021, Part II. LNCS (LNAI), vol. 12713, pp. 29–41 (2021)
14. Roy, K.K., et al.: Mining weighted sequential patterns in incremental uncertain databases. Inf. Sci. **582**, 865–896 (2022)
15. Eom, C.S., et al.: Effective privacy preserving data publishing by vectorization. Inf. Sci. **527**, 311–328 (2020)
16. Leung, C.K., et al.: Theoretical and practical data science and analytics: challenges and solutions. Int. J. Data Sci. Anal. **16**(4), 403–406 (2023)

17. Yoo, K., et al.: Big data analysis and visualization: challenges and solutions. Appl. Sci. **12**(16), 8248:1–8248:5 (2022)
18. Lac, L., et al.: Computational frameworks integrating deep learning and statistical models in mining multimodal omics data. J. Biomed. Inform. **152**, 104629 (2024)
19. Leung, C.K.: Biomedical informatics: state of the art, challenges, and opportunities. BioMedInformatics **4**(1), 89–97 (2024)
20. Leung, C.K., et al.: Big data science on COVID-19 data. In: IEEE BigDataSE 2020, pp. 14–21 (2020)
21. Leung, C.K., et al.: Explainable data analytics for disease and healthcare informatics. In: IDEAS 2021, p. 12:1–12:10 (2021)
22. Zhou, X., et al.: Deep learning-empowered big data analytics in biomedical applications and digital healthcare. IEEE/ACM Trans. Comput. Biol. Bioinform. **21**(4), 516–520 (2024)
23. Choudhery, D., Leung, C.K.: Social media mining: prediction of box office revenue. In: IDEAS 2017, pp. 20–29 (2017)
24. D'Souza, R.R., et al.: Discovery of patent influence with directed acyclic graph network analysis. In: IDEAS 2023, pp. 17–24 (2023)
25. Elsner, J.B., et al.: Statistical models for tornado climatology: long and short-term views. PLOS ONE **11**(11), e0166895:1–e0166895:20 (2016)
26. Trafalis, T.B., et al.: Active learning with support vector machines for tornado prediction. In: ICCS 2007. LNCS, vol. 4487, pp. 1130–1137 (2007)
27. Trafalis, T.B., et al.: Machine-learning classifiers for imbalanced tornado data. Comput. Manage. Sci. **11**(4), 403–418 (2013)
28. Barajas, C.A., et al.: Performance benchmarking of data augmentation and deep learning for tornado prediction. In: IEEE Big Data 2019, pp. 3607–3615 (2019)
29. Lagerquist, R., et al.: Deep learning on three-dimensional multiscale data for next-hour tornado prediction. Mon. Weather Rev. **148**(7), 2837–2861 (2020)
30. McGuire, M.P., Moore, T.W.: Prediction of tornado days in the United States with deep convolutional neural networks. Comput. Geosci. **159**, 104990 (2022)
31. Soni, S., et al.: Deep learning based weather forecast: a prediction. In: ICOEI 2021, pp. 1736–1742 (2021)
32. Lakshmanan, V., et al.: A spatiotemporal approach to tornado prediction. In: IJCNN 2005, pp. 1642–1647 (2005)
33. Coccomini, D.A., Zara, G.: Predicting tornadoes days ahead with machine learning. arXiv: 2208.05855 (2022)
34. McGovern, A., et al.: Enhancing understanding and improving prediction of severe weather through spatiotemporal relational learning. Mach. Learn. **95**(1), 27–50 (2013)

Siren Federate: Bridging the Gap Between Document and Relational Data Systems for Efficient Exploratory Graph Analysis

Stéphane Campinas[ID], Matteo Catena[✉][ID], and Renaud Delbru[ID]

Siren, Galway, Ireland
{stephane.campinas,matteo.catena,renaud.delbru}@siren.io
https://siren.io/

Abstract. Investigative workflows requires interactive exploratory analysis on large heterogeneous knowledge graphs. Current databases show limitations in enabling such task. This paper discusses the architecture of Siren Federate, a system that efficiently supports exploratory graph analysis by bridging the document-oriented and relational models. Technical contributions include distributed join algorithms, adaptive query planning, query plan folding, and semantic caching. Experiments show that Siren Federate exhibits low latency and scales well with the amount of data, the number of users, and the number of computing nodes.

Keywords: Exploratory Graph Analysis · Knowledge Graph · Database and Information System Architecture · Distributed Join Algorithms · Document-oriented Database

1 Introduction

Siren provides its Investigative Intelligence platform to Law Enforcement, National Security and Cyber-threat investigators. Investigative intelligence is a specialized area of data analytics with the goal of uncovering threats and criminal activities [1,3] through the analysis of inter-connected data.

Knowledge graphs [13] are fundamental elements in investigation systems, as they integrate diverse data into a unified graph for data analysis. In fact, investigations often involve connecting the dots across large amount of structured (such as database table records), semi-structured (such as XML, JSON, or logs), and unstructured data (such as text or multimedia content). Malicious actors exploit the increasing volume and complexity of these data to blend in and operate. Therefore, it is essential that exploratory graph analysis works at scale to uncover complex chains of dependencies hidden within massive volumes, thus allowing to trace incidents, detect vulnerabilities, and manage risks. Additionally, investigative systems must have fast response times: investigators often interact with the system through an explorative and iterative process, which can be hindered by large system latency [18].

R. Chbeir et al. (Eds.): IDEAS 2024, LNCS 15511, pp. 186–200, 2025.
https://doi.org/10.1007/978-3-031-83472-1_13

Current database systems face scalability and flexibility challenges implementing these requirements, as detailed in Sect. 2. Siren Federate addresses these challenges by integrating relational and graph analytics into Elasticsearch, a distributed Information Retrieval (IR) system [10]. Its extension enables efficient analysis of massive knowledge graphs while retaining searching and relevance ranking capabilities of IR systems.

In this paper, we discuss how Siren Federate bridges the gap between document-oriented and relational models, we illustrate its architecture, and finally we evaluate its performance on a large synthetic dataset.

2 Motivations

By mapping data sources to entities (vertices) and their relationships (edges) in a graph structure, knowledge graphs provide a flexible and dynamic framework for data integration and retrieval, crucial in rapidly evolving domains. Investigative workflows on knowledge graphs often involves exploratory analysis [17] for discovering patterns and generating new leads. An investigator may start with limited information and iteratively expand their search within the graph to uncover new evidences. The system must guide users in searching, filtering, and drilling down through this data to pinpoint potential entities of interest. Additionally, graph analytical capabilities such as path finding, centrality, or community detection, can help discovering relevant subgraphs to investigators.

Siren's platform supports this workflow by combining several data interaction paradigms – search, analytic dashboards, set-to-set navigation[1], and graph visualization – into a coherent model. For instance, individuals, cellphone data, calls, texts, and network cells are linked together in a Signals Intelligence scenario to form a complex graph. Set-to-set navigation guides investigators in connecting together those different datasets. Applying filters to one set impacts all (in)directly connected sets, allowing exploration of the relevant information. The investigator can move from cellphones to related records, like locations visited by cellphone's owners whom received crime-related texts. Graph visualization helps visualizing the inter-connected sets, identifying patterns or clusters, and answering questions like *"Which people own which cellphone? Do they connect to the same network cell? Do they meet with other groups of users at other times?"*.

Effective exploratory graph analysis must handle diverse query workloads: (a) searching textual documents (e.g., social media, open web, mobile forensic data) and arbitrary records with relevance ranking and highlighting; and (b) relational and graph database workloads [4] such as OLTP queries (processing only a localized part of the graph), OLAP queries (spanning large portions of the graph), neighborhood queries, traversal queries over long graph paths, and global graph analytics (e.g., global pattern matching, graph search, community detection, path finding, centrality). The challenge is to maintain such a mixed workload at large scale and interactive speed, with response times ranging from sub-seconds to seconds to not impact the investigator's workflow.

---

[1] Set-to-set navigation is a type of relational faceted navigation [23].

Despite the capabilities of graph and relational databases, they often fall short in meeting investigative workflows requirements [4]. Native graph stores struggle with large-scale data due to limited sharding and replication. Relational databases handle well structured data and complex queries, but lack flexibility for heterogeneous data and have limited graph, text processing, and search capabilities [7]. Multi-model databases attempt to unify various data models, but their origins in specific models often lead to inefficiencies in handling diverse workflows. Achieving optimal performance across relational, graph, and full-text search remains difficult due to challenges in query processing, schema design, and indexing [20]. Finally, polyglot architectures do not fully resolve these issues, as they require data duplication and movement across multiple specialized backends [21]. This led us to use Elasticsearch as it supports a flexible data model, scales well on commodity machines, and offers advanced text processing and search capabilities. Elasticsearch offers interactive speed by leveraging inverted indices for search, columnar storage for fast analytics, and effective caching strategies.

However, document-oriented databases like Elasticsearch have limitations in joining data. Joins must be pre-planned at indexing time, storing documents to be joined on the same index shard.[2] Without resorting to data duplication, such a mechanism is only suitable for hierarchical relationships but not for more complex ones like networks. In fact, joins are needed to implement a competent graph analytics system [32]. Therefore, a distributed architecture that combines the strengths of graph databases, relational databases, and IR systems is needed. It must scale to massive heterogeneous graphs and support efficient relational and graph operations, to enable the iterative exploratory analysis of intelligence workflows. Siren Federate addresses these challenges by incorporating query-time distributed join capabilities between different indices into Elasticsearch.

3   Bridging the Document-Oriented and Relational Models

In a document-oriented store, one approach to model a graph is to map vertices and edges to documents [4]. IR systems provide flexible data modeling and advanced search capabilities, but lack the necessary relational join operations required by exploratory graph analysis (e.g., to find adjacent vertices). Siren Federate bridges this gap by allowing joins within the document-oriented model, thus supporting the analysis of knowledge graphs.

Several works attempt to bridge between document-oriented and relational models by mapping the first into the second [6,14,27,31]. However, these approaches are limited to what relational databases propose and miss optimizations offered by IR systems for processing and searching documents. Siren Federate takes the opposite approach, mapping from a relational data model to a document-centric data model in order to fully leverage what IR systems offer.

In the relational model, a join combines rows from multiple tables into a new table. Instead, in the document-oriented model, queries are applied to documents

---

[2] https://www.elastic.co/guide/en/elasticsearch/reference/8.13/joining-queries.html

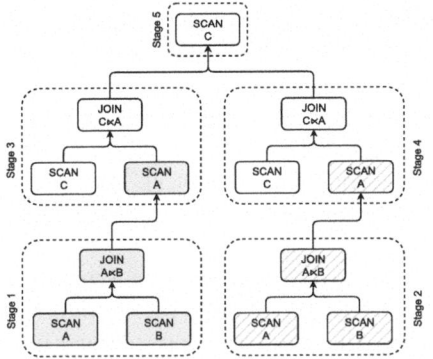

Fig. 1. A staged logical query plan

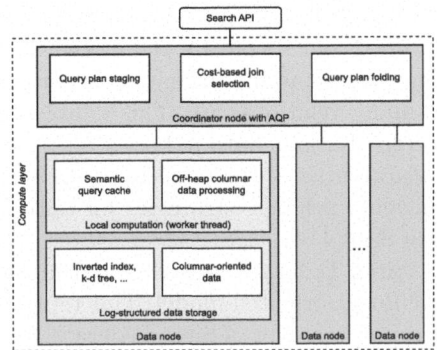

Fig. 2. Siren Federate's architecture

in an index, and returns matching documents. Siren Federate expresses a join operation ⋈ within this model as the process of finding documents from an index (the parent set) that are related to documents from another index (the child set) according to specific conditions, like field equality. Siren Federate implements (a) the *semi-join* ⋉ for filtering the parent set's documents based on the child set's documents; and (b) the *inner-join* ⋈ for extending the parent set's documents with fields from the matching child set's documents. These mechanisms are well suited for the iterative exploration needed in investigative intelligence, allowing to refine a target set of documents with results from previous investigation stages.

The logical steps for Siren Federate to join two document sets from indices A and B are highlighted in gray in Fig. 1. Two of the steps involve a SCAN operation, searching over the parent set A and child set B to retrieve subsets of documents to be joined. These documents may need to be exchanged across the computing cluster according to one of the different strategies described in Sect. 4.3. The parent and child subsets are then locally joined on the cluster's nodes by a JOIN operation, using data structures like hash tables, inverted indices, or k-d trees. The join results are tuples (in the relational sense) representing documents from the parent index that have fulfilled the join conditions. These tuples are then used by another SCAN operation to filter the parent index, retrieving the parent documents that meet the join conditions. This model also supports multi-join operations, with multiple child sets joined with the parent set. This is represented using a non-binary tree structure, where each SCAN operation can be associated with more than one child JOIN operation (see Fig. 1).

Siren Federate follows a *late materialization* approach, scanning only fields from parent documents to evaluate a join operation and avoid manipulating entire documents. Each document is associated with a global ID (see Sect. 4.1), to uniquely identify it across the system. Tuples produced by the join operation include this ID, rather than the entire document content. Upstream operations can use this ID to materialize necessary fields. This strategy is used for various operations, such as filtering, sorting, aggregating, and retrieving document content. These operations are delegated to the underlying Elasticsearch engine, which is optimized for handling such tasks efficiently.

Given the document-centric model, tuples produced by the join must be grouped and sorted by the global document ID, as the join may produce scattered tuples about the same parent document, for example in the case of many-to-many relationships. This enables the parent SCAN operation to efficiently merge the join output based on ordered document IDs, which align with the natural order of the underlying log-structured storage (see Sect. 4.1). We employ efficient exchange strategies for optimizing the grouping and sorting operations (see Sect. 4.3).

Siren Federate uses this logical model to integrate relational joins into the document-oriented model. This representation drives the architecture design and runtime behavior of Siren Federate. For instance, the adaptive query planner uses it to stage the query plan execution. The semantic information embedded within this model is used by the semantic caching, but also for folding the query plan. Finally, this model retains the search engine's capabilities to efficiently execute filters, sorting, and aggregations.

4   Siren Federate Architecture

This section introduces the core architectural components of Siren Federate, shown in Fig. 2. Siren Federate serves as the compute layer of an investigative system, leveraging the distributed computing and storage architecture of Elasticsearch for scalability. The application layer of the investigative system relies on Siren Federate's relational and analytical capabilities via its search API.

The distributed IR system consists of a cluster of computing nodes. Each node plays a different role in the cluster: coordinator nodes are responsible for planning the execution of a request received from the search API, while data nodes are responsible for storing data and executing operations dictated by the coordinator's query plan. This architecture ensures sub-second to seconds response time at scale, as computational load is distributed across data nodes, which can independently process the log-structured data storage to produce results (Sect. 4.1). Data nodes execute scan and join operations using a columnar data processing model (Sect. 4.2) and different join algorithms (Sect. 4.3). The query plan defined by the coordinator is divided into multiple stages (Sect. 4.4). Redundant operations of the query plan are folded to avoid unnecessary computation (Sect. 4.5). The logical query plan is processed iteratively, stage-by-stage, interleaving its physical planning with its execution. At each iteration, a cost-based query optimizer checks the semantic cache to reuse existing join results (Sect. 4.6) or selects the most efficient join algorithm.

4.1   Log-Structured Distributed Data Store

Siren Federate leverages Elasticsearch's distributed data store, which horizontally partitions data across nodes using document sharding. An index is partitioned into shards, and each document is routed to a shard. A shard is a Lucene index [11], based on a log-structured model [22], and composed of one or more index segments. The log-structured model adopts an append-only update strategy and consists in creating a file-based data structure called index segment. Segments are immutable and get merged over time or when a size threshold is reached. This append-only model allows for (1) implementing a lightweight read-lock mechanism to guarantee data consistency during the execution of distributed joins, enabling the concurrent execution of queries and real-time data updates; and (2) dynamically generating a global ID for documents by combining shard and segment IDs with the document's insertion order, thanks to the immutability of segments. This global ID enables the quick location of a document's physical position in the cluster, and is leveraged to achieve late materialization during the computation of join operations, as explained in Sect. 3.

4.2   Columnar In-Memory Processing

Siren Federate stores data for intermediate join computation into off-heap main memory using a columnar layout and leverages compression algorithms optimized for specific data types. During join operations, Siren Federate processes only the relevant fields, such as join key fields and global document IDs. The data exhibits a tabular structure, with tuples corresponding to documents and columns to their fields. There are two approaches for processing tabular data: row-at-a-time and column-at-a-time.

The row-at-a-time approach reads whole tuples even if only a few columns are needed, leading to CPU cache misses and negatively impacting the performance. Following best practices from [15], Siren Federate adopted the column-at-a-time approach, improving the query performance by a factor of 2 compared to the row-at-a-time implementation.[3]

The column-at-a-time approach uses a batch-processing pipeline. Each batch stores a fixed number of tuples, stored in a columnar fashion. The size of a batch is optimized to fit within the CPU cache line to avoid cache misses. Although batches are currently processed sequentially by a worker thread, future plans include parallel batch processing to increase throughput. A profiling tool[4] showed an increase of the CPU cache usage with the column-at-a-time approach: Siren Federate ver. 27.5 increased "cache-references" by 25% compared to ver. 22.6 that uses the row-at-a-time approach, demonstrating enhanced CPU cache utilization.

4.3   Distributed Join Algorithms

Siren Federate implements join techniques that leverage the intrinsic data structures of the underlying IR system to ensure scalability and high performance. An example with two distributed indices A and B is shown in Fig. 3. Both indices are partitioned into three shards, whose data needs to be exchanged across the computing nodes in order to be joined. The available join strategies are:

Broadcast Hash Join. Data from the child index is forwarded to all computing nodes hosting shards of the parent index (see Fig. 3, left). Local hash tables,

---

[3] https://info.siren.io/content/siren-benchmark-whitepaper.
[4] https://github.com/async-profiler/async-profiler.

created from the received data, are probed while scanning the parent index's columnar storage. Worker threads process segments in parallel, and local hash tables are shared across these worker threads.

Broadcast Index Join. This strategy utilizes Lucene's inverted indexes (akin to burst tries [12]) for binary values, and Bkd-trees [26] for numerical values. Data are exchanged like the broadcast hash join (see again Fig. 3, left), but the child set data is used for index lookups over the parent set, eliminating exhaustive scans of the columnar storage. Worker threads process segments and probe the index with the received data. This is effective for graph expansion or path finding tasks, where the objective is to incrementally expand relationships from a group of records.

Partitioned Hash Join. Inspired by [30], it leverages the columnar storage to scan data from the parent and child indices, partitioning data across computing nodes, and creating localized hash tables for each partition (see Fig. 3, middle). This method employs morsel-driven parallelism and involves a two-step partitioning to create fixed-sized work units: an initial node partitioning at the scan level (sender side) and a second partitioning at the join level (receiver side). This method achieves better parallelism and reduced memory and network overheads compared to strategies like the broadcast hash join. In Fig. 3 only three computing nodes are shown for the sake of space. However, this join strategy can leverage all cluster nodes regardless the number of shards.

Routing Join. Similar to the broadcast hash join, it leverages the document sharding to reduce network traffic. It reuses the sharding routing function of the parent index to partition and exchange the child set's tuples to the corresponding parent set's shards [5] (see Fig. 3, right). Each worker thread employs either a hash table-based strategy (like the broadcast hash join) or an inverted index-based strategy (like the broadcast index join) to compute the results. Preliminary experiments (not presented in this work) indicate a 30% reduction in response times compared to the broadcast hash join strategy.

These strategies optimize specific scenarios. The role of the query planner (Sect. 4.4) is to select the most cost-effective join strategy by considering factors such as shard topology and set cardinality to optimize the cluster's utilization.

4.4 Adaptive Query Planner

Accurate join cardinality estimation is crucial for planning the most effective join algorithm. Unfortunately, this is challenging with complex query plans with deeply nested joins, common in investigative scenarios. Traditional static methods, based on histograms and simple cardinality estimation formulas, often yield inaccurate estimations due to assumptions like attribute independence and distribution uniformity, resulting in sub-optimal selection of join algorithms. These inaccuracies are exacerbated as the complexity of the query plan increases [16]. Index-based join sampling, while more accurate, is computationally expensive, especially in distributed systems where it requires data shuffling across the network, and also suffers from inaccuracies with long sequence of joins.

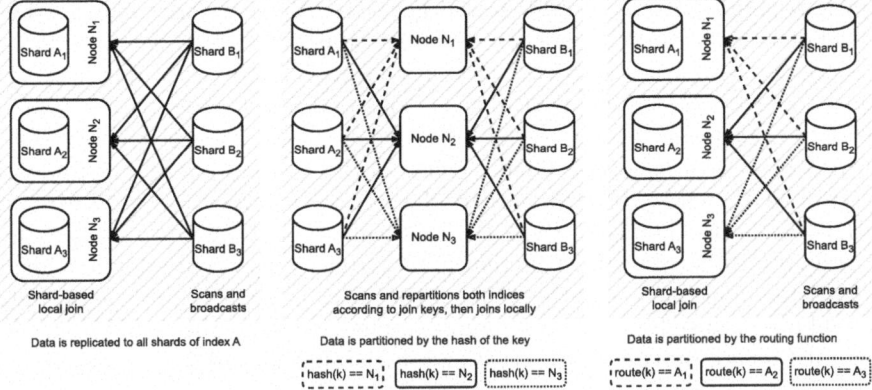

Fig. 3. Federate's distributed join algorithms: (left) Broadcast Hash/Index Join, (middle) Partitioned Hash Join, (right) Routing Join. Arrows represent data exchange between computing nodes

To address this, Siren Federate implements an adaptive query planner (AQP) that interleaves planning and execution via stages [9]. This approach collects runtime statistics during execution, allowing more accurate cardinality estimation compared to static methods, especially for long sequence of joins. This enables to dynamically adjust the query plan based on real-time feedback. AQP operates in several key phases:

Logical Plan Generation. The planner generates a logical query plan divided into stages. Each stage corresponds to a materialization point where an intermediate result is fully created before proceeding further. Typically, it comprises a logical join and two logical scans.

Physical Optimization. The planner gathers statistical information, computes costs for various join strategies, and selects the optimal one. This is repeated for each stage, leveraging runtime cardinality estimates from already computed nested joins (stages).

Execution. The physical sub-graph of each stage is executed, materializing intermediate results before proceeding.

Parallelization. The query plan enables parallel execution of independent stages. Independent stages are executed concurrently, while dependent stages must wait for predecessors to complete.

To illustrate how AQP works, consider a dataset with three indices. An index A of documents representing cellphones, with fields containing the phone number, the operator, reference to the person who subscribed the contract, etc. An index B of documents representing the online activity of a person such as social media posts, with fields like person's identity and textual content. An index C of documents representing each call detail record (CDR) with fields such as time, duration, completion status, source and destination numbers of a call [29]. Imagine we want to find all CDRs related to phones used by people involved in suspicious online activities. AQP would generate a logical query plan in stages (logical plan generation) as shown in Fig. 1. Assume filters (e.g., keyword matching or vector search) applied to B identifies crime-related posts, then the set of "phones used by people involved in suspicious online activities" is the result of $A \ltimes B$ (Stages 1 and 2). The set of CDRs where these phones are the caller is returned by $C \ltimes A$ using the CDR's caller as the join key (Stage 3). Similarly, the set of CDRs where these phones are the callee is returned by $C \ltimes A$ using the CDR's callee as the join key (Stage 4). The disjunction of these two sets produces the desired results (Stage 5). Different join strategies may be used depending on the statistical information gathered from the previous stages (physical optimization and execution). Since Stages 1 and 2 are independent, they can be executed in parallel before moving to Stages 3 and 4 (parallelization).

4.5   Query Plan Folding

User queries often contain redundant operations that negatively impact query processing, such as repeated searches or joins. Redundancies commonly occur when investigating related entities through various graph topologies, boolean expressions, or batched requests targeting the same entities with diverse filters or aggregations. Redundant operations affect also SQL query processing [19,28].

To address this challenge, Siren Federate adopts a query plan folding that uses the semantic definition of query operators to detect and merge redundant operators across one or more logical query plans. The semantic definition of an operator captures its logical meaning, structure, dependencies, and state of the data tables it involves [8].

Siren Federate handles not only the folding of selection and scan operations [28], but also of join operations. The folding strategy consolidates redundant operations into a unified shared operator. In the previous AQP example, the operators from Stage 1 and its subsequent SCAN A are folded with those from Stage 2 and SCAN A (highlighted by a hatched background in Fig. 1) since they represent the same join. However, Stage 3 and 4 are not folded as they have different join conditions: the CDR's caller number is used in Stage 3, while the CDR's callee number is used in Stage 4. Similarly, SCAN C is not folded between the two stages because it scans different fields.

4.6   Semantic Caching

In exploratory graph analysis, iterative analysis often results in recurrent execution of the same join operations. By caching these, the system can optimize subsequent related queries, reducing the latency and computational load.

Semi-joins are well-suited for caching compared to other join types, as their outputs can be represented as sets of document IDs. Exploiting this, Siren Federate employs semantic caching, relying on the semantic definition of query operators (Sect. 4.5). Compared to conventional caching methods which operate at the query syntactic level [24], semantic caching [8] indexes cache entries according to query operator semantics, guarantying data consistency and resilience to changes in the underlying data, even when query operators depend on multiple data sources derived from descendant query operators. Unlike traditional caching strategies that focus on raw results which can lead to large memory overhead, the semantic caching strategy uses compact bitset representations to efficiently encode semi-join outputs. This reduces memory consumption and increases the potential amount of cached operations, enhancing the overall system efficiency.

To illustrate this, take the example from Sect. 4.4. In a subsequent iteration, the investigator wants to identify, from a new index D, people owning phones involved in previously found CDRs. This adds a new join $D \ltimes C$, with C being the results of the previous iteration. The AQP generates a logical plan including the subtree from Fig. 1. With semantic caching, the results of the subtree can be reused, meaning that only the additional join $D \ltimes C$ must be computed.

This strategy benefits graph analytics, particularly path finding algorithms which can be represented as sequences of semi-joins. Semantic caching of semi-joins reduces redundant computations, minimizing the number of operations and associated I/O, and resulting in a more efficient execution of graph queries [25].

4.7   Principles for a Competent Graph Analytics System

In [32], ten Wolde et al. identify eight core features necessary for a competent graph analytics system. Siren Federate leverages and extends the capabilities of Elasticsearch to meet these principles:

Fast Scans on Elements with Schema. Siren Federate uses Elasticsearch's dynamic schema capabilities and column-oriented storage for fast attribute scanning. Dynamic schema-awareness offers flexibility in handling knowledge graph variability and optimizes query processing by adapting to the data structure.

Skippable Compressed Columnar Storage. Elasticsearch supports fast columnar scans with data skipping via pushed-down predicates, combining columnar storage ordered by document IDs with an inverted index or k-d trees.

Vectorized or Data-Centric Execution. Siren Federate adopts column-at-a-time (vectorized) processing for its data pipelines during scan and join operations, and an in-memory vector format with compression and data skipping.

Morsel-Driven Multi-core. Siren Federate uses morsel-driven parallelism during scan and join operations to distribute constant-sized work units (morsels) across worker threads, reducing load imbalance and optimizing CPU cache usage. Dynamic index segment splitting is critical for better parallelism during scans of large segments, and reducing query execution latency.

State-of-the-Art Query Optimization. Siren Federate's AQP splits query plans into stages and employs runtime estimation to avoid the overhead of traditional table sampling, ensuring efficient query execution.

Bulk APIs/Algebras. Elasticsearch's boolean algebra operates as a bulk API for predicate evaluation by enabling manipulation of document sets efficiently. Siren Federate integrates relational algebra that also functions as a bulk API.

Out-of-Core Buffer Manager. While performing in-memory join operations, Siren Federate leverages Elasticsearch's ability to handle out-of-core data sizes efficiently during scan. Frequently accessed columns are cached at the operating system level, ensuring scan operations in RAM.

Explicit Control over Memory Locality. Siren Federate uses off-heap memory management to reduce garbage collection overhead when handling gigabytes of data in memory for short durations, and optimizes memory locality through columnar storage, morsel-driven parallelism, and effective data partitioning.

To further enhance performance in a distributed environment, Siren Federate adopts three additional core features:

Data Locality. Leveraging data locality minimizes data movement across the network, a common bottleneck as it requires additional intermediate serialization and copy of the data. Siren Federate performs late materialization of documents using global IDs to quickly locate documents in the cluster and co-locates data by reusing existing data routing coming from document sharding.

Data Exchange. Effective data exchange operators exploit the physical distribution and structure of the storage to maximize memory locality without data reorganization. Preserving the implicit order, even partially, of materialized tuples during scan and join operations improves the performance of the exchange operator, which must group and sort tuples based on the document ID as explained in Sect. 3. Radix partitioning [33] is a highly efficient method for clustering tuples from a range of documents together, improving document sorting.

Caching. Implementing compact caching strategies for intermediate results allows reuse across queries and users, reducing redundant computations and improving overall system efficiency, especially in incremental exploratory scenarios.

5   Evaluation

This section evaluates the scalability and performance of Siren Federate in computing semi-joins[5] across various dimensions, such as the number of cluster nodes, concurrent users and data volume.

Dataset. We use a synthetic dataset[6] with 15.6 billion documents tracking the positional information of cell phones to mimic scenarios where analysts monitor phone calls. This dataset covers 100 days, with 156 million documents per day, 6.5 million unique phone identifiers, and 2,400 positions per phone. One Elasticsearch index per day is created, with 8 primary shards with no replicas. The total size of the data is 2.7 TB.

---

[5] We focus on semi-joins as they are more suitable than inner-joins for large datasets. For semi-joins, the output size is linear with the cardinality N of the parent set, compared to $O(N \times M)$ for inner-joins (M being the cardinality of the child set). Semi-joins are used extensively for exploratory graph analysis tasks like set-to-set navigation, graph expansion, and pathfinding, due to their efficiency and scalability.

[6] https://gist.github.com/scampi/07e7bd556fe016a5cba6c092c3f418fb.

Setup. The benchmark tests use AWS "i3.4xlarge" machines with Broadwell processors (16 vCPUs), a local NVME drive for Elasticsearch data, a gp2 drive for the OS, and a 10 Gbps network link. We use Java Virtual Machine 20.0.1, Elasticsearch 8.7.1, and Siren Federate 31.1. Elasticsearch is configured with 30 GB heap memory, and Siren Federate with 16 GB off-heap memory.

Experiments. We evaluate Siren Federate with varying cluster sizes (12 to 36 nodes), data volumes, and concurrent users (1, 5 and 10). The system is setup to serve the maximum number of concurrent users. We use the following queries with different complexity:

Q1 joins phone numbers in a given area on one day with those in another area on another day (78 million documents per set); **Q2** is similar to Q1 but over a week (546 million documents per set); **Q3**, given a phone number, finds other phones at the same location over 90 days (14 billion documents filtered with 2,160 documents); **Q4** finds phones at the same location on two different days (156 million documents each); **Q5** is similar to Q4 but over a week (more than 1 billion documents per set); **Q6** is similar to Q4 but over two weeks (more than 2 billion documents per set).

Queries use the *partitioned hash join*, except Q3, which uses the *broadcast index join* due to the small cardinality of its child set. We measure the execution time for a randomly-selected query with a fixed number of concurrent users, bypassing query caches. The benchmark runs until at least 100 measurements per query are produced, reporting the 90th percentile processing time (P90).

Results. Figure 4 reports P90 for queries Q1 to Q3. With one concurrent user, Q1 joins a total of 156 million documents in subsecond time. However, P90 does not decrease as the number of nodes increases because the amount of data joined is too small. The latency of the join phase represents an insignificant part of the response time, while the scan phase is tied to a limited number of shards and cannot be further distributed.

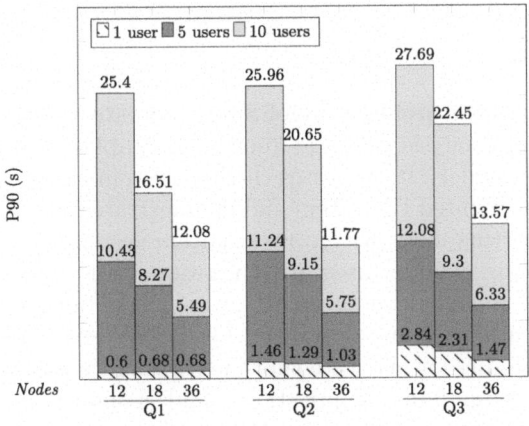

Fig. 4. Query times for queries Q1 to Q3 with varying number of nodes and users

Q2 joins over 1 billion documents ($\times 7$ more data than Q1), with P90 increasing by at most $\times 2.5$, indicating a better usage of the computing resources. However, we notice that response time does not decrease as expected as the number of nodes increases. In fact, increasing the number of nodes by $\times 3$ only decreases P90 by 30%. We assume this is because the latency of Q1 and Q2 are dominated by some fixed overhead during the query planning and pipeline execution. This requires further analysis.

Q3, which filters 14 billion documents using the *broadcast index join* strategy, shows P90 decreasing by $\times 2$ as nodes increases by $\times 3$. This strategy avoids partitioning and shuffling such a large parent set while still distributing load across the full cluster as the number of nodes increases.

With 5 concurrent users, P90 of Q1 increases by $\times 12.5$ – on average across cluster's configurations – when compared to response times with 1 user; Q2 increases by $\times 6.8$; Q3 by $\times 4.2$. With 10 concurrent users, the increase of P90 is by at most $\times 2.3$ w.r.t. 5 users. Differently than moving from 1 to 5 concurrent users, doubling the number of users doubles the response times in this case. This significant increase in latency is particularly evident for Q1, and it may underline as previously noted an overhead in query planning and execution for short queries. Such overhead may impact scaling, and needs to be investigated in future work. Nonetheless, these results show that Siren Federate scales well with the number of users and nodes, achieving subsecond to second response times over large datasets.

For Q4, Q5, and Q6, the aim is to further assess the ability of the system to scale with the amount of data, joining parent and child sets containing hundreds of millions to billions of documents each. With 36 nodes, the reported P90 is 1.92, 5.00, and 7.58 seconds for Q4, Q5, and Q6 respectively. We can observe a sub-linear scaling factor with the size of the join operation. Between Q4 and Q5 the size of the join operation increases by $\times 7$, but P90 only increases by 2.2 s. Between Q5 and Q6, the size of the join operation increases by $\times 2$ but P90 only increases by 1.5 s. These results confirm that Siren Federate scales well with the amount of data.

6   Conclusion

This paper presented the architecture of Siren Federate, a system that supports efficient exploratory analysis on large knowledge graphs by bridging document and relational data models. It addresses challenges in integrating graph analytics within a document-oriented IR system and demonstrates the effectiveness of this approach for supporting data-driven investigation systems.

Key contributions include integrating relational join operations within the document-oriented model, leveraging IR system capabilities, and implementing distributed join algorithms optimized for IR systems. We also introduced adaptive query planning for accurate runtime cardinality estimation, query plan folding to reduce redundant computations, and semantic caching to enhance iterative query performance. Columnar in-memory processing optimizations and Elasticsearch's log-structured distributed architecture were also highlighted.

Performance evaluations using a synthetic dataset with billions of documents demonstrated the efficiency and scalability of Siren Federate. The results show the capability to achieve sub-second to second response times on large datasets, which is critical for supporting the iterative workflow of investigators.

Our architecture retains the search and relevance ranking capabilities of the IR system while introducing efficient relational operations and graph analytics at scale. This demonstrates how a combination of document and relational system features can enhance scalability and analytical capabilities for exploratory analysis of large knowledge graphs.

Future efforts will focus on optimizing parallel processing of segments, incorporating spatial joins for enhanced analytics, expanding graph analytics, and developing new adaptive optimization techniques. Finally, we also plan to evaluate our system on additional, standard datasets, such as the LDBC Social Network Benchmark [2].

Acknowledgments. We used ChatGPT to improve writing in some parts of this otherwise original work. All AI-generated text has been carefully checked for correctness.

References

1. Akhgar, B., Bayerl, P.S., Sampson, F. (eds.): Open Source Intelligence Investigation. Springer, Heidelberg (2016)
2. Angles, R., et al.: The LDBC social network benchmark (2024)
3. Atzenbeck, C., Ozgul, F., Hicks, D.L.: Linking and organising information in law enforcement investigations. In: Proceedings of the 13th International Conference on Information Visualisation (IV) (2009)
4. Besta, M., et al.: Demystifying graph databases: analysis and taxonomy of data organization, system designs, and graph queries. ACM Comput. Surv. **56**(2) (2023)
5. Campinas, S., Catena, M., Delbru, R.: Method to reduce network communication for distributed join on shared-nothing databases. International (PCT) Application PCT/EP2023/059801 (2023)
6. Chasseur, C., Li, Y., Patel, J.M.: Enabling JSON document stores in relational systems. In: Proceedings of the 16th International Workshop on the Web & Databases (WebDB) (2013)
7. Davoudian, A., Chen, L., Liu, M.: A survey on NoSQL stores. ACM Comput. Surv. **51**(2) (2018)
8. Delbru, R., Campinas, S.: Semantic caching of semi-join operators in shared-nothing and log-structured databases. U.S. Patent Application 17/597923 (2020)
9. Deshpande, A., Ives, Z., Raman, V.: Adaptive query processing. Found. Trends Databases **1**(1) (2007)
10. Gormley, C., Tong, Z.: Elasticsearch: The Definitive Guide: A Distributed Real-Time Search and Analytics Engine. O'Reilly (2015)
11. Hatcher, E., Gospodnetic, O.: Lucene in Action (In Action Series). Manning (2004)
12. Heinz, S., Zobel, J., Williams, H.E.: Burst tries: a fast, efficient data structure for string keys. ACM Trans. Inf. Syst. **20**(2) (2002)
13. Hogan, A., et al.: Knowledge graphs. ACM Comput. Surv. **54**(4) (2021)

14. Karpathiotakis, M., Alagiannis, I., Ailamaki, A.: Fast queries over heterogeneous data through engine customization. Proc. VLDB Endow. **9**(12) (2016)
15. Kersten, T., et al.: Everything you always wanted to know about compiled and vectorized queries but were afraid to ask. Proc. VLDB Endow. **11**(13) (2018)
16. Leis, V., et al.: Cardinality estimation done right: index-based join sampling. In: Proceedings 7th Conference on Innovative Data Systems Research (CIDR) (2017)
17. Lissandrini, M., et al.: Knowledge graph exploration systems: are we lost? In: Proceedings of the 12th Conference on Innovative Data Systems Research (CIDR) (2022)
18. Liu, Z., Heer, J.: The effects of interactive latency on exploratory visual analysis. IEEE Trans. Vis. Comput. Graph. **20**(12) (2014)
19. Lu, A., Fang, Z.: SQL2FPGA: automatic acceleration of SQL query processing on modern CPU-FPGA platforms. In: Proceedings of the 31st IEEE Annual International Symposium on Field-Programmable Custom Computing Machines (FCCM) (2023)
20. Lu, J., Holubová, I.: Multi-model databases: a new journey to handle the variety of data. ACM Comput. Surv. **52**(3) (2019)
21. Noel, S., Bodeau, D.J., McQuaid, R.: Big-data graph knowledge bases for cyber resilience. In: Proceedings of the NATO IST-153/RWS-21 Workshop on Cyber Resilience (2017)
22. O'Neil, P., et al.: The log-structured merge-tree (LSM-tree). Acta Inform. **33**(4) (1996)
23. Oren, E., Delbru, R., Decker, S.: Extending faceted navigation for RDF data. In: Proceedings of the 5th International Semantic Web Conference (ISWC) (2006)
24. Papailiou, N., et al.: Graph-aware, workload-adaptive SPARQL query caching. In: Proceedings of the ACM International Conference on Management of Data (SIGMOD) (2015)
25. Pini, D., Tummarello, G., Delbru, R.: Optimization of database sequence of joins for reachability and shortest path determination. U.S. Patent 11720564 (2022)
26. Procopiuc, O., et al.: BKD-tree: a dynamic scalable KD-tree. In: Proceedings of the 8th International Symposium on Advances in Spatial & Temporal Databases (SSTD) (2003)
27. Roijackers, J., Fletcher, G.H.L.: On bridging relational and document-centric data stores. In: Proceedings of the 29th British National Conference on Databases (BNCOD) (2013)
28. Sahal, R., Khafagy, M., Omara, F.: Big data multi-query optimisation with apache flink. Int. J. Web Eng. Technol. **13** (2018)
29. Sammons, J.: The Basics of Digital Forensics: The Primer for Getting Started in Digital Forensics. Syngress (2014)
30. Schuh, S., Chen, X., Dittrich, J.: An experimental comparison of thirteen relational equi-joins in main memory. In: Proceedings of the ACM International Conference on Management of Data (SIGMOD) (2016)
31. Ágnes Vathy-Fogarassy, Hugyák, T.: Uniform data access platform for SQL and NoSQL database systems. Inform. Syst. **69** (2017)
32. ten Wolde, D., et al.: DuckPGQ: efficient property graph queries in an analytical RDBMS. In: Proceedings of the 13th Conference on Innovative Data Systems Research (CIDR) (2023)
33. Zhang, Z., Deshmukh, H., Patel, J.M.: Data partitioning for in-memory systems: myths, challenges, and opportunities. In: Proceedings of the 9th Conference on Innovative Data Systems Research (CIDR) (2019)

Query Processing and Applications

Outlier-Weighted Traffic Flow Prediction Using Online Autoencoders

Himanshu Choudhary[1] , Ahmad B. Alkhodre[2], and Marwan Hassani[1(✉)]

[1] Eindhoven University of Technology, Eindhoven, The Netherlands
m.hassani@tue.nl
[2] Islamic University of Madinah, Madinah, Saudi Arabia
aalkhodre@iu.edu.sa

Abstract. In today's urban landscape, traffic congestion poses a critical challenge, especially during outlier scenarios. These outliers can indicate abrupt traffic peaks, drops, or irregular trends, often arising from factors such as accidents, events, or roadwork. Moreover, given the dynamic nature of traffic, the need for real-time traffic modeling also becomes crucial to ensure accurate and up-to-date traffic predictions. To address these challenges, we introduce the Outlier Weighted-Autoencoder Modeling (OWAM) framework. OWAM employs autoencoders for local outlier detection and generates correlation scores to assess neighboring traffic's influence. These scores serve as weighted factors for neighboring sensors, before fusing them into the model. This integration enhances the traffic model's performance and supports effective real-time updates, a crucial aspect for capturing dynamic traffic patterns. OWAM demonstrates a favorable trade-off between accuracy and efficiency, rendering it highly suitable for real-world applications. The findings contribute directly to an industrial application but also to the development of more efficient and adaptive traffic prediction systems. The code and datasets of our framework are publicly available at https://github.com/himanshudce/OWAM.

1 Introduction

Traffic congestion is a well-known problem impacting public health, economic productivity, and contributing to environmental issues. It also contributes to environmental issues due to increased emissions and air pollution [19]. As increasing road capacity is often not feasible, optimizing traffic controllers becomes crucial for enhancing traffic flow. These systems can help alleviate congestion, reduce travel times, and enhance overall traffic flow.

Traffic modeling has advanced considerably, resulting in effective traffic prediction models in general. Yet, challenges arise when these models encounter outlier scenarios. In the realm of traffic, outliers refer to situations where data significantly deviates from normal traffic patterns. These outliers can indicate abrupt traffic peaks, drops, or irregular trends, often arising from factors such as accidents, events, or roadwork. Furthermore, these traffic prediction many traffic prediction models lack the ability to generalize effectively for real-time predictions. Traffic is inherently dynamic, experiencing continuous fluctuations due to factors like accidents and special events. Thus, reducing training and evaluation time while ensuring the model's robustness for real-time updates and efficiency in handling real-world conditions is essential.

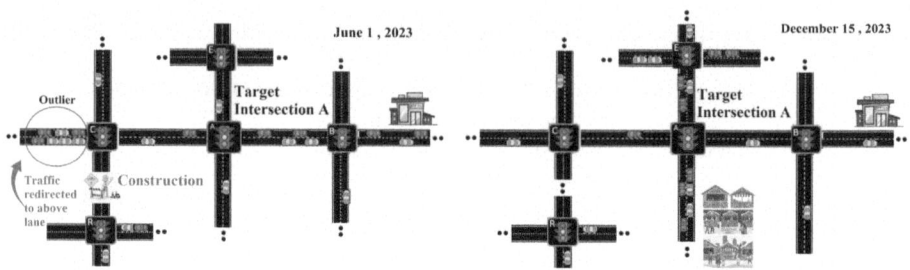

(a) Traffic condition on June 1 after an ac-   (b) Traffic condition on December 15, cident.   around Christmas time.

Fig. 1. Traffic conditions of the city on different dates.

In urban traffic prediction, abrupt changes in traffic patterns e.g. at intersection A (see Fig. 1) pose challenges, exemplified in Figs. 1a and 1b. Such scenarios, marked by outliers and shifts in dynamics, lead to congestion if not predicted accurately. Addressing this, we introduce the Outlier Weighted Autoencoder Modeling (OWAM) framework, aiming to improve predictions in dynamic and outlier traffic scenarios. Leveraging Autoencoders (AE) in real-time, we detect outliers at each sensor, integrating Earth Mover's distance for improved performance. These outlier scores facilitate correlation calculations between the target sensor and its neighbors, allowing for precise influence determination. Additionally, OWAM incorporates outlier-based model updates to handle both outlier scenarios and dynamic changes effectively. We validate our approach on real-world datasets: Hague, METR-LA, and PEMS-BAY.

The OWAM framework demonstrates superior performance compared to previous outlier-based approaches and the LSTM baseline. Although it falls short of surpassing the heavy graph-based techniques in terms of RMSE, it notably stands out in terms of runtime efficiency, rendering it highly suitable for real-world traffic management systems. Our main contributions are:

1. an innovative Outlier Weighted Autoencoder Modelling (OWAM) framework, that enhances traffic flow prediction in both outlier and real-time scenarios in an industrial application,
2. novel outlier detection architecture that integrates Earth mover's distance with AEs to improve the accuracy of outlier identification,
3. an extensive experimental evaluation over three real-world datasets to demonstrate the effectiveness of our proposed architecture on traffic prediction tasks utilizing traffic flow and traffic speed,
4. providing an open-source implementation of OWAM framework for reproducibility and further research.

This paper extends the short paper recently published in [6] with an extensive explanation of the theoretical concepts behind OWAM and a thorough experimental evaluation results and discussion. The remainder of this paper is structured as follows: In Sect. 2, we review relevant literature on outlier detection techniques and traffic modeling. Section 3 provides essential background and mathematical formulation. Section 4 presents the OWAM framework's design. Section 5 describes datasets and metrics. Section 6 discusses our findings on traffic prediction. Section 7 concludes the paper.

2   Related Work

This section reviews relevant literature on outlier detection techniques and traffic modeling which are central to our research. Outlier detection focuses on identifying data points deviating significantly from the norm, with methods such as OC-SVM [23], LOF [4], and Isolation Forests [15]. However, these approaches often struggle with real-time traffic flow forecasting due to high response times, scalability issues, and challenges in handling high-dimensional real-time data. To address dynamic tasks, notable contributions in online anomaly detection for data streams include HST [24], RRCF [10], and an extension of LOF for data streams [20].

AEs have proven highly effective in both offline and online scenarios. Kit-NET [18] efficiently identifies network intrusions using an ensemble of AEs, while MemStream [2] combines denoising AEs with a memory module to address concept drift and memory poisoning in real-time outlier detection for streaming data. Probability-Weighted Autoencoders (PW-AE) [5] are proposed to handle burst anomalies and higher proportions of outliers in data streams by optimizing the loss function based on anomaly probability. Building upon this promising work [5], we further modify it for enhanced outlier detection, as detailed in subsequent sections.

In traffic modeling, approaches aim to improve prediction accuracy and efficiency, falling into three main groups: parametric techniques, machine learning techniques, and deep learning techniques [16]. Parametric methods like Kalman filters and ARIMA models use statistical parameters for time series analysis, but often struggle with complex patterns. Machine learning common methods including SVM and KNN incorporate spatial characteristics, often integrating dynamic distance measures and state matrices for better accuracy. Deep learning employs neural networks with multiple layers to extract space-time dependencies from diverse data sources. For instance, CNNs extract spatial features from traffic images, LSTMs capture temporal dependencies in flow data, encoders reduce dimensionality, and GCNs model regional relationships. Hybrid models combining these techniques also enhance performance.

Outlier-based hybrid approaches offer valuable insights into traffic modeling by focusing on outlier detection in traffic flow data. Polson et al. [21] demonstrated the effectiveness of neural networks in handling outlier scenarios in traffic data. On the other hand, LOF-based models have been proposed to detect outliers within Probability Distributions of traffic flows (FPDs) [7,8].

Prior outlier-based approaches were constrained by traditional outlier detection methods, predominantly tested on similar datasets, and were limited in their ability to intelligently integrate information. Additionally, researchers largely overlooked the real-time aspect of traffic, where traffic consistently evolves and adapts to patterns in response to various ongoing events, a critical aspect demanding further investigation. Regular model updates are vital for ensuring accurate and up-to-date traffic models that reflect real-time changes in traffic flow. These models should strike a balance between accuracy and efficiency to facilitate real-world scenarios.

3 Preliminaries and Problem Formulation

Traffic controllers (TC) manage traffic signal timing at intersections, optimizing various traffic movements. They rely on data from multiple traffic monitoring and management (TM) systems, which collect real-time traffic information at each sensor I, including vehicle volume, speed, and occupancy. Our objective is to enhance the prediction accuracy of the traffic management systems TM_I, assisting TCs in decision-making and optimizing traffic flow. We trained a model M with both offline and online objectives for comparison and to assess real-time applicability.

Let $L^+ = \{(x(i), y(i)) \,|\, \forall i \in S\}$ be a sequence of $|S|$ tuples, where each tuple consists of a feature-vector $x(i) \in R^d$ and a corresponding real value to be predicted $y(i) \in R$. Additionally, let $L_{\text{train}} = \{(x(i), y(i)) \,|\, \forall i \in \{1, \dots, P\}\}$ be a sub-sequence used for training the model M. Let $L_{\text{test}} = \{(x(i), y(i)) \,|\, \forall i \in \{1, \dots, m\}\}$ be the sub-sequence used for testing the model.

Here, L_{train} and L_{test} are disjoint subsets of L^+. Given a function $f : R^d \to R$ that was trained on L_{train} with P samples, the objective can be formulated as $min(z_1)$, representing the sum of losses of each pair $(x(i), y(i))$:

$$z_1 = \frac{1}{N} \sum_{i=1}^{N} (\mathcal{L}_r, L_{\text{test}}, f(L_{\text{train}})) \tag{1}$$

Where N represents the total number of sensors and $\mathcal{L}_r$ represents a regression loss.

In addition to the traditional (offline) settings, we incorporated a real-time incremental updating strategy with time windows T of varying lengths. Let L_{train} contain the data up to time point t, and $L_{test} = \{(x(i), y(i)) \,|\, \forall i \in \{t + 1, \dots, t_n\}\}$ represent infinite sub-sequences with interval length T. For a given validation protocol $\mathcal{V}$, the problem can be formulated by $min(z_2)$ such that -

$$z_2 = \frac{1}{N} \sum_{i=1}^{N} \sum_{j=1}^{t_n} \mathcal{V} (\mathcal{L}_r, L_{\text{test}}, f(L_{\text{train}})) \tag{2}$$

For the validation protocol $\mathcal{V}$, we employ the test-then-train approach. In this evaluation method, models are first validated based on a performance metric and subsequently trained on each instance within the time window T of L_{test} at its arrival time. To achieve this objective, we introduce the OWAM framework and outline its key components in the following sections.

Flow Probability Distribution (FPD). Flow Probability Distributions (FPDs) [17] represent traffic occurrence probabilities within specific time intervals, derived from real traffic data and encapsulating traffic distribution patterns. FPDs can be constructed for various traffic parameters such as vehicle flow or average speed to provide a comprehensive understanding of traffic dynamics.

Let $FPD(HB)$ represent the FPD for a specific sensor B during time period H. To compute $FPD(HB)$, we begin with a set of aggregated traffic flow values, X_{HB}: $X_{HB} = (x_{hB1}, x_{hB2}, \ldots, x_{hBH})$. Here, x_{hB} signifies the aggregated traffic flow or speed value x_h for intersection B during the h-th time interval, where h ranges from 1 to H. To calculate the FPDs, we apply the formula as created by Zimek and Djenouri in [4]:

$$FPD(HB) = \frac{\sum_{x_h \in X_H} P(x_h)}{P(x_{hB})}, \quad \forall x_{hB} \in X_{HB}, \quad h = 1, \ldots, H \qquad (3)$$

Outlier Detection using Autoencoders. AEs, commonly used for anomaly detection, aim to reconstruct $\hat{D}$ of input data D by encoding it into a lower-dimensional representation and then decoding it back to its original format (see Equation 4). Anomalies, deviating significantly from the training data, result in higher reconstruction errors/loss (see Equation 5). As shown in Fig. 2, our research includes training AE on continuous traffic stream of FPDs in real-time and if faced with an anomaly, say, characterized by a left-skewed distribution

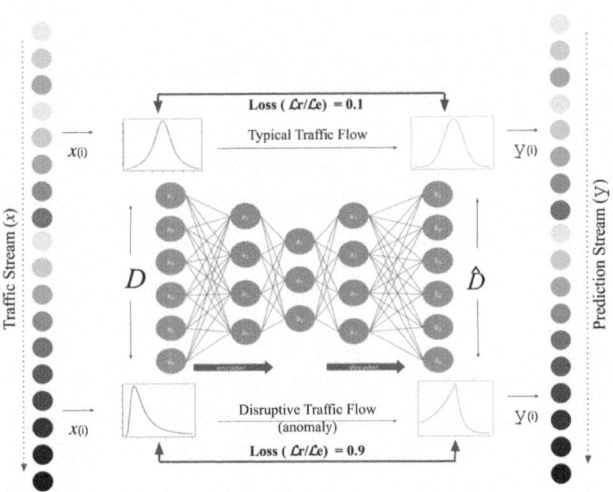

Fig. 2. Online autoencoder architecture for outlier detection.

with substantial magnitude (see Fig. 2), the model generates a higher reconstruction loss, indicating the presence of an outlier.

$$\hat{D} = Decoder(Encoder(D)) \tag{4}$$

and

$$Reconstruction\ Loss(\mathcal{L}) = \mathcal{L}_{r/e}(\hat{D}, D) \tag{5}$$

To train the autoencoder, we aim to minimize the reconstruction loss ($\mathcal{L}_r$), typically measured using Root Mean Square Error (RMSE). Building upon the previous work by Bazan et al. [1], where they found that Earth Mover's distance (EMDs) exhibited superior performance in comparing two distributions, we utilized this knowledge and used EMD as a loss function $\mathcal{L}_e$ in AEs as can be depicted in the Fig. 2, where loss can be used either as RMSE ($\mathcal{L}_r$) or EMD ($\mathcal{L}_e$). Earth's Mover Distance (EMD) is well-suited for capturing complex differences between entire distributions, in contrast to Euclidean distance (RMSE), which emphasizes point-wise errors. EMD quantifies the minimum effort required to transform one distribution into another, considering their overall structure, and is less influenced by individual extreme points. Finally, this research incorporates the Long Short-Term Memory (LSTM) model [12], a widely used sequence prediction model, in its existing form. It is important to highlight that our proposed framework can be integrated with any traffic modeling framework, as we do not alter the original model but enhance its performance by passing only crucial information.

4   Methodology

This section presents a comprehensive overview of our novel framework, Outlier Weightage Autoencoder Modeling (OWAM), along with an elaboration on the rationale behind its design choices.

Figure 3 outlines our comprehensive traffic modeling approach. Initially, traffic data is collected at 5-minute intervals. Subsequently, Flow Probability Distributions (FPDs) are generated from this data, and these distributions undergo online AE-based outlier detection to compute local anomaly scores for each sensor. A Pearson's correlation coefficient is then employed to calculate the correlation score between the target sensor and its neighboring sensors to determine their influence. These correlation scores are used as weights to adjust the importance of the neighboring sensor information before integrating it with the data from the target sensor. This integrated information is subsequently fed into an LSTM model for efficient traffic prediction. It is important to note that not only LSTM but any other modeling framework can benefit from the weighted information generated by OWAM. To address real-time changes, we continually recalculate outlier information using online AE for every T time window and incrementally update the LSTM model with this revised information. Each component is discussed below to provide a comprehensive understanding of the OWAM framework.

Creating Flow Probability Distribution (FPDs). The first step involves creating Flow Probability Distributions (FPDs) from the collected data (see Eq. 3), which offers several advantages. FPDs help to smooth out short-term fluctuations and noise, simplify analysis, and capture higher-level traffic patterns and trends. They enhance the identification of peak hours, daily variations, and weekly trends, providing a better understanding of traffic dynamics. Additionally, FPDs are more robust to outliers compared to individual 5-min intervals, improving analysis reliability. Choosing the right time window for FPDs is crucial, balancing pattern capture and noise reduction. The authors in [17] indicated that very short intervals can introduce excessive noise, while very long intervals may over smooth data, potentially missing patterns.

Autoencoders for Outlier Detection. In the second step, the continuous stream of Flow Probability Distributions (FPDs) undergoes online anomaly detection to identify outliers at each sensor. Utilizing online AEs, which adapt to incoming data in real-time, allows continuous updates of weights and parameters to capture evolving patterns and anomalies without the need for retraining on the entire dataset. Probability-weighted AE (PW-AE) were primarily employed for their superior performance, as detailed in Sect. 2. Additionally, we enhance outlier detection by introducing a novel PW-AE-EMD model, integrating EMD loss ($\mathcal{L}_e$) to measure dissimilarity between the input FPD (D) and the reconstructed FPD ($\hat{D}$) (see Equation 4 and 5). We preserve sensor outlier scores without additional threshold filtering. By maintaining the raw outlier scores, we gain a comprehensive understanding of the dynamics of outliers and their effects on nearby sensors. This approach enhances our traffic pattern analysis.

Weighted Correlation Analysis for Relevant Sensor Identification. To proceed, we need target sensors for correlation analysis and prediction. Due to

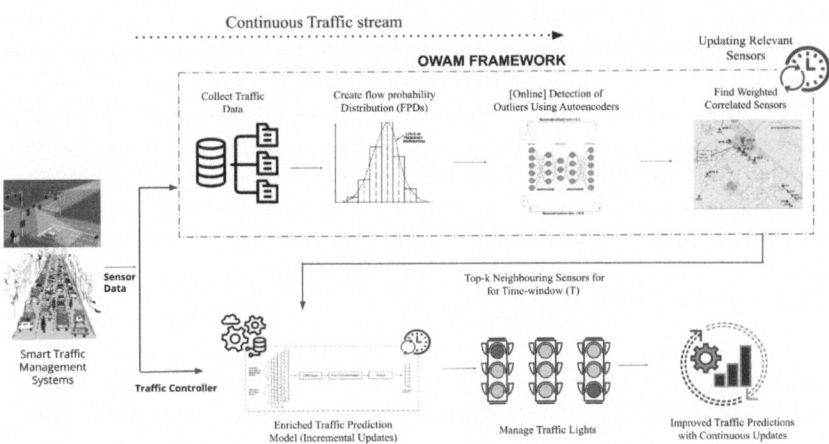

Fig. 3. Overview of OWAM framework with real-time updates.

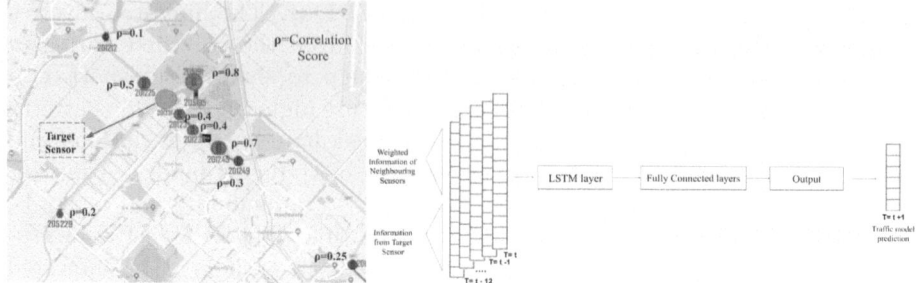

(a) Weights of correlated sensors.   (b) LSTM overview for traffic prediction.

Fig. 4. Correlated sensor and a standard LSTM overview.

computational constraints, we select representative samples from each dataset. For the Hague dataset, we use same target sensors as [17], while for METR-LA and PEMS-BAY, we randomly select five due to the unavailability of exact sensor locations. We compute correlation scores between target sensors and others using Pearson's correlation coefficient. Since sensors are closely situated, we skip introducing time lag for traffic propagation. Utilizing correlation scores, we introduce outlier-weighted correlation integration, assigning weights based on correlation coefficients. As shown in Fig. 4a, the green circle represents the target sensor, while the neighboring sensors are depicted as red circles with varying intensities corresponding to their correlation scores. This mechanism allows the model to emphasize on more influential sensors based on their outlier characteristics, enhancing traffic predictions.

Traffic Prediction. After obtaining weighted correlation data from neighboring sensors, we integrated it with the target sensor's data and fed it into the LSTM model (see Fig. 4b) for traffic flow prediction. This weighted information applies to any traffic prediction model, not just LSTM. Our experiments involved varying sensor subsets, determined by a threshold (θ). For example, in the METR-LA dataset with 207 sensors, choosing the top 5% ($\theta = 0.05$) meant selecting the top 10 correlated neighboring sensors. In offline settings, we split the data into an 80/20 ratio for training and evaluation. The LSTM model processed sequential data, with input vectors (x_i) representing the previous 1-hour window and output vectors (y_i) predicting the subsequent 5-minute traffic situation. We conducted comparative analysis offline due to the lack of online versions for the evaluated models.

Real-Time Traffic Modelling. We've implemented an incremental updating strategy (see Fig. 3) for addressing traffic dynamics. Correlated sensors are refreshed at time windows (T) for incremental updates to the LSTM model. After initially training the base model with the first 50% of data, we update it with the next t_n time windows of interval T. For each subsequent window at $t + 1$, outlier scores are obtained using online AE, correlation scores are recalculated, and neighboring sensors' significance is adjusted with incoming data. Before updating the LSTM model, its performance is evaluated and then trained with the current window's data following the test-then-train methodology [3]. We explore various time windows, from 1 h to 1 month, comparing outlier-weighted updates (incremental updates using OWAM), standard real-time updates (incremental updates without OWAM, keeping neighboring sensors fixed), and an offline model with no updates post-training, aiming to demonstrate the benefits of real-time updates and outlier-weighted correlations in traffic prediction.

5   Experimental Setup

This section presents the following key components: performance metrics, an overview of the datasets along with their statistics, and system details. An open source implementation of OWAM together with the datasets and results is available under: https://github.com/himanshudce/OWAM.

We employed the Root Mean Squared Error (**RMSE**) **metric**, and model convergence and prediction time, to evaluate the performance of the system. RMSE quantifies the average difference between predicted values $(\hat{y}_i)$ and true values (y_i) in a dataset. It is computed as: $RMSE = \sqrt{\frac{1}{n}\sum_{i=1}^{n}(\hat{y}_i - y_i)^2}$.

This research utilizes three datasets to investigate traffic patterns: The Hague dataset, METR-LA dataset, and PEMS-BAY dataset, each providing valuable insights into traffic flow dynamics.

The **Hague dataset** comprises traffic flow data from 23 intersections in The Hague, Netherlands, spanning from January 1, 2018, to March 31, 2020, with 5-minute intervals. Minimal missing values (less than 0.01%) were imputed using linear interpolation. Sensor distances and average travel times obtained from the Google Maps API complement the dataset, enhancing its utility for modeling. The **METR-LA dataset**, a benchmark in traffic modeling, consists of speed measurements from 207 sensors along the highways of Los Angeles. It covers the period from March 1st, 2012, to June 30th, 2012, providing a diverse set of traffic patterns for analysis. The **PEMS-BAY dataset**, similar to METR-LA, serves as another benchmark dataset for traffic modeling. Collected by California Transportation Agencies (CalTrans), it contains speed measurements from 325 sensors in the Bay Area, spanning from January 1st, 2017, to May 31st, 2017. In order to get higher-level insights into these datasets, we also computed the correlations among all pairs of sensors and examined their distributions. The results revealed significant inter-node (spatial) correlations in both METR-LA and Hague datasets, while the correlations in PEMS-BAY dataset are notably weaker. This observation is depicted in Fig. 5d, 5e and 5f.

By analyzing the velocity distributions presented in Fig. 5c, 5b, and 5a, notable differences can be observed among the datasets. PEMS-BAY dataset demonstrates a more uniform and closer-to-free-flow velocity distribution compared to METR-LA. This observation aligns with the lower inter-node correla-

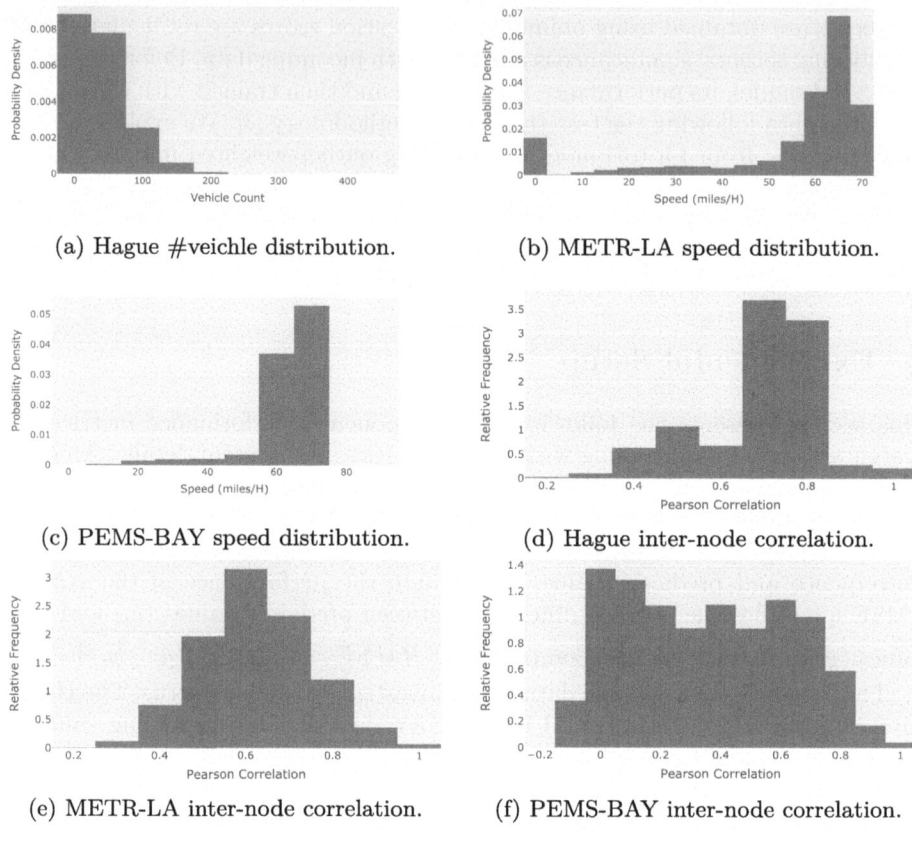

(a) Hague #veichle distribution.

(b) METR-LA speed distribution.

(c) PEMS-BAY speed distribution.

(d) Hague inter-node correlation.

(e) METR-LA inter-node correlation.

(f) PEMS-BAY inter-node correlation.

Fig. 5. Distributions of speed, number of vehicles, and inter-node correlations.

tions depicted in Fig. 5f, suggesting simpler and less-correlated traffic conditions in PEMS-BAY as compared to METR-LA. Furthermore, the vehicle distribution in Hague dataset reveals that most intersections experience low volumes of vehicles at any given time, indicating light and smooth traffic conditions in the area. The experiments were performed on a Mac M1 machine with 16 GB RAM running on macOS 13. To leverage GPU capabilities, the experiments utilized the MPS backend.

6   Results

To comprehensively assess OWAM's performance, we evaluated several aspects. Firstly, we examined its impact on the dimensionality of the traffic prediction model and explored the integration of Earth's Mover Distance loss in AEs to enhance predictions. Secondly, we compared OWAM with an LSTM Baseline and established techniques to gauge its effectiveness. Thirdly, we scrutinized

Table 1. Some stats of utilized datasets.

Dataset	Hague	METR-LA	PEMS-BAY
#Samples	236,736	34,272	52,116
#Sensors	23	207	325
Sample Rate	5 min	5 min	5 min
Time Range	2 Years	4 months	6 months
Traffic Indicator	#Vehicles	Speed (mi/h)	Speed (mi/h)

OWAM's adaptability to dynamic traffic changes in real-time scenarios. For this we used three real datasets that are actively utilized in the literature: Hague, METR-LA and PEMS-BAY. Some statistics of these datasets are explained in Table 1, the distribution of some of their attributes is visualized in Fig. 5.

In analyzing dimensionality, we aimed to identify the optimal number of sensors for predicting traffic flow at the target sensor. Figure 6b illustrates RMSE scores obtained from the LSTM-based OWAM framework for three datasets, revealing optimal threshold values for peak performance. While the Hague dataset showed consistent performance with 50% or all neighboring sensors, METR-LA and PEMS-BAY datasets exhibited performance degradation with all sensors, emphasizing the dataset-specific nature of sensor selection. Notably, using only the target sensor ($\theta = 0$) or all available sensor information ($\theta = 1$) resulted in lower prediction performance across all datasets, underlining the importance of selecting an optimal number of correlated sensors for accuracy and efficiency.

Regarding training time (see Fig. 6a), there was no consistent trend observed across datasets, though an increase in training time was noticeable for the higher-dimensional PEMS-BAY dataset.

Table 2 highlights the advantages of using EMD as the loss function. The OWAM model with EMD loss consistently outperforms the model relying on the RMSE loss (Euclidean Distance) function. While the improvements may not be substantial in the Hague scenario, an overall enhancement is still evident.

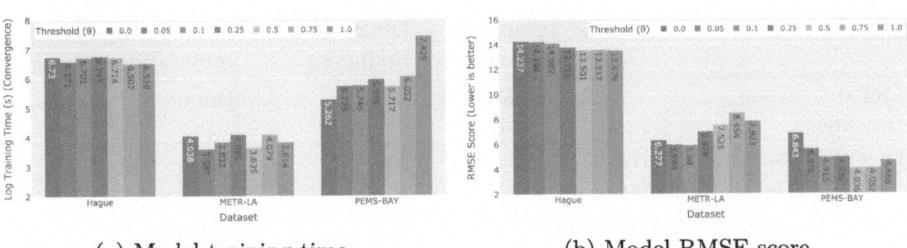

(a) Model training time.   (b) Model RMSE score.

Fig. 6. Model training time and RMSE score for different thresholds for Hague, METR-LA, and PEMS-BAY datasets.

Table 2. Comparison of EMD loss and RMSE loss on integration with autoencoders.

Model	Dataset	Train Time	RMSE
OWAM (EMD Loss)	Hague	677.739	**13.479**
	METR-LA	**37.903**	**7.525**
	PEMS-BAY	313.028	**4.915**
OWAM (RMSE Loss)	Hague	**245.465**	13.539
	METR-LA	81.914	7.918
	PEMS-BAY	**289.274**	5.537

The RMSE score decreases by 5% and 11% for METR-LA and PEMS-BAY, respectively. Although no clear pattern emerges in training times, it's important to note that RMSE is computationally more efficient, featuring $O(n)$ time complexity, whereas EMD involves solving a linear programming problem, often less efficient than $O(n)$, especially for large datasets.

To evaluate the advantages of incorporating EMD loss into the outlier model, we measured the distribution of outlier score for RMSE and EMD for a random sensor. The results for **EMD** show a mean loss of 0.70 with a standard deviation of 0.13. And for **RMSE** a mean loss of 0.81 and a standard deviation of 0.11. When EMD loss is used, it leads to a higher standard deviation in outlier scores compared to RMSE loss. This increased variance represents the model's capacity to differentiate highly relevant sensors from less relevant ones, allowing it to accurately capture complex data patterns. Conversely, a lower standard deviation indicates less differentiation among sensors, potentially resulting in the oversight of important correlations and data patterns.

Table 3 provides a comprehensive overview of the features associated with the compared approaches. The "LSTM Baseline" represents the standard modeling approach without any specific preprocessing steps. "DGCRN" (Diffusion Graph Convolution Recurrent Network) [14] stands as a representative state-of-the-art

Table 3. Feature comparison of traffic prediction models.

Model	Spatial Dependency	Temporal Dependency	Dimensionality Reduction	Neighboring Sensor Integration
LSTM Baseline	None	LSTM	No	Binary
DGCRN	Distance Matrix + Graph Convolutions	GRU	No	Binary
OBIS	Outlier Correlation (LOF)	LSTM	Yes	Binary
OWAM	Outlier Correlation (AE)	LSTM	Yes	Weighted

Table 4. Results for the Hague dataset: best in **bold**, second best *italicized*.

Model Name	RMSE	Train Time (s)	Instance Pred Time (ms)	Eval Time (s)
OWAM	*13.479*	677.74	**1.006**	**47.617**
HST [24]	13.631	**244.31**	2.14	101.32
Kit-Net [18]	13.540	*305.37*	*1.019*	*48.267*
DGCRN [14]	**13.110**	5043.31	5.91	280.00
OBIS [17]	13.547	682.81	1.08	51.30
LSTM Baseline	13.547	682.81	0.63	29.83

graph-based approach in traffic analysis. The choice of DGCRN is motivated by its recognition in the literature and the availability of open-source code which offers valuable insights into the contributions of outlier-based research compared to cutting-edge graph-based techniques in traffic management. This comparison helps in assessing the practical potential of our framework in intelligent traffic management systems. "OBIS" [17] based in [9], represents the state-of-the-art in outlier-based approaches. Furthermore, we also compared OWAM to other online outlier-based methods (HST and Kit-Net), where sensors are selected similarly based on weighted correlation with only differentiation being the outlier technique. It should be noted that all state-of-the-art and other methods, including DGCRN, were applied using the same settings as OWAM to ensure a fair comparison. This ensures the reproducibility of results and provides a clear basis for evaluating OWAM's performance relative to other models.

Tables 4, 5, and 6 summarize the comparative analysis among different models. OWAM outperforms both the previously developed outlier-based model and the Baseline LSTM in terms of RMSE scores, showing an average improvement of 10% over OBIS and 12% over Baseline LSTM. Furthermore, OWAM achieves similar or better training and evaluation times compared to OBIS while reducing training time by 15% compared to the baseline LSTM. Additionally, OWAM surpasses both HST and Kit-Net in terms of RMSE and prediction time, which

Table 5. Results for the METR-LA dataset: best in **bold**, second best *italicized*.

Model Name	RMSE	Train Time (s)	Instance Pred Time (ms)	Eval Time (s)
OWAM	*5.880*	**46.18**	**1.18**	**8.09**
HST [24]	6.815	*47.53*	3.99	27.37
Kit-Net [18]	5.883	55.44	1.69	11.58
DGCRN [14]	**4.161**	11263.23	45.23	310.04
OBIS [17]	7.421	91.37	*1.57*	*10.78*
LSTM Baseline	8.831	106.06	1.62	11.11

Table 6. Results for the PEMS-BAY dataset: best in **bold**, second best *italicized*.

Model Name	RMSE	Train Time (s)	Instance Pred Time (ms)	Eval Time (s)
OWAM	*4.036*	**303.94**	**14.23**	**97.56**
HST [24]	4.816	372.86	16.15	110.73
Kit-Net [18]	4.312	*356.96*	15.23	104.393
DGCRN [14]	**1.591**	25900.67	113.79	780.04
OBIS [17]	4.375	377.49	*14.48*	*99.27*
LSTM Baseline	5.434	405.05	16.30	111.74

can be attributed to its advanced outlier detection techniques and the weighted inclusion of top neighboring sensors, improving both prediction accuracy and computational efficiency.

OWAM competes closely with DGCRN on the Hague dataset but lags behind in RMSE for METR-LA and PEMS-BAY datasets, highlighting DGCRN's strength in handling complex traffic networks. Despite this, DGCRN's longer training times and higher memory consumption make OWAM a more suitable choice for real-time traffic applications. In summary, OWAM's superior performance and faster processing times compared to DGCRN make it the preferred

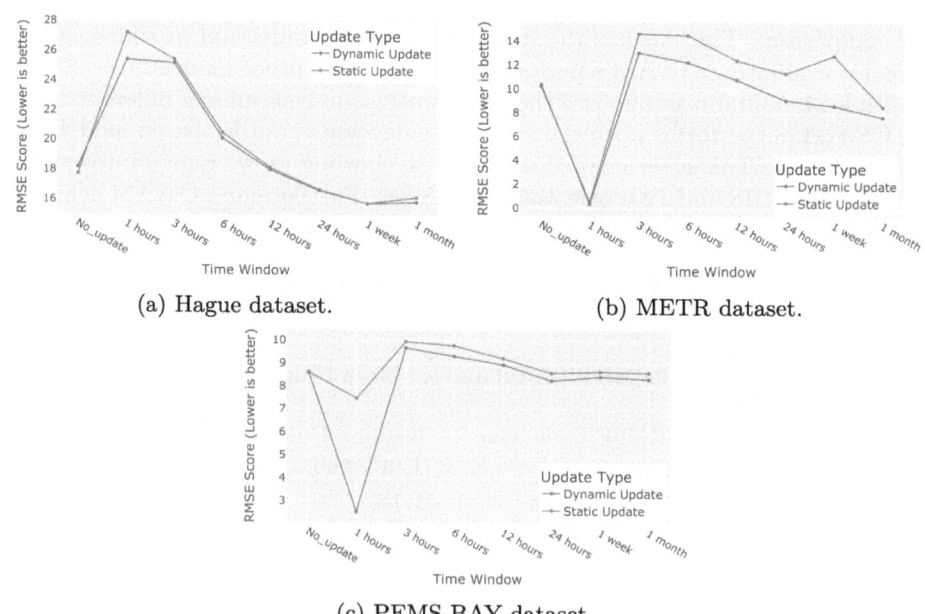

(a) Hague dataset.   (b) METR dataset.

(c) PEMS-BAY dataset.

Fig. 7. Comparison between real-time dynamic and static updates.

option for real-world scenarios, offering scalability and efficiency in large-scale, real-time traffic modeling.

In real-time settings, dynamic updates using outlier-weighted correlations significantly enhance model performance compared to static and non-updated models, particularly for short time windows (see Figs. 7b and 7c) or long time windows (see Fig. 7a). Optimal update window selection is critical, influenced by specific requirements and computational constraints. While smaller time windows do not always guarantee better performance, a careful analysis of data patterns is essential before determining the appropriate update frequency. Although OWAM does not strictly require this, we recommend setting time window size through a validation step. Overall, OWAM presents a promising solution for traffic flow prediction, showing competitive performance against baseline and state-of-the-art models. Its advanced outlier detection techniques and weighted inclusion of top neighboring sensors enhance accuracy and computational efficiency. Model selection should be customized based on application-specific requirements. While GNNs like DGCRN excel in accuracy-driven tasks, OWAM offers a more efficient choice for real-time applications.

7 Conclusion

We presented the OWAM framework, which advances outlier techniques and EMD loss to enhance outlier detection and seamlessly incorporate anomaly information into traffic prediction models. Additionally, OWAM provides dynamic model updates for efficient real-time traffic modeling. Experimental results demonstrate OWAM's significant improvement in prediction accuracy, outperforming other models by 10% compared to OBIS and 12% compared to the baseline LSTM model. While Graph Neural Networks (GNNs) exhibit more robust performance, OWAM excels in its efficiency for real-time applications. In the future, we plan to extend OWAM's applicability to various domains that require anomaly detection and real-time responsiveness after immediate concept drift detection [11,13,22], such as environmental monitoring, fraud detection [25], and healthcare.

References

1. Bazan, E., Dokládal, P., Dokladalova, E.: Quantitative analysis of similarity measures of distributions. In: Proceedings of the 30th British Machine Vision Conference (BMVC) (2019)
2. Bhatia, S., Jain, A., Srivastava, S., Kawaguchi, K., Hooi, B.: MemStream: memory-based anomaly detection in multi-aspect streams with concept drift. In: Proceedings of the ACM Web Conference (WWW) (2022)
3. Bifet, A., Holmes, G., Pfahringer, B., Pfisterer, A.: Machine Learning for Data Streams: With Practical Examples in MOA. MIT Press (2018)
4. Breunig, M.M., Kriegel, H.P., Ng, R.T., Sander, J.: LOF: identifying density-based local outliers. In: Proceedings of the ACM International Conference on Management of Data (SIGMOD), pp. 93-104 (2000)

5. Cazzonelli, L., Kulbach, C.: Detecting anomalies with autoencoders on data streams. In: Proceedings of the European Conference on Machine Learning & Knowledge Discovery in Databases (ECML/PKDD), vol. 1, pp. 258–274 (2023)
6. Choudhary, H., Hassani, M.: Autoencoder-based continual outlier correlation detection for real-time traffic flow prediction. In: Proceedings of the 39th ACM/SIGAPP Symposium on Applied Computing (SAC), pp. 218–220 (2024)
7. Djenouri, Y., Zimek, A.: Outlier detection in urban traffic data. In: Proceedings of the of the 8th International Conference on Web Intelligence, Mining & Semantics (WIMS) (2018)
8. Djenouri, Y., Zimek, A., Chiarandini, M.: Outlier detection in urban traffic flow distributions. In: Proceedings of the IEEE International Conference on Data Mining (ICDM), pp. 935–940 (2018)
9. Fitters, W., Cuzzocrea, A., Hassani, M.: Enhancing LSTM prediction of vehicle traffic flow data via outlier correlations. In: Proceedings of the 45th IEEE Annual Computers, Software & Applications Conference (COMPSAC), pp. 210–217 (2021)
10. Guha, S., Mishra, N., Roy, G., Schrijvers, O.: Robust random cut forest based anomaly detection on streams. In: Proceedings of the 33nd International Conference on Machine Learning (ICML), pp. 2712–2721 (2016)
11. Hassani, M.: Concept drift detection of event streams using an adaptive window. In: Proceedings of the 33rd International ECMS Conference on Modelling & Simulation, pp. 230–239 (2019)
12. Hochreiter, S., Schmidhuber, J.: Long short-term memory. Neural Comput. $9(8)$, 1735–1780 (1997)
13. Huete, J., Qahtan, A.A., Hassani, M.: PrefixCDD: effective online concept drift detection over event streams using prefix trees. In: Proceedings of the 47th IEEE Annual Computers, Software & Applications Conference (COMPSAC), pp. 328–333 (2023)
14. Li, F., Feng, J., Yan, H., Jin, G., Jin, D., Li, Y.: Dynamic graph convolutional recurrent network for traffic prediction: benchmark and solution. ACM Trans. Knowl. Discov. Data $17(1)$, 1–21 (2022)
15. Liu, F.T., Ting, K., Zhou, Z.H.: Isolation forest. In: ICDM, pp. 413–422 (2009)
16. Medina-Salgado, B., Sánchez-DelaCruz, E., Pozos-Parra, P., Sierra, J.E.: Urban traffic flow prediction techniques: a review. Sustain. Comput.: Inform. Syst. 35 (2022)
17. Mertens, T., Hassani, M.: Can we learn from outliers? Unsupervised optimization of intelligent vehicle traffic management systems. In: Proceedings of the European Conference on Machine Learning & Knowledge Discovery in Databases (ECML/PKDD), pp. 521–537 (2022)
18. Mirsky, Y., Doitshman, T., Elovici, Y., Shabtai, A.: Kitsune: an ensemble of autoencoders for online network intrusion detection. In: Proceedings of the 5th Annual Network & Distributed System Security Symposium (NDSS) (2018)
19. Pasquale, C., Papamichail, I., Roncoli, C., Sacone, S., Siri, S., Papageorgiou, M.: Two-class freeway traffic regulation to reduce congestion and emissions via nonlinear optimal control. Transp. Res. Part C: Emerg. Technol. 55, 85–99 (2015)
20. Pokrajac, D., Lazarevic, A., Latecki, L.J.: Incremental local outlier detection for data streams. In: Proceedings of the IEEE Symposium on Computational Intelligence & Data Mining (CIDM), pp. 504–515 (2007)
21. Polson, N.G., Sokolov, V.O.: Deep learning for short-term traffic flow prediction. Transp. Res. Part C: Emerg. Technol. 77, 1–19 (2017)

22. Scharwächter, E., MÜller, E., Donges, J., Hassani, M., Seidl, T.: Detecting change processes in dynamic networks by frequent graph evolution rule mining. In: ICDM, pp. 1191–1196 (2016)
23. Schölkopf, B., Williamson, R.C., Smola, A., Shawe-Taylor, J., Platt, J.: Support vector method for novelty detection. In: Proceedings of the 13th Conference on Advances in Neural Information Processing Systems (NIPS), pp. 582–588 (1999)
24. Tan, S.C., Ting, K.M., Liu, F.T.: Fast anomaly detection for streaming data. In: Proceedings of the 22nd International Joint Conference on Artificial Intelligence (IJCAI), pp. 1511–1516 (2011)
25. Tariq, H., Hassani, M.: Topology-agnostic detection of temporal money laundering flows in billion-scale transactions. In: Machine Intelligence and Data Science Applications (to appear)

Speculative Query Support for RDBMS with Flexible Query Order and Benefit Verification

Anna Sasak-Okoń$^{(\boxtimes)}$ [ID]

Maria Skodowska-Curie University, Pl. M. Curie-Skodowskiej 5,
20-031 Lublin, Poland
anna.sasak@umcs.pl

Abstract. This paper discusses the implementation and performance of speculative query execution for relational databases through dynamic analysis of incoming user query streams. A middleware known as the Speculative Layer employs a particular multigraph representation for groups of successive input queries called the Speculation Window which is used to create and execute speculative queries in advance. The speculatively obtained results are applied when executing user queries. This paper introduces a new strategy for managing queries within the Speculation Window. Depending on the availability of results from executed speculative queries, the order of user queries within the Speculation Window can be adjusted. When a user query is to be executed without speculative support, we choose to postpone it in favour of executing one of the subsequent user queries. This decision is made with the anticipation that the speculative results obtained soon will benefit the delayed query. Experiments conducted in a multithreaded environment with SQLite database, show that the new flexible execution strategy reduces the number of user queries executed without the speculative support. Additional series of experiments verifies that the certain parameters describing the speculative support system such as Speculation Window size or length of allowed query execution delay are properly chosen.

Keywords: Speculative Computations · Speculative Database Queries · Graph Query Modelling

1 Introduction

Speculative parallelization, optimistic parallelisation [1] or thread-level speculation is a technique which executes in parallel parts of code originally intended to run sequentially. As a consequence, in certain cases, speculative threads may deliver data subsequently updated by the predecessor thread, causing dependence violation. Thus, all results obtained by speculative threads must be validated. If positive validation is not possible, obtained results are rejected and speculative threads are restarted with the new input data. If, however the assumptions made speculatively for values or control flow are correct, the considerable efficiency improvement can be obtained [7–9]. According to the literature there are three main types of speculation techniques [10,11]: control speculation - an execution of an instruction before the instruction it is control dependent, data dependence speculation - an assumption that the input data value for a particular instruction has been already computed and stored in the corresponding memory location, data values speculation - a prediction of the data value that is going to be produced to use it earlier as an input data for another instruction.

The speculative execution framework outlined in our prior papers [2,4–6] focuses on parallelised speculative assistance for execution of user query streams. It incorporates a multi-threaded middleware called the Speculative Layer positioned between user applications and the RDBMS. The objective of the Speculative Layer is to select and execute speculative queries. These speculative results stored in the main memory structures referred to as the Speculative DB (SDB) form a quickly accessible data set. The selection process for speculative queries to execute is called speculative analysis. The speculative analysis is ongoing for groups of successive user queries termed Speculation Window (SW). For each SW distinct graph representations of each user query are generated and subsequently merged into a single multigraph following a defined set of rules. SW shifts along the queue of user queries advancing one query at a time. Up until now, for each SW iteration, the first user query in line was executed nonspeculatively, regardless of speculative results availability. In this study we present a novel approach to nonspeculative query execution contingent upon speculative results availability. If the first query in the SW can't make use of already prepared speculative results, we allow it to be replaced by the next query in the SW which has at least one executed speculative query assigned. We anticipate that one of the speculative results obtained soon will prove useful for the postponed query. After executing the nonspeculative query, the SW advances, and we retry executing the delayed user query.

The remaining text of the paper is composed of four parts. In the first part the related work is described. Next, an overview of the Speculative Layer structure is provided. We describe guidelines for generating query graphs and the procedure of speculative analysis resulting in optimally defined set of speculative queries ready for use. Section 4 presents the new, flexible strategy for the nonspeculative query execution. Section 5 contains results of two series of experiments. First series of experiments presents effectiveness comparison between the Speculative Layer execution with old and new strategies for the nonspeculative query execution. The second series of experiments verifies the validity of the parameter describing the size of the SW chosen for the previous versions of the speculative algorithm. This Section also includes a discussion of strength and weaknesses of delayed execution of user queries. We also propose a necessary constraint in the user query execution delay preventing the advantage off loss over gain.

2   Related Work

Speculative execution is an active research field in relational databases. A speculative support for SQL query processing was presented in [12]. Authors use database idle time to execute asynchronous anticipated database transformations in a parallel way. They are initiated based on fragmentary actions of users formulating their final queries. If the lookahead performed queries were useful for the final query, it was executed in a shorter time. A proposal based on data hints is described in [13,14]. Authors speculatively support data integration for separate heterogeneous data sources (data gathering plans) performing some operations ahead of their normal schedule. If predictions based on received earlier tips are correct, significant plan implementation speedup is provided.

A speculative support for transaction protocols was proposed in [15,16]. A speculative protocol was introduced providing a faster access to data locked due to ongoing transaction. Two transaction images were used - old and recent - for two simultaneous speculative executions. Finally based on the blocking transaction real result, one of the speculative results was validated. Paper [17] describes how the speculative computations can be used in record analysis for ranked queries which aim in returning records from different data sources according to some preference function. The original method includes creating speculative version of the ranking algorithm. It speculatively assume the degree of query results variation as negative if the data source is slower. Due to this, the query results are returned faster but with a risk of some inaccuracy.

There are also papers [18,19] which focus on multiple query optimization. Common sub-expressions determined for a group of queries can be used either to choose more appropriate execution plan or to form and execute a set of materialized views. The total query execution time is decreased as common tasks are performed only once. These articles, however, do not contain speculative execution concept.

Graph structures are often used as a formalism to aid in representing and analysing queries as they naturally depict and conceptualize entities and their relationships [20,21]. Query optimization methods based on cashing the results of already executed queries are also very popular [22–25]. This approach is based on an assumption that the same query will be requested again. In such case, instead of evaluating this query, the cached results can be reused. The proposed algorithms attempt to find reusable components of earlier query plans and then to develop a new plan which can reuse them. This way only data not available in cache is loaded from server. Even though the new query execution plan may not be optimal, the execution time is shorter due to data reuse. Most of the described optimization methods are parts of a general concept of query folding [26]. Query folding is the process of determining whether and how a query can be answered using a given set of resources. These resources can be materialized views, cached results of previous queries, or even queries answerable by other databases.

It's important that none of the optimization techniques described above target speculative support, which covers need of multiple future queries at a time.

Saving the results obtained speculatively resembles caching methods, however, rather than relying on previous query execution history, we prefer to analyse queries which are queued for imminent execution and support them with data prepared in advance. Our method's key original aspects include executing specially defined Speculative Queries alongside a multigraph-oriented approach.

3 The Speculative Layer

The rapid growth of online shopping has induced the development of database applications tailored to its needs. These applications are intensively used as tools for browsing products and handle numerous queries with similar structures. Shop users create these queries by specifying their search criteria, directly impacting the attributes and conditions within the SELECT and WHERE clauses. Consequently, successive queries include certain component operations, whose results, if obtained speculatively, can be utilized multiple times.

Based on these insights, a dynamic speculative approach for executing SQL user queries was proposed. This model is implemented as an additional multithreaded middleware known as the Speculative Layer, situated between user shop applications and the RDBMS. User queries form a stream that is continuously monitored and analysed by the Speculative Layer. This speculative analysis focuses on consecutive user queries grouped within a SW structure, which moves along the query stream. Each user query in SW is represented by its query graph, generated according to a set of predefined rules. They are then combined to form a collective representation of the entire query group, in the form of a *Query Multigraph* (Q_M). Common operations within user queries are identified and leveraged to generate new queries termed speculative queries. To facilitate this, an extended version of Q_M is developed, known as the *Speculative Query Multigraph* (SQ_M). It incorporates speculative edges, a unique type of edge absent in Q_M denoting potential speculative queries. Selected speculative queries are executed concurrently with the original user queries, and their results are stored in a dedicated RAM database structure named the Speculative DB (SDB). Data from the SDB is accessible during the execution of user queries, thus reducing the need for disk database reads.

User query executed within each SW (so far always the first query in SW) is referred to as the nonspeculative query. If there are speculative queries assigned to the nonspeculative query, it is modified to incorporate them; otherwise, it is executed without speculative support. Any user query can use the speculative results prepared for each relation listed in its FROM clause. More details about combining results from multiple speculative queries for the execution of a single user query are provided in [3]. After the results of the nonspeculative query are returned to the user, the SW shifts by one query. Consequently, the representation of the executed user query in Q_M is substituted with the representation of the next user query from the queue, newly entering the SW. The SW move is succeeded by the speculative analysis process to generate a new set of speculative queries for execution. This operational sequence is reiterated as long as there are pending user queries awaiting execution. Figure 1 presents the general scheme of the RDBMS cooperating with the Speculative Layer.

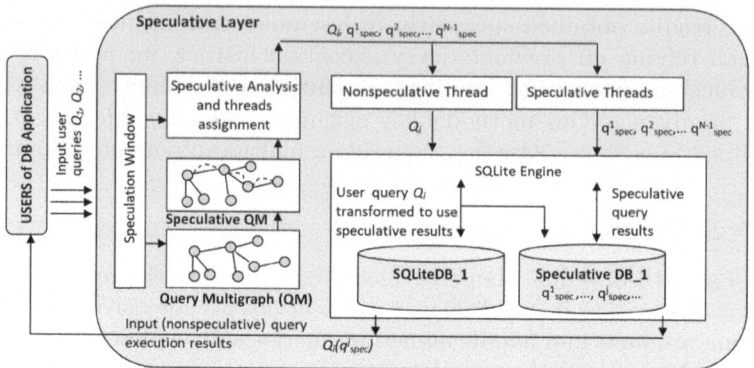

Fig. 1. General scheme of the Speculative Layer

The Speculative Layer is designed to support conjunctive queries with arithmetic comparisons (CQAC) with IN and LIKE accepted for sets and string comparisons. A nested query is allowed in a WHERE clause returning a value for the attribute condition (SELECT...WHERE...AND attribute<(SELECT... FROM... WHERE)). Single query graphs (Q_M) are created according to rules similar to these proposed in [27]. There are three types of vertices used: *relation*, *attribute* and *value*. A value vertex is used to represent both single values and sets of values appearing in a user query WHERE clause. These vertices are adjacent to edges which are representing where, in a user query, a particular relation, attribute or value appeared. Thus there are: *membership* edges (μ) for the SELECT clause, *predicate* (θ) edges for the WHERE clause joining the relation vertex with an attribute vertex and *selection* (σ) edges also for the WHERE clause joining an attribute vertex with a value vertex. Due to the possibility of the modifying query appearance in the user query queue, there are three more types of edges used to represent different types of modifying queries: *delete* (δ), *insert* (η), and *update* (υ) edges.

Q_M constitutes a joint representation of user queries from the SW. Q_M vertices set is a union of vertices and edges set is a multiset of edges of all component query graphs. Figure 2 presents the Q_M created for the following component queries:

(Q_1) SELECT $A_{2,2}, A_{2,3}, A_{1,3}$ FROM R_1, R_2 WHERE $A_{1,4} = A_{2,1}$ AND $A_{2,2} > 7$
(Q_2) SELECT $A_{1,2}, A_{2,2}$ FROM R_1, R_2 WHERE $A_{1,4} = A_{1,2}$ AND $A_{1,3}$ LIKE '%xx' AND $A_{2,2} < 2$
(Q_3) SELECT $A_{2,2}, A_{2,3}$ FROM R_2 WHERE $A_{2,2} > 5$

Figure 2 also includes speculative edges (marked by a red dashed line) representing one of possible speculative queries to be executed for this SW:

– sq_1: SELECT $A_{2,1}, A_{2,2}, A_{2,3}$ FROM R_2 WHERE $A_{2,2} > 5$.

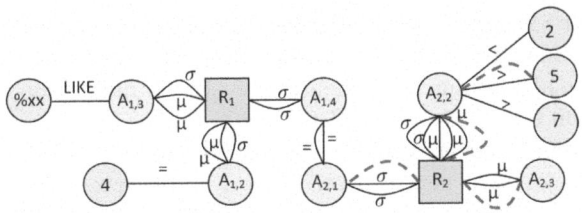

Fig. 2. Q_M representing four user queries.

Such speculative query results, if executed, can be used while executing two user queries from the SW: Q_1 and Q_3. In general, for this particular Q_M three more awaiting speculative queries would be generated - one for each value vertex.

- sq_2: SELECT $A_{1,3}, A_{1,2}, A_{1,4}$ FROM R_1 WHERE $A_{1,3}$ LIKE %xx
- sq_3: SELECT $A_{2,1}, A_{2,2}, A_{2,3}$ FROM R_2 WHERE $A_{2,2} > 7$
- sq_4: SELECT $A_{2,1}, A_{2,2}, A_{2,3}$ FROM R_2 WHERE $A_{2,2} < 2$

Speculative edges inserted in Q_M indicate for which graph elements the respective speculative queries are to be generated. Depending on the speculation result we introduce two types of speculative queries: *speculative parameter* or *speculative data* queries. Speculative parameter edges are inserted for the nested queries components in Q_M. If a nested query was executed as a speculative parameter query, its results can be used as a parameter value in its parent user query. The aim of speculative data queries is to obtain and save in the SDB a subset of records or/and attributes of a database relation. The goal is to generate such speculative query (its SELECT and WHERE clauses) so as its results could be used for execution of as many input queries as possible.

As there is allowed possibility of the modifying query occurrence in the query stream, and thus in the SW, we introduce additional type of speculative edges called *speculative state*. The presence of a modifying query in SW means that both already executed and awaiting speculative queries may refer to invalid data. Its results can't be used without validation process following the modifying query execution. Depending on the modification type (delete, insert or update) we verify which relation and attributes are being referred to. If the results of the speculative queries marked with the *speculative state* consist of all required attributes we can execute the modifying query on them (positive validation). In other case, we delete this speculation due to missing attributes (negative validation).

The algorithm of inserting speculative edges is executed for each attribute vertex incident to a selection edge. First, nested queries are identified resulting in new speculative parameter queries. Then, all combinations of the value vertices adjacent to the analysed attribute vertex are considered to form a WHERE clause of generated speculative data queries. Appropriate (adjacent) relation vertex and additional attribute vertices are selected to form SELECT and FROM clauses of the aforementioned speculative data query. Finally, if a modifying query appears in SW, then speculative edges corresponding to succeeding user queries must be marked with the additional speculative state edges. Detailed description of the algorithms implemented for Q_M manipulation and speculative queries generation can be found in [4,6].

Based on earlier experiments reported in [2], the SW size was set to 5, and each SW was assigned 3 active threads. It was observed that increasing the SW size or the number of threads didn't improve user query execution. One thread is always allocated for executing the nonspeculative user query, leaving the remaining two threads available for executing selected speculative queries. From the pool of pending speculative queries generated during the Q_M speculative analysis, 2 are selected for execution. The highest execution priority is given to those speculative queries that can benefit the most user queries. We take into account two defined size reduction metrics for speculative queries: *vertical selectivity* and *horizontal selectivity* (see [5] for definitions). These metrics represent the decrease in the number of rows and columns resulting from the query execution. To avoid creating full copies of database relations in RAM, we prioritize speculative queries with low values of aforementioned metrics. If a new speculative query cannot be selected for execution within a particular SW, the speculative thread reports a no-job status. This status usually occurs if the change in the Q_M structure is insufficient to produce new and unique speculative queries after SW move. Executed speculative queries are recorded on a list and linked to user queries that can use their results.

4   A New Strategy for the Speculation Window Execution

The previous execution strategy for the SW was rigid, meaning the nonspeculative query always occupied the first position in it. This approach maintained the execution order of user queries, but it posed a risk that speculative results might not be available in time for certain nonspeculative queries.

In the new strategy, the nonspeculative query is not limited to the first position in SW. If the initial query in SW would proceed without relying on speculative results, the subsequent query in SW, capable of using existing speculative query results, is executed instead. This approach allows the speculative algorithm to generate and execute new speculative queries that will benefit the delayed user query. The anticipated reduction in execution time through the use of speculative results is expected to outweigh any potential delay in execution.

Figure 3 presents the new nonspeculative query execution strategy for three consecutive SWs represented by blue arrays. Below each blue user query square, there are orange circles containing ids of the executed speculative queries assigned to a particular user query for the potential use. We can see, that for the SW1 there is no need for a query order change, as the Q_1 has the sq_1 speculative query assigned. Thus, the Q_1 query is executed nonspeculatively using the results of sq_1. Then, SW moves by one user query, Q_1 is replaced by Q_6. For the SW2 the first query in the group has no executed speculative queries assigned (no circles below). Then, we look for the next query in the SW which

could be executed instead. Thus, Q_3 would be executed nonspeculatively for this
SW using selected speculative results (sq_2, sq_3 or both). SW moves again but
the non-executed Q_2 remains in it, while the Q_3 is replaced by the next query
from the user query queue Q_7. For the SW3 we can see that the parallel speculative query execution provided a speculative query sq_6 assigned to previously
omitted Q_2. Thus, for the SW3 the nonspeculative query is Q_2 which, due to a
new strategy, would be executed with the use of the speculative query results
sq_6.

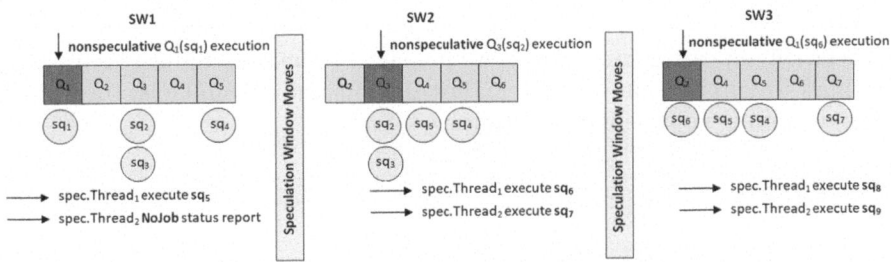

Fig. 3. Flexible query execution strategy for 3 consecutive Speculation Windows.

5   Experimental Results

5.1   Test Environment and Queries

The Speculative Layer is coded in C++ and employs multithreaded execution
using the Pthread library. It utilizes the SQLite 3.8.11.1 engine as its DBMS.
For our experiments, we employed the database structure and data (8 relations,
1GB data) from the widely recognized TPC-H database benchmark [28]. Additionally, we created a set of 8 query templates referring to 2–5 database relations
and used them to generate 3 sets of 1000 input queries each, incorporating random values in the attribute conditions within WHERE clauses. Template T8
represents modifying queries. Modifying queries are handled according to the
strategy mentioned in Sect. 2 and described in [6] and don't interfere with the
new new strategy of nonspeculative query execution. Each test queries set contains approximately 4% of modifying queries and 96% of select queries with the
uniform density for each of T1–T7 templates.

5.2   A New Nonspeculative Query Execution Strategy

In our initial comparison, we evaluated how the revised nonspeculative query
execution approach for the SW influences user query execution. We contrasted
the outcomes between the former strategy (where the nonspeculative query consistently led the SW) and the new method (where the nonspeculative query is

chosen as the first query in the SW that can use already obtained speculative query results). The experiment was carried out with a SW size of 5 and 3 active threads. We have also limited the number of nonspeculative query execution delay steps to the size of SW to prevent user query starvation. Figure 4 illustrates the average execution times obtained for each query template, depending on the number of speculative query results used (ranging from 0 - red bars - to 3 - yellow bars) for one user query. It is evident that each speculative query use leads to further reductions in user query execution time. The most substantial speedups are observed for T5 and T7 templates. These user queries show significantly longer execution times when executed without speculative support (approximately 163 and 181 s). By employing those three sets of speculative query results (one for each relation appearing in a query), we managed to notably reduce their execution times (by up to 9 times shorter). For the remaining templates, the average time reduction is approximately 2 times.

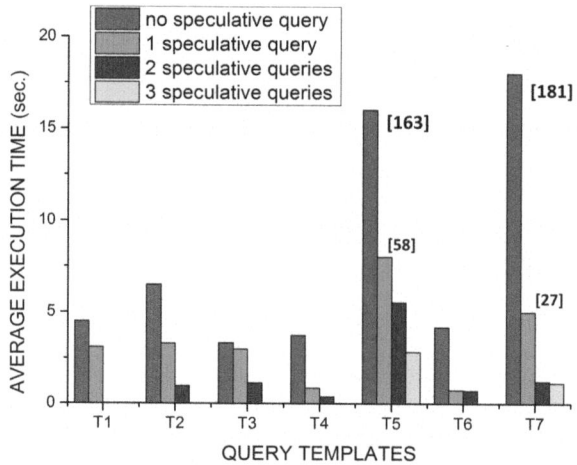

Fig. 4. The average execution time for 7 query templates with and without use of the speculative query results.

We then analysed the effects of the new approach on the overall number of queries executed, both with and without speculative support. While executing the user queries test sets with the speculative support and the old strategy for the nonspeculative query choice, we obtained approximately 90 no-job reports from speculative threads for 1000 user queries. It means that approximately for as much as 90 SWs, there weren't two unique speculative queries to be executed, which would meet our requirements (result size and possible use). The total number of 1716 speculative queries were executed for each test set and 113 user queries were executed without the use of the speculative results.

The second execution - with the new strategy which allows for the nonspeculative query choice based on potential use of speculative results - resulted in 58 user queries (for 1000 user queries) executed without the use of the speculative result, which is almost two time less than before. We also got 70 no-job reports and a total number of 1748 speculative queries were executed, each of which was assigned approximately to 1,98 user queries for potential use. The query order change resulting from the new execution strategy occurred 132 times involving 58 user queries, with an average delay across the entire set being 2.59.

5.3 Speculation Window Size

Due to the substantial decrease in the number of user queries executed without speculative support following the adoption of the new strategy, we opted to conduct a new set of experiments to explore whether modifying the SW size could offer further advantages.

Table 1 illustrates the outcomes obtained for SW sizes ranging from 5 to 15. It is noticeable that as the SW size increases, there is a significant rise in the number of "no job" reports, reaching up to 26% for SW size = 15 (we assume that the maximal number of executed speculative queries is 2 for each user query). This is primarily due to the fact that with larger SW and Q_M sizes, each movement of the SW only replaces one user query in the multigraph structure which is insufficient to generate new distinct speculative queries for execution. Consequently, we observe a decrease in the total number of executed speculative queries for the test set and stabilization of the number of user queries executed without speculative support for bigger SWs. The bottom part of Table 1 displays outcomes specific to the new approach for executing nonspeculative queries. We observe that the number of user queries facing execution delays due to the absence of speculative query results hovers around 60 and doesn't appear to correlate with the SW size. However with bigger SWs the allowed execution delay is also growing (up to 14 for the SW size 15) posing a threat the wait time outweighs the potential benefits of utilizing speculative query results.

We have already established that each speculative query results used while executing user query provides further execution time reduction. Thus, the general goal of the Speculative Layer execution optimization is not only to reduce the number of the user queries executed without the speculative support, but also to maximize the number of speculative results used for one user query. We have investigated how, for each query template, the SW size affects the numbers of the user queries executed without the speculative support or with one or more speculative queries results used. SW size 5 tends to provide the biggest reduction in number of queries executed without speculative support. Further changes in the SW size provide minor changes in the number of user queries executed with the use of speculative results, for example increasing the number of user queries executed with the use of two speculative results but reducing the number of user queries executed with the use of one speculative query results. The T1 template shows very uniform results as these are the user queries with a nested query in their WHERE clauses and even without optimization we manage to execute all of them with the use of the speculative results. Also T3 and T5 templates are worth noticing as they show the biggest reduction in the number of user

Table 1. Execution results obtained for different sizes of the SW Size

Speculation Window Size										
5	6	7	8	9	10	11	12	13	14	15
% of nojob reports										
7.04%	8.55%	10.67%	13.60%	14.33%	17.07%	22.49%	20.95%	20.86%	22.57%	26.21%
% of user queries executed without speculative results used										
5.60%	4.21%	3.1%	3.5%	3.5%	3.4%	3.9%	3.6%	3.4%	3.2%	3.6%
Total number of executed speculative queries										
1748	1717	1667	1594	1592	1539	1496	1480	1481	1445	1419
Number of user queries executed with changed order										
58	61	61	66	66	70	70	66	64	59	66
Total number of queries order change in SW										
132	180	226	274	300	335	372	383	406	373	402
Average delay in nonspeculative query execution										
2,59	2,95	3,70	4,15	4,55	4,78	5,46	5,80	6,34	6,32	6,09

queries executed without the speculative support. Both these templates also show the highest fluctuations in presented data values. SW size = 5 also provides the smallest delay due to new strategy for nonspeculative query execution (see Table 1 bottom rows).

5.4 Benefit Verification

The speculative support system is designed for user query streams executed one after another. If a user query is executed using the speculative result prepared in advance, its execution time is shorter and user's waiting time is also shorter. On the other hand, if the query execution is delayed, the benefit from anticipated speculative results use is reduced in terms of user's waiting time. We have thus analysed what is the balance of profit from the use of speculative results and loss from delayed execution. For the SW size = 5, 58 user queries were executed with changed order. The delay was from 1 to 4 (assumed delay limit). In the group of delayed queries almost 78% were executed with the use of acquired speculative results providing considerably shorter execution time, especially if we consider long running user queries, such as T5 and T7 templates (see Fig. 4 for execution times for each template). On the other hand as much as 14 user queries (22%) were delayed and still executed without the speculative support. What's more all of them were delayed for the maximal number of 4 SW moves generating average delay of 40.5 s.

We have then examined the delayed queries executed with the use of speculative result. In this group 16 (which is almost 40% of the group) were delayed for 3 or 4 SW moves generating average delay of 35 s. Such delay in case of short running queries such as T1, T3, T4 or T6 doesn't seem to be beneficial. On the other hand if a user query is executed with the use of speculative results and additionally ahead of its proper turn the benefit from the speculative support is growing (the average additional speed-up due to earlier execution was 17,2 s), especially if the omitted query is the long running one.

Based on the above conclusions we decided to further limit the accepted delay first to 3, then to 2 and finally to 1 SW move. Table 2 presents the obtained results. We can see that the strictest limitation on the number of SW moves during which the nonspeculative query can be delayed provides the least benefits. In a group of 58 delayed user queries only 13 were executed with use of the speculative results. On the other hand the average delay of 6.87 s for queries not executed with speculative support is acceptable. Each one more SW move of possible delay of nonspeculative query execution increases the number of queries executed with the speculative support while the time of average delay is also growing. The reasonable choice seems to be 2 SW moves of acceptable delay. In such case almost half of delayed queries were executed with the speculative support and the demonstrated delay time of 12–13 s doesn't prevail benefits from the use of speculative results. With shorter delays there is also less risk that long-delayed query would still be executed without speculative results. Longer delays (3 or more SW moves) could be considered if the query set mostly comprised long-running queries showing the greatest reduction in execution time due to speculative support.

Considering the last set of parameters for the Speculative Layer i.e. SW size 5 and maximal allowed user query delay 2, only about 7.8% of user queries were executed without speculative results, whereas our previous experiments [2,3,5,6] shown approximate 15% of user queries executed without speculative support. Even though 36 delayed queries did not benefit from speculative results, 101 user queries were executed ahead their normal schedule intensifying the benefit from speculative support.

Table 2. Results for speculative execution with limited user query delay.

	Delay limit (SW moves)			
	4	3	2	1
No of delayed user queries executed with spec. support	44	35	22	13
Avg. delay of user queries executed with spec. support	28.57	16.12	12.04	4.74
No of delayed user queries executed without spec. support	14	24	36	45
Avg. delay of user queries executed without spec. support	40.5	26.6	13.43	6.86

6   Conclusion

The paper presents a novel approach to executing non-speculative queries within the speculative query execution support system named the Speculative Layer.

In the basic algorithm of the Speculative Layer, the nonspeculative query was always the first query in the SW. In this paper, we propose a new, more flexible strategy of the nonspeculative query choice. If a user (nonspeculative) query is to be executed without speculative support, we allow a query order change and instead we execute the first user query in the SW which can use already prepared speculative results.

In the first series of experiment we allowed the maximal user query delay equal SW size. For a test database and three sets of randomly generated queries each, the new adopted strategy helped to execute almost 95% of user queries with the use of speculative results and the user query order change was reported for approximately 58 queries for each set. Then, we investigated whether the SW size established earlier, set to 5, remains an optimal choice for the updated execution strategy. The obtained results prove that the size choice is still valid. Larger SW sizes resulted in minor changes in the number of queries executed using speculative results, however causing longer delays in nonspeculative query execution. Finally we have analysed the balancing of profit form the use of speculative results and the loss from delayed execution. The reasonable allowed delay seems to be 2 SW moves which provide almost half of delayed queries to be executed wit the speculative support. Longer delays could be consider for specific query sets containing a lot of long-running queries.

Further efforts will focus on development of even more adaptable user query execution strategy for the SW. We will explore the concept of a variable shift speculation window, where multiple user queries are executed as nonspeculative queries within each SW, depending on the available speculative results. This strategy will lead to larger changes in the multigraph structure, potentially resulting in a broader range of speculative queries.

Disclosure of Interests. The authors have no competing interests to declare that are relevant to the content of this article.

References

1. Estebanez, A., Llanos, D.R., Gonzales-Escribano, A.: A Survey on thread-level speculation techniques. ACM Comput. Surv. **49**(2), 1–39 (2017)
2. Sasak-Oko, A.: Speculative query execution in relational databases with graph modelling. In: Proceedings of the FEDCSIS 2016. ACSIS, vol. 8, pp. 1383–1387 (2016)
3. Sasak-Oko, A., Tudruj, M.: Graph-based speculative query execution for RDBMS. In: Proceedings of the 13th International Conference in Parallel Processing & Applied Mathematics (PPAM), pp. 303–313 (2018)
4. Sasak-Oko, A., Tudruj, M.: Graph-based speculative query execution in relational databases. In: Proceedings of the 16th International Symposium on Parallel & Distributed Computing (ISPDC), pp. 122–131 (2017)
5. Sasak-Oko, A., Tudruj, M.: Speculative query execution in RDBMS based in analysis of query stream multigraphs. In: Proceedings of the 24th Symposium on International Database Engineering & Applications (IDEAS), pp. 208–218 (2020)

6. Sasak-Oko, A.: Modifying queries strategy for graph-based speculative query execution for RDBMS. In: Proceedings of the 13th International Conference on Parallel Processing & Applied Mathematics (PPAM), pp. 408–418 (2020)
7. Silc, J., Ungerer, T., Robic, B.: Dynamic branch prediction and control speculation. Int. J. High Perform. Syst. Archit. **1**(1), 2–13 (2007)
8. Pan, S., So, K., Rahmeh, J.T.: Improving the accuracy of dynamic branch prediction using branch correlation. In: Proceedings of the International Conference on Architectural Support for Programming Languages & Operating Systems (ASPLOS), pp. 76–84 (1992)
9. Moshovos, A., Breach, S.E., Vijaykumar, T.N., Sohi, G.S.: Dynamic speculation and synchronization of data dependences. In: Proceedings of the 24th Annual International Symposium on Computer Architecture (ISCA), pp. 181–193 (1997)
10. Kaeli, D., Yew, P.: Speculative Execution in High Performance Computer Architectures. Chapman Hall/CRC (2005)
11. Kejariwal, A., et al.: On the performance potential of different types of speculative thread-level parallelism. In: Proceedings of the 20th Annual International Conference on Supercomputing (ICS), p. 24 (2006)
12. Polyzotis, N., Ioannidis, Y.: Speculative query processing. In: Proceedings of the 1st Biennial Conference on Innovative Data Systems Research (CIDR), pp. 1–12 (2003)
13. Barish, G., Knoblock, C.A.: Speculative plan execution for information gathering. Artif. Intell. **172**(4–5), 413–453 (2008)
14. Barish, G., Knoblock, C.A.: Speculative execution for information gathering plans. In: Proceedings of the 6th International Conference on Artificial Intelligence Planning Systems (AIPS), pp. 184–193 (2002)
15. Reddy, P.K., Kitsuregawa, M.: Speculative locking Protocols to improve performance for distributed database systems. IEEE Trans. Knowl. Data Eng. **16**(2), 154–169 (2004)
16. Ragunathan, T., Krishna, R.P.: Improving the performance of read-only transactions through asynchronous speculation. In: Proceedings of the Spring Simulation Multiconference (SpringSim), pp. 467–474 (2008)
17. Hristidis, V., Papakonstantinou, Y.: Algorithms and applications for answering ranked queries using ranked views. VLDB J. **13**(1), 49–70 (2004)
18. Ge, X., et al.: LSShare: an efficient multiple query optimization system in the cloud. Distrib. Parallel Databases **32**(4), 593–605 (2014)
19. Chaudhari, M.B., Dietrich, S.W.: Detecting common subexpressions for multiple query optimization over loosely-coupled heterogeneous data sources. Distrib. Parallel Databases **34**, 119–143 (2016)
20. Preti, G., Lissandrini, M., Mottin, D., Velegrakis, Y.: Mining patterns in graphs with multiple weights. Distrib. Parallel Databases **39**(2), 281–319 (2019). https://doi.org/10.1007/s10619-019-07259-w
21. Goonetilleke, O., Koutra, D., Liao, K., Sellis, T.: On effective and efficient graph edge labeling. Distrib. Parallel Databases **37**, 5–38 (2019)
22. Faisal, H.M., et al.: A query matching approach for object relational databases over semantic cache. In: Application of Decision Science in Business and Management. IntechOpen (2020)
23. Ahmad, M., Qadir, M.A., Sanaullah, M.: Query processing over relational databases with semantic cache: a survey. In: Proceedings of the IEEE International Multitopic Conference, pp. 558–564 (2008)

24. Wang, F., Agrawal, G.: Query reuse based query planning for searches over the deep web. In: Proceedings of the International Conference on Database & Expert Systems Applications (DEXA), pp. 64–79 (2010)
25. Cybula, P., Subieta, K.: Query optimization by result caching in the stack-based approach. In: Proceedings of the 3rd International Conference on Objects and Databases (ICOODB), pp. 40–54 (2010)
26. Gryz, J.: Query folding with inclusion dependencies. In: Proceedings of the 14th International Conference on Data Engineering (ICDE), pp. 126–133 (1998)
27. Koutrika,G., Simitsis, A., Ioannidis, Y.: Explaining structured queries in natural language. In: Proceedings of the International Conference on Data Engineering (ICDE), pp. 333–344 (2010)
28. TPC benchmarks. http://www.tpc.org/tpch/default.asp. Accessed 01 Apr 2024

Made to Measure: Towards Approximability of Query Evaluation Engines

Daniel Flachs[✉][iD] and Guido Moerkotte[iD]

University of Mannheim, 68131 Mannheim, Germany
{flachs,moerkotte}@uni-mannheim.de

Abstract. Accurate, efficiently computable cost functions are important for cost-based query optimizers. We investigate how to automatically find the best approximate cost functions minimizing the q-error for a hash join on measurements from extensive experiments. We systematically analyze which factors influence the approximation error, which input parameters are beneficial, and how this relates to plan quality in the big picture. From this, we derive suggestions for improvements for the query evaluation engine (QEE) that benefit both the hash join's runtime and the approximation error.

Keywords: query evaluation · hash join · cost function · cost model

1 Introduction

The cost model is a significant component of a cost-based query optimizer. The process to build a cost model is usually as follows:

1. Start with an existing query evaluation engine (QEE).
2. Analyze the QEE (and the hardware) and build a parameterized cost model, which typically contains (hardware) coefficients that need to be calibrated.
3. Calibrate the coefficients (e.g., via microbenchmarks).
4. Run (usually limited) experiments to validate the model.

There exist numerous examples of this approach in the literature [4,14,30]. Alternatively, some modern approaches apply learning to combine steps 2 and 3 into a single step [27].

We suggest a rather different, iterative approach (called the *loop*), where improvements to the precision of the cost approximations can not only be achieved by improving the approximation itself, but predominantly by improving the QEE:

1. Start with an existing QEE.
2. Run extensive experiments to obtain runtime measurements.
3. Generate best approximations for the measurements.
4. If the approximation errors are too large for certain cases, improve the QEE by applying optimization techniques, or even by introducing new specialized physical operators, and go to step 2.

The original version of the chapter has been revised. A correction to this chapter can be found at https://doi.org/10.1007/978-3-031-83472-1_27

R. Chbeir et al. (Eds.): IDEAS 2024, LNCS 15511, pp. 235–250, 2025.
https://doi.org/10.1007/978-3-031-83472-1_16

Thereby, one should have an eye on the following *requirements* for a cost function. It should be (R1) accurate and (R2) efficiently computable [3], (R3) it should be frugal concerning its parameters, and (R4) it should guarantee modest error propagation. Clearly, if there is a large discrepancy between the estimated and the true cost of a query evaluation plan (QEP), the query optimizer might choose a bad (suboptimal) plan. Efficiency is required as the cost function is invoked for every plan alternative the query optimizer considers (see Sect. 2.1). The necessity to consider (R3) (Sect. 2.3) and (R4) (Sect. 2.4) will become clear later on.

The usage of the *best* approximation in step 3 of the loop is fundamental to our approach. Consider a set of approximation functions $\mathbb{F}$ and the best approximation $f \in \mathbb{F}$ that minimizes the error for a set of measurements $\mathcal{M}$. If the error is too large, there is only a limited number of *possibilities to improve the error* apart from changing $\mathbb{F}$:

(A1) Add parameters: Runtime behavior that cannot be explained by the current set of function input parameters may lead to large approximation errors.

(A2) Consider (meaningful) subsets of the measurements: Let $\mathcal{M}' \subseteq \mathcal{M}$ denote a subset of the set of measurements, and let $q_\mathcal{M}, q_{\mathcal{M}'}$ denote the maximum approximation error on $\mathcal{M}$ and $\mathcal{M}'$, respectively. Then, it is straightforward to see that $q_{\mathcal{M}'} \leq q_\mathcal{M}$.

(A3) Improve the approximability of the measurements: This boils down to improving the QEE.

This limited number of possibilities helps in drawing conclusions from errors of best approximations.

To exemplify our proposed loop, we focus on predicting the single-threaded execution time of single key/foreign key main memory hash joins under no load, similar to Cheng et al. [4]. To ensure *extensive experiments* as stated in step 2 of *the loop*, relation and attribute domain sizes for our experiments vary between 1 and 2^{30}, and attribute values are generated with and without skew.

We make the following contributions: We present necessary fundamentals on error metrics, plan quality, and best approximations, and consider error propagation through cost functions (Sect. 2). Next, we illustrate the loop (Sect. 3): We first consider the best approximations for the standard hash join, and, as one improvement, for the 3D hash join [5]. Our analysis factors in the influence of certain cost function parameters as well as possible QEE improvements onto the precision of the best approximation. Furthermore, we argue how cost-based optimization can be designed more effectively and efficiently based on our results (Sect. 4). We also review other related work and compare it to ours (Sect. 5). Finally, Sect. 6 concludes the paper.

2   Fundamentals

This section introduces fundamentals regarding the q-error as the error metric of choice, relationships between errors and plan quality, methods to obtain the best approximation under the q-error, and the propagation of cardinality estimation errors through cost functions.

Algorithm 1. Create Join Tree Procedure [17, p. 62]

```
 1: function CREATEJOINTREE(T_1, T_2)
 2:     Input: two join trees T_1, T_2
 3:     Output: the best join tree for joining T_1 and T_2
 4:     BestTree ← null
 5:     for each impl ∈ Implementations do
 6:         Tree ← T_1 ⋈^impl T_2
 7:         if BestTree = null ∨ COST(BestTree) > COST(Tree) then
 8:             BestTree ← Tree
 9:         end if
10:         Tree ← T_2 ⋈^impl T_1
11:         if COST(BestTree) > COST(Tree) then
12:             BestTree ← Tree
13:         end if
14:     end for
15:     return BestTree
16: end function
```

2.1 The Big Picture: Where Cost Functions Live

Cost functions are called from the innermost loop of the query optimizer. To generate optimal join trees for a query, a dynamic programming join ordering algorithm like *DPccp* [18] can be used. It must iterate all *csg-cmp-pairs* (connected subgraphs of the query graph and their complements) in a bottom-up order valid for dynamic programming. For each csg-cmp-pair, the CREATEJOINTREE procedure from Algorithm 1 [17, p. 62] is called with the two optimal join trees associated with the pair. The number of csg-cmp-pairs depends both on the type of query graph and the number of relations, and grows fast [18,22], which justifies requirement (R2). Within CREATEJOINTREE, cost functions are repeatedly evaluated for different join implementations, and, in case of non-symmetric cost functions, for both possible commutations. In contrast to cost functions, cardinality estimation must only be called once per *plan class*, i.e., once per connected subgraph of the query graph.

2.2 Error Metrics and Plan Quality

To evaluate the quality of a cost function, we require an error metric that gives a notion of deviation between the true (measured) cost and the estimated cost given by the cost function. A popular error metric in the database context for both cardinality and cost estimations is the *q-error* [19]. Let $x > 0$ be some true value and $\hat{x} > 0$ be its estimate. Then, $\text{qerr}(x, \hat{x}) := \max\left(\frac{x}{\hat{x}}, \frac{\hat{x}}{x}\right)$, the factor by which the estimate deviates from the true value. The q-error is a relative error measure that is symmetric for both under- and overestimates, e.g., $\text{qerr}(0.5, 2) = \text{qerr}(2, 0.5) = 4$, and has a minimum of 1. If $q := \text{qerr}(x, \hat{x})$, then it holds that $(1/q)x \leq \hat{x} \leq qx$.

The following relationship between the estimation error of the cost function and the quality of the plan chosen by the optimizer justifies the use of the q-error: If the query optimizer chooses a plan based on the estimated cost given by a cost function with an estimation error of at most q, the query optimizer might choose a suboptimal plan. However, this plan can be at most a factor of q^2 from the optimal plan w.r.t. the true cost. This motivates why we should always strive for a cost function with a small maximum estimation error. The following theorem captures this formally [11]. Let $\mathcal{P} := \{P_1, ..., P_n\}$ be a set of query evaluation plans, e.g., those for a query Q from which the optimizer can choose.

Theorem 1. *Let $f(P_i)$ and $\hat{f}(P_i)$ denote the true and the estimated cost for $P_i \in \mathcal{P}$, respectively. The optimal plan $P_{opt} \in \mathcal{P}$ minimizes $f(P)$, whereas the query optimizer chooses $P_{best} \in \mathcal{P}$, which minimizes $\hat{f}(P)$.*
If, for all $P_i \in \mathcal{P}$,

$$\mathrm{qerr}(f(P_i), \hat{f}(P_i)) \leq q$$

for some $q \geq 1$, then

$$\mathrm{qerr}(f(P_{best}), f(P_{opt})) \leq q^2.$$

Observe that the opposite is not true: large errors in the cost function do not necessarily imply bad plans. This fact is sometimes misunderstood. To clarify, we present the following new theorem:

Theorem 2. *Let $f(P_i)$ denote the true cost for $P_i \in \mathcal{P}$, and let $P_{opt} \in \mathcal{P}$ be the optimal plan chosen under cost function f. Let $g : \mathbb{R} \to \mathbb{R}$ be a strictly increasing function, and let $\hat{f} := g \circ f$. Let $P_{best} \in \mathcal{P}$ be the plan chosen under $\hat{f}$. Then $f(P_{opt}) = f(P_{best})$.*

Observe that the error of $\hat{f}$ is unbounded. A simple example would be $g(x) = ax$ ($a > 0$), which results in $\mathrm{qerr}(f(P), \hat{f}(P)) = a$ for any $P \in \mathcal{P}$. However, the optimizer still chooses a plan with minimal true cost under $\hat{f}$. This is due to the fact that the total order on $\mathcal{P}$ implied by f is maintained by $\hat{f}$, i.e., $f(P) \leq f(P') \Leftrightarrow \hat{f}(P) \leq \hat{f}(P')$ for any two plans $P, P' \in \mathcal{P}$.

2.3   Approximation Toolbox

Notation and Best Approximation. We will first fix some notation and formally define the problem of finding the best approximation for a set of points.

Given a set of pairs $\mathcal{M} = \{(\boldsymbol{x}_i, y_i) \mid 1 \leq i \leq m, \; \boldsymbol{x}_i \in \mathbb{R}^n, \; y_i \in \mathbb{R}\}$ and a set of functions $\mathbb{F}$, we want to find a function $\hat{f}(\boldsymbol{x}) \in \mathbb{F}$ that minimizes $\max_{1 \leq i \leq m} \mathrm{qerr}(y_i, \hat{f}(\boldsymbol{x}_i))$. We call the problem one-dimensional if $n = 1$, and multidimensional otherwise. In our cost function scenario, $\boldsymbol{x}_i$ is a vector of input parameter values (e.g., cardinalities), and y_i is a scalar measured cost value. Let us denote by f the true cost function with $f(\boldsymbol{x}_i) = y_i$, whereas $\hat{f}$ is the approximate cost function. Note that the true cost function f is typically unknown and can only be measured experimentally at a finite number of discrete points.

We restrict ourselves to a subset $\mathbb{F}$ of all possible functions. Let $\mathbb{P}_k, k \in \mathbb{N}_0$, denote the set of all polynomials of degree k, i.e.,

$$p(\boldsymbol{x}) = p(x_1, ..., x_n) = \sum_{\substack{i_1,...,i_n \\ i_1+...+i_n \leq k}} a_{i_1,...,i_n} \cdot x_1^{i_1} \cdot ... \cdot x_n^{i_n}. \tag{1}$$

Let $\mathbb{E}_k := \{e^{p(\boldsymbol{x})} \mid p(\boldsymbol{x}) \in \mathbb{P}_k\}$. Then, we consider $\hat{f}$ from $\mathbb{F} := \bigcup_k \mathbb{F}_k$ where $\mathbb{F}_k := \mathbb{P}_k \cup \mathbb{E}_k$. Note that a polynomial of degree k in n variables has $\binom{n+k}{k}$ monomials. This motivates requirement (R3) from Sect. 1.

Approximation Algorithms. There exist methods to find the best approximation minimizing the maximum q-error using linear, polynomial, or exponential functions for the one- and multidimensional case [19,24]. The best approximation can be found by either using an iterative algorithm [19] for the one-dimensional case or by reducing the problem of finding the best approximation to a Second Order Cone Programming (SOCP) problem and using a solver like the MOSEK software package [20,24] in the multidimensional case. We implemented both approaches to obtain the best approximations. Our implementations not only return the best approximation's maximum q-error, but also generate the resulting cost functions as C-functions. To the best of our knowledge, we are the first to consider best approximations under the q-error for the construction of cost functions.

2.4 Error Propagation

We now consider error propagation through cost functions, as captured by requirement (R4). This topic has been neglected in the literature thus far. This is different from the error propagation through the operators of a query plan [10]. To this end, we must first distinguish between errors in the *cost function itself* (f vs. $\hat{f}$) and errors in the *inputs* to the cost function. The latter can originate, e.g., from imprecise cardinality estimates. Let us denote the true, error-free inputs by $\boldsymbol{x}$ and the estimates by $\hat{\boldsymbol{x}}$. This results in four combinations: true costs on true inputs $f(\boldsymbol{x})$; true costs on estimated inputs $f(\hat{\boldsymbol{x}})$; approximate costs on true inputs $\hat{f}(\boldsymbol{x})$; approximate costs on estimated inputs $\hat{f}(\hat{\boldsymbol{x}})$. The latter is reality and needs to be compared to the true $f(\boldsymbol{x})$ to apply Theorem 1.

Let us consider how errors from the input parameters propagate through a polynomial cost function depending on the degree. Let $\boldsymbol{x} = (x_1, ..., x_n)$, $\hat{\boldsymbol{x}} = (\hat{x}_1, ..., \hat{x}_n) \in \mathbb{R}^n$ be the true and estimated input values (e.g., cardinalities) to a cost function $f \in \mathbb{P}_k$ of degree k. Fix the input parameter with the largest q-error: $s := \arg\max_{1 \leq i \leq n} \mathrm{qerr}(x_i, \hat{x}_i)$, $q_s := \mathrm{qerr}(x_s, \hat{x}_s)$. Observe that f contains a monomial of the form $a\hat{x}_s^k$ with all other exponents being 0. Since $\hat{x}_s$ is bounded by $[(1/q_s)x_s, q_s x_s]$, $\hat{x}_s^k$ is bounded by $[(1/q_s)^k x_s^k, q_s^k x_s^k]$, and further,

$\mathrm{qerr}(x_s^k, \hat{x}_s^k) = (\mathrm{qerr}(x_s, \hat{x}_s))^k = q_s^k$. All other monomials have q-errors of at most q_s^k. Furthermore, the sum of all monomials, i.e., the whole polynomial, still has a q-error bounded by q_s^k [17, p. 481]. The same bound holds for $\mathbb{E}_k$ and log-transformed inputs. As a consequence, higher-degree polynomials amplify the input q-error more severely than lower-degree polynomials, making lower degrees desirable. Further, considering (additional) inputs places more burden on the cardinality estimation component and potentially introduces additional errors.

Next, consider the error introduced by the cost function itself, i.e., $\hat{f}$ in contrast to f, denoted by $q_{\hat{f}}$. Then, the overall error of $\hat{f}(\hat{x})$ is bounded by $q_s^k \cdot q_{\hat{f}}$. Hence, all of $q_s, q_{\hat{f}}$, n and k should be as small as possible to minimize the overall error.

3   Q-Errors of Best Approximations

Next, we illustrate *the loop*. Therefore, we analyze the q-errors of the best approximations of our extensive experiments (see Appendix). Since we consider *best* approximations, the resulting q-errors can be used to determine cases where improvements to the QEE are a necessity. We begin our analysis with the standard chaining hash join (*CH-join* for short) and propose optimizations applicable to it to improve upon its approximability. As a major QEE improvement (the implementation of a new hash join), we analyze the 3D hash join (*3D-join* for short) [5]. Both hash join implementations resolve collisions by chaining. The CH-join uses flat collision chains, such that build attribute values colliding under the hash function are inserted at the head of a linked list. In contrast, the 3D-join uses hierarchical collision chains which group the same build attribute values and store them together. This has been shown to work particularly well for build attributes containing duplicates [5].

In the first subsection, we introduce the possible input parameters for our cost functions. We then analyze the CH-join. Starting with the complete CH-join, we identify cases in which the approximation yields errors deemed too large. Thereby, we consider q-errors of at most 2 to be acceptable, since Theorem 1 tells us that this yields a maximum loss factor in plan quality of 4. Further, cardinality estimates, which exhibit much larger q-errors, are used as inputs to the cost functions. Thus, adding more precision to the cost functions most probably does not help.

We reduce the approximation error by either (1) improving the approximation (by A1, A2) in Sect. 3.3, or (2) improving the QEE (A3) in Sect. 3.4. The overall goal is to have a cost function with a small error, small degree and as few input parameters as possible.

3.1   Parameters Considered

We divide the considered cost function input parameters into three groups, summarized in Table 1: (1) *Fundamental, compulsory parameters* are input (and, if

Table 1. Cost function input parameters.

Category	Input Parameters
(1) fundamental + compulsory	$c_{\mathrm{bld}}, c_{\mathrm{prb}}, c_{\mathrm{res}}$
(2) fundamental + optional	$nodv, c_{\mathrm{sj}}$
(3) derived + optional	mac

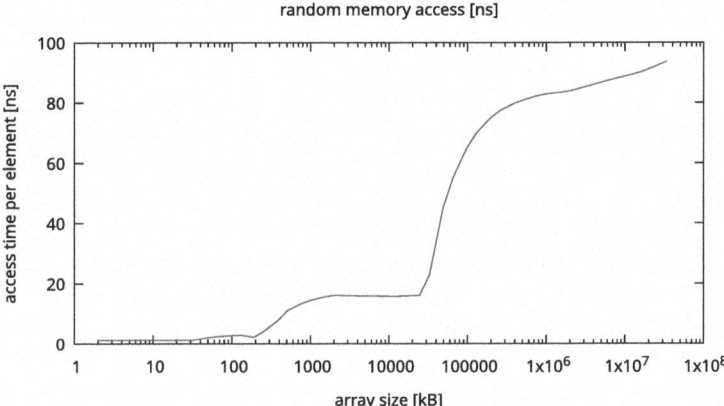

Fig. 1. Random memory access cost per element.

applicable) output cardinalities. Every approximation considers at least these inputs. We refer to them by $c_{\mathrm{bld}}, c_{\mathrm{prb}}, c_{\mathrm{res}}$ (build, probe, and result, respectively). They should usually be available from the query optimizer's cardinality estimation component. (2) *Fundamental, optional parameters* are used to improve the approximations by providing additional information that cannot be derived from compulsory parameters. The number of distinct values of the foreign key attribute $nodv$ and the semi-join cardinality c_{sj} of the probe side fall into this category. The latter denotes the number of probe tuples that find at least one match. These parameters are also part of the cost models in previous work [4]. (3) The third category are *derived parameters* that can be calculated from the fundamental parameters. They are always optional. They are often found in machine learning-based cost estimation approaches [27]. The only derived parameter we consider is mac, the cost of one random memory access, which depends on the memory consumption of the hash join. Figure 1 illustrates this non-linear, not easy to approximate relationship for an Intel Xeon E5-2690 v3 CPU. Note that in order to apply the functions in $\mathbb{E}_k$, all parameters must be log-transformed, e.g., c_{bld} becomes $\ln(c_{\mathrm{bld}})$.

Our cost functions output the wall-clock execution time of the complete join or a join phase (build, probe), or a number related to the execution time. In general, cost values can be arbitrary numerical values, but must be consistent across different join implementations to be comparable in Algorithm 1.

Table 2. Approximation q-errors for the complete CH-join.

Complete CH-Join			$\mathbb{E}_1$	$\mathbb{E}_2$	$\mathbb{E}_3$	$\mathbb{E}_4$	$\mathbb{E}_5$	$\mathbb{P}_1$	$\mathbb{P}_2$
$c_{\text{bld}}, c_{\text{prb}}, c_{\text{res}}$			72.19	24.70	24.41	23.78	23.66	39.49	33.49
$c_{\text{bld}}, c_{\text{prb}}, c_{\text{res}}$	$nodv$		69.72	21.02	20.47	20.47	20.47	39.39	—*
$c_{\text{bld}}, c_{\text{prb}}, c_{\text{res}}$		c_{sj}	37.47	2.74	2.21	2.17	2.17	36.98	—*
$c_{\text{bld}}, c_{\text{prb}}, c_{\text{res}}$	$nodv$	c_{sj}	35.70	2.69	1.94	1.66	1.60	36.89	—*

* Numerical overflow during approximation.

Recall the function classes $\mathbb{P}_k$ and $\mathbb{E}_k$, where k denotes the degree of the polynomial. In each of the following subsections, we will consider the best linear and quadratic approximations (i.e., cost functions from $\mathbb{P}_k, \mathbb{E}_k, k \in \{1, 2\}$) with input parameters limited to category (1) (fundamental compulsory). Afterwards, we will drop the restriction on the function's degree and consider $1 \leq k \leq 5$. Further, we will examine the impact of adding optional parameters from categories (2) and (3). This will show that the approximation quality usually increases with a higher degree and/or more input parameters. We refer to the number of input parameters as the *dimensionality* of the approximation. Recall that we would like to keep this number as small as possible while maintaining an approximation with small degree and low error. We show an interesting subset of the approximation q-errors for different cases in Tables 2 to 4. As none of the polynomial approximations with $k \geq 3$ yield usable results due to numerical overflows, we restrict the tables to $\mathbb{P}_1$ and $\mathbb{P}_2$.

3.2 The Complete CH-Join

We start by presenting the q-errors of the best approximations for the complete CH-join. By 'complete join', we mean the whole join comprising the build and the probe phase. The results are shown in Table 2. With only the fundamental compulsory parameters $c_{\text{bld}}, c_{\text{prb}}$, and c_{res} as inputs, the best linear approximation is from $\mathbb{P}_1$ and has a q-error of 39.5, the best quadratic approximation is from $\mathbb{E}_2$ with a q-error of 24.7. Both are clearly not satisfactory. The best approximation from $\mathbb{E}_5$ without additional parameters yields an error of 23.7. Obviously, cardinalities alone are not sufficient for a good approximation of the runtime of the complete CH-join.

Adding the number of distinct values as the fourth parameter reduces the error to 20.5 ($\mathbb{E}_3$), adding the semijoin size yields a q-error of 2.21 ($\mathbb{E}_3$). Adding both *nodv* and c_{sj} yields errors of 1.94, 1.66 and 1.60 (exponential with degrees 3, 4 and 5, respectively). The latter are below 2 and perfectly acceptable only if the parameters can be estimated well and error propagation (cf. Sect. 2.4) is not a concern. Next, we are interested in finding a good approximation without *nodv* or c_{sj} and of lower degree by dissecting the CH-join into sub-cases.

Table 3. Approximation q-errors for the CH-join, dissected by the uniqueness of the build attribute.

CH-Join				$\mathbb{E}_1$	$\mathbb{E}_2$	$\mathbb{E}_3$	$\mathbb{E}_4$	$\mathbb{E}_5$	$\mathbb{P}_1$	$\mathbb{P}_2$
build unique										
$c_{bld}, c_{prb}, c_{res}$				33.97	2.47	2.20	2.17	2.17	2.65	2.17
$c_{bld}, c_{prb}, c_{res}$	*nodv*			33.97	2.42	1.78	1.63	1.55	2.47	1.90
$c_{bld}, c_{prb}, c_{res}$		c_{sj}		33.97	2.47	2.20	2.17	2.17	2.65	2.17
$c_{bld}, c_{prb}, c_{res}$	*nodv*	c_{sj}		3.90	2.25	1.73	1.66	1.60	36.55	27.71
$c_{bld}, c_{prb}, c_{res}$	*nodv*		*mac*	33.46	2.24	1.62	1.54	1.51	2.43	1.83
build non-unique										
$c_{bld}, c_{prb}, c_{res}$				46.99	21.21	19.95	19.62	18.84	37.38	28.31
$c_{bld}, c_{prb}, c_{res}$	*nodv*			25.20	2.76	1.76	1.66	1.60	37.18	28.23
$c_{bld}, c_{prb}, c_{res}$		c_{sj}		3.97	2.54	1.74	1.66	1.60	36.55	27.67
$c_{bld}, c_{prb}, c_{res}$	*nodv*	c_{sj}	*mac*	2.84	1.77	1.33	1.24	1.24	35.99	17.47

3.3   Dissecting the CH-Join

We subdivide the CH-join analysis based on the uniqueness of the build attributes and the join phases (build, probe).

Uniqueness of the Build Attribute. We did not limit our experiments to building the hash table for the key/foreign-key join on the unique key attribute. Instead, we also consider building on the foreign key attribute containing duplicates. Note that the uniqueness of the build attributes is inferrable, e.g., by applying the known functional dependencies from the SQL standard. This allows us to generate separate best approximations for the 'build unique' and the 'build non-unique' case, as summarized in Table 3.

Unsurprisingly, the *build unique* case exhibits better approximability. The best linear and quadratic approximation with no additional parameters are polynomials with an error of 2.65 and 2.17, respectively. Higher degrees or using exponential functions do not improve the errors. Considering additional parameters, we achieve an error of 1.78 with *nodv* and 1.62 with *nodv* and *mac* with a function from $\mathbb{E}_3$. Recall that this is an approximation for the complete join, so although we consider 'build unique' here, *nodv* is meaningful for the foreign-key probe side.

The *build non-unique* case is harder to approximate as it introduces more variability due to duplicate build keys leading to longer hash table collision chains. This is especially true for the measurements with high skew. The best linear and quadratic approximations without additional parameters both have errors above 20. Even the best approximation from $\mathbb{E}_5$ yields an error of still 18.8. Adding *nodv* or c_{sj} reduces the error to about 1.76 and 1.74 with degree 3, respectively, and 1.77 when combining both parameters and *mac* (degree 2).

Join Phases. An obvious possibility to dissect the CH-join (or any hash join, for that matter) further, is to separately analyze the two join phases build and probe. In this section, we consider separate cost functions for them. In most cases, it is sufficient to have a cost function for the complete join without subdividing it into phases, as there is usually one build and one probe, and their costs sum up, even over multiple joins. One case where the cost of the individual phases is of interest is when a hash table can be beneficially reused, as, for example, for TPC-H Q9.

Starting with *build*, we find that the best linear and quadratic approximations for all builds using c_{bld} are never below 2.30, even if we add *mac*. We can further split the builds into unique and non-unique, as above. *Unique builds* can be approximated nicely by a quadratic function with an error of 1.64 or 1.13, without or with *mac*, respectively. Approximating *non-unique builds* using the same restrictions as for unique results in errors of 2.32 (without *mac*) and 2.19 (with *mac*).

For the *probe* phase, there are larger differences between 'unique' and 'non-unique'. A linear function using only $c_{\mathrm{bld}}, c_{\mathrm{prb}}, c_{\mathrm{res}}$, and *mac* approximates probing on a unique build side with an error of 2.65. In contrast, a non-unique build side cannot be approximated better than with an error of 18.7 under the same conditions with a quadratic function.

3.4   Improving the CH-Join/QEE

From the previous subsections, it became clear that duplicates on the build side are the weak spot of the CH-join in terms of how well the measurements can be approximated. Adding *nodv* and c_{sj} as parameters to cost functions is one remedy to decrease errors. However, these are not easy to estimate. Thus, instead of using them, we suggest to improve the query evaluation engine so that its runtime behavior is more amenable to approximation. To that end, we suggest three optimizations: prefetching, partitioning, and the 3D-join.

It is known that prefetching techniques like AMAC [12] improve the runtime of hash joins. Further, preliminary experiments indicate that using AMAC also improves the approximability by reducing the variability that memory accesses introduce if data does not fit into the cache.

It is also well known that partitioning can improve the runtime of main memory hash joins [15,26]. We thus analyze the best approximations if we limit the cardinalities of both join relations to a maximum value, which corresponds to the maximum partition size. We call this case *lim*, and choose 2^{23} tuples as the maximum partition size. We also introduce *mil* as an interesting opposite case to 'lim', where either join argument has a cardinality of 2^{22} or above[1].

Unfortunately, the best approximation for 'lim' without additional parameters still has an error of over 20. Again, additional parameters like c_{sj} can push the error below 2.

---

[1] 'Lim' and 'mil' overlap to allow for a smooth transition from 'lim' to 'mil' in the overall cost function.

Table 4. Approximation q-errors for the 3D-join for build non-unique.

3D-Join *build non-unique*			$\mathbb{E}_1$	$\mathbb{E}_2$	$\mathbb{E}_3$	$\mathbb{E}_4$	$\mathbb{E}_5$	$\mathbb{P}_1$	$\mathbb{P}_2$
$c_{bld}, c_{prb}, c_{res}$			34.11	2.63	2.32	1.83	1.71	3.06	2.52
$c_{bld}, c_{prb}, c_{res}$	*nodv*		31.75	2.49	1.86	1.56	1.38	3.01	2.43
$c_{bld}, c_{prb}, c_{res}$		c_{sj}	28.74	2.49	1.80	1.56	1.38	2.97	2.40
$c_{bld}, c_{prb}, c_{res}$	*nodv*	c_{sj}	11.62	2.49	1.75	1.56	1.38	2.97	2.40
$c_{bld}, c_{prb}, c_{res}$	*nodv*	c_{sj} *mac*	11.59	2.29	1.51	1.38	1.38	2.70	1.74

We can also combine the uniqueness of the build key from Sect. 3.3 with limiting the partition size. Then, linear and quadratic approximations for the 'build unique' case with only the fundamental compulsory parameters have errors as low as 1.67 and 1.51, respectively.

Contrarily, for the problematic case 'build non-unique' (Sect. 3.3), the gains by partitioning joins are insufficient to push the q-error below 2: The best approximation without additional parameters has an error of 17.3. It decreases to 1.43 when using *nodv* and c_{sj}. As these are difficult to estimate, we investigate the 3D-join [5], which has been shown to be superior to the CH-join for non-unique builds, as a third possible QEE improvement in the next subsection.

3.5   Best Approximations for the 3D-Join

This subsection addresses cost function approximations for the 3D-join as a possible improvement for the CH-join in certain cases. We will describe an interesting subset of the approximations, focusing on the cases where the CH-join cannot be approximated well.

The complete 3D-join can be approximated with an error of 3.06, 2.52, and 2.06 without extra parameters for degrees 1, 2, and 5, which is a lot better than the same case for the CH-join, but still not satisfactory. If we restrict ourselves to the Achilles' heel of the CH-join, 'build on non-unique', where the build side contains duplicates while still considering the complete join, the approximation error remains the same for degrees 1 and 2, but improves to 1.83 and 1.71 for degrees 4 and 5 (see also Table 4). Recall that the same conditions could only be approximated with an error of 18.8 for the CH-join. If we add *mac*, we achieve an error 1.98 with a quadratic function. If we use partitioning joins, a quadratic approximation without additional parameters approximates 'lim' with an error of 1.59 and 'mil' with 1.92.

3.6   Take Home Message

In summary, we found that QEE improvements have a positive effect on both the runtime and the approximation quality. In other words: The query runs faster *and* the cost function becomes more precise. As a take home message, we formulate that QEE development and cost function development must go hand

Fig. 2. Plans considered by CREATEJOINTREE. The key relation is <u>underlined</u>. p is an equality join predicate.

in hand. Further, we observe that our set of functions $\mathbb{F}$ seems to be suitable to approximate our measurements well. Otherwise it would note have been possible to achieve an improvement by increasing the degree or the number of parameters.

4 Smart Cost-Based Optimization

Figure 2 shows the four plans enumerated by Algorithm 1 during plan generation, where underlining marks the key relation R. We now briefly investigate if Algorithm 1 can be modified to reduce the number of plans considered. Our extensive experiments allow us to calculate

$$\max_{(R,S)} \frac{\min(f(P_b(R,S)), f(P_c(R,S)))}{\min(f(P_a(R,S)), f(P_d(R,S)))} \tag{2}$$

with quite some certainty, where $f(P_x(R,S))$ denotes the measured execution time of plan P_x evaluated for some instances R and S. The result of this term is 1.063. Thus, if we restrict the plans considered by CREATEJOINTREE to plans $P_b \equiv \left(S \bowtie_p^{\text{ch}} \underline{R}\right)$ and $P_c \equiv \left(\underline{R} \bowtie_p^{\text{3d}} S\right)$, we lose at most 6.3% in execution time by not considering the other two plans. Note that this number is smaller than the q-error of any best approximation from Sect. 3, especially if taking into account that the cardinality estimates which are used as inputs to the cost functions typically contain errors. Thus, we can safely prune two plans without looking at their costs. As a side effect, we additionally save two cost function evaluations.

5 Related Work

We focus on cost functions developed to predict the execution time for joins. They are used to predict a join's I/O time (for disks [4,6,7,25] or SSDs [1,2,23]), CPU time [4,21,29,30] and/or main memory access time [13,14,29,30].

All approaches fall into two broad categories. The first category includes approaches where explicit cost formulas are designed. They contain coefficients that need to be calibrated. This step is non-trivial, and there even exist papers which solely concentrate on calibration [31]. Papers in the second category do not come up with explicitly designed cost formulas, neither do they need any calibration. Instead, they combine steps (2) and (3) of the traditional process (Sect. 1) by directly deriving some approximation of a set of measurements. This has been achieved by using multivariate regression [32] or learning techniques [8,16,27,28]. Most of the existing work falls into the first category. Clearly, our work falls into the second category. In the following, we pick a few typical papers from each category and present them as examples.

Manegold et al. [14] provide a comprehensive set of formulas to model main memory access costs. Their cost function is inherently non-linear. As parameters, the sizes of the input and output relations are considered, but neither the semijoin size nor the number of distinct values. Unfortunately, the evaluation is underspecified and rather weak. They consider only joins where the two input relations and the output relation are of equal size and in the range of 128 kB to 128 MB. No more details are given. Further, the reader has to extract the errors from a log-scaled graph. Visually, the q-error of their model looks quite high for cases where the relations consume less than 1 MB.

The paper by Siddiqui et al. [27] falls into the second category. From an observed workload, they extract common subexpressions to derive several possible models and a meta-model to choose between them. As input parameters they consider the input and output sizes of each operator as well as a set of 22 derived parameters, but neither the semijoin size nor the number of distinct values. The first level of derivation is to use square roots and logarithms of the parameters. Then, the resulting terms are combined using multiplication and/or division to calculate all 22 derived parameters. In the evaluation, they report the median squared-log-error $\ln\left(\frac{\hat{c}+1}{c+1}\right)$, where c is the true cost and $\hat{c}$ is the estimate cost. The result is 19%. It is unclear what conclusions can be drawn from this number.

Cheng et al. [4] propose a cost model based on CPU and disk I/O costs to predict the running time of three join implementations. Notably, their model has 43 parameters out of which more than half are hardware and system constants that need to be calibrated. The data- and query-dependent input parameters include input and output cardinalities, selectivities, distinct value counts, and semijoin sizes. Though their formulas include basic cardinality estimation, e.g., multiplying cardinalities and selectivities, the cost functions themselves are linear. Their validating experiments are limited to a single 1:1 join of two relations with 100 K and 1 M tuples, where join attribute values are sampled uniformly from a domain with 10 M values. No cost function estimation errors are given. Instead, it is only performed visually on the runtime plots.

Zhu and Larson [32] present a statistical approach to create linear cost models using multivariate linear regression for single-table filter and two-table join queries. They aim to predict query execution time (consisting of CPU and disk I/O time) while minimizing the L_2 norm between true and estimated costs. Similar to our work, they divide their parameters into basic and additional ones. Basic parameters are input, intermediate and result cardinalities, and additional ones are, e.g., the physical tuple size. They neither consider c_{sj} nor $nodv$. Using an iterative, rule-based approach, they start with a basic model and try to add and remove parameters to find the best regression model w.r.t. the standard error. Their experiments are, however, not well described. The number of experiments, the size of the join inputs and the output, the data distribution, and the join predicates remain unclear. They report the quality of their regression models for different query classes in terms of the standard error and the coefficient of multiple determination. Notably, the final cost function only contains a subset of the basic parameters and no additional parameters.

6   Conclusion

We investigated the approximability of the execution times of two hash join implementations to generate cost functions for them. Our analysis suggests that cost functions with small estimation error, small degree, and few input parameters are desirable. Further, we draw the conclusion that QEE development and cost function development must go hand in hand to improve both runtime and cost function accuracy. Since our approach is very general, we plan to examine the approximability of other algebraic operators and parallel variants as future work. Further, we intend to investigate how imprecise cost function inputs and cost function inaccuracy influence plan quality in practice when used in a plan generator.

Acknowledgments. We would like to thank Simone Kehrberg for proofreading the paper, and Nazanin Rashedi for helpful comments and discussions. We would also like to thank the anonymous reviewers for their suggestions to improve the paper.

Appendix: Experimental Setup

This section describes the experimental setup. For the implementation details for the two join methods and the hardware specification see [5].

We consider a single key/foreign key join between a key relation R and a foreign key relation S. The build cardinalities are varied between 2^0 and 2^{30} in steps of powers of 2. As probe cardinalities, we choose 2^{10}, 2^{15}, 2^{20}, and 2^{25}. To generate the foreign keys, we draw $|S|$ random samples according to a (1) uniform and (2) standard Zipf distribution from a domain $[1, |S|/2^d]$ ($0 \leq d \leq 10$). We evaluate the four plans shown in Fig. 2 and measure their runtimes for build and probe in terms of timestamp counter clock ticks, a proxy for the wall-clock time [9, Chap. 18.17].

References

1. Bausch, D., Petrov, I., Buchmann, A.: On the performance of database query processing algorithms on flash solid state disks. In: Proceedings of the 22nd International Workshop on Database and Expert Systems Applications (DEXA), pp. 139–144 (2011). https://doi.org/10.1109/DEXA.2011.60
2. Bausch, D., Petrov, I., Buchmann, A.: Making cost-based query optimization asymmetry-aware. In: Proceedings of the 8th International Workshop on Data Management on New Hardware (DaMoN), pp. 24–32 (2012). https://doi.org/10.1145/2236584.2236588

3. Chaudhuri, S.: An overview of query optimization in relational systems. In: Proceedings of the 17th ACM SIGACT-SIGMOD-SIGART Symposium on Principles of Database Systems (PODS), pp. 34–43 (1998). https://doi.org/10.1145/275487.275492

4. Cheng, J., et al.: An efficient hybrid join algorithm: a DB2 prototype. In: Proceedings of the 7th International Conference on Data Engineering (ICDE), pp. 171–180 (1991). https://doi.org/10.1109/ICDE.1991.131464

5. Flachs, D., Müller, M., Moerkotte, G.: The 3D hash join: building on non-unique join attributes. In: Proceedings of the 12th Conference on Innovative Data Systems Research (CIDR) (2022). https://www.cidrdb.org/cidr2022/papers/p18-flachs.pdf

6. Haas, L., Carey, M., Livny, M., Shukla, A.: Seeking the truth about ad hoc join costs. VLDB J. **6**(3), 241–256 (1997). https://doi.org/10.1007/S007780050043

7. Harris, E., Ramamohanarao, K.: Join algorithm costs revisited. VLDB J. **5**(1), 64–84 (1996). https://doi.org/10.1007/S007780050016

8. Hilprecht, B., Binnig, C.: Zero-shot cost models for out-of-the-box learned cost prediction. Proc. VLDB Endowment **15**(11), 2361–2374 (2022). https://doi.org/10.14778/3551793.3551799

9. Intel Corporation: Intel 64 and IA-32 Architectures Software Developer Manual, vol. 3B, June 2024. https://www.intel.com/content/www/us/en/developer/articles/technical/intel-sdm.html

10. Ioannidis, Y., Christodoulakis, S.: On the propagation of errors in the size of join results. In: Proceedings of the 1991 ACM SIGMOD International Conference on Management of Data (SIGMOD), pp. 268–277 (1991). https://doi.org/10.1145/115790.115835

11. Kastrati, F., Moerkotte, G.: Optimization of conjunctive predicates for main memory column stores. Proc. VLDB Endowment **9**(12), 1125–1136 (2016). https://doi.org/10.14778/2994509.2994529

12. Kocberber, O., Falsafi, B., Grot, B.: Asynchronous memory access chaining. Proc. VLDB Endowment **9**(4), 252–263 (2015). https://doi.org/10.14778/2856318.2856321

13. Liu, F., Blanas, S.: Forecasting the cost of processing multi-join queries via hashing for main-memory databases. In: Proceedings of the 6th ACM Symposium on Cloud Computing (SoCC), pp. 153–166 (2015). https://doi.org/10.1145/2806777.2806944

14. Manegold, S., Boncz, P., Kersten, M.: Generic database cost models for hierarchical memory systems. In: Proceedings of the 28th International Conference on Very Large Databases (VLDB), pp. 191–202 (2002). https://doi.org/10.1016/B978-155860869-6/50025-1

15. Manegold, S., Boncz, P., Kersten, M.: Optimizing main-memory join on modern hardware. IEEE Trans. Knowl. Data Eng. **14**(4), 709–730 (2002). https://doi.org/10.1109/TKDE.2002.1019210

16. Marcus, R., Papaemmanouil, O.: Plan-structured deep neural network models for query performance prediction. Proc. VLDB Endowment **12**(11), 1733–1746 (2019). https://doi.org/10.14778/3342263.3342646

17. Moerkotte, G.: Building Query Compilers. Technical Report, University of Mannheim (2023). https://pi3.informatik.uni-mannheim.de/~moer/querycompiler.pdf

18. Moerkotte, G., Neumann, T.: Analysis of two existing and one new dynamic programming algorithm for the generation of optimal bushy join trees without cross products. In: Proceedings of the 32nd International Conference on Very Large Data Bases (VLDB), pp. 930–941 (2006). https://dl.acm.org/citation.cfm?id=1164207

19. Moerkotte, G., Neumann, T., Steidl, G.: Preventing bad plans by bounding the impact of cardinality estimation errors. Proc. VLDB Endowment **2**(1), 982–993 (2009). https://doi.org/10.14778/1687627.1687738

20. Mosek ApS: Mosek (2023). https://www.mosek.com/

21. Nam, Y.M., Han, D., Kim, M.S.: SPRINTER: a fast n-ary join query processing method for complex OLAP queries. In: Proceedings of the ACM International Conference on Management of Data (SIGMOD), pp. 2055–2070 (2020). https://doi.org/10.1145/3318464.3380565

22. Ono, K., Lohman, G.: Measuring the complexity of join enumeration in query optimization. In: Proceedings of the 16th International Conference on Very Large Databases (VLDB), pp. 314–325 (1990). https://www.vldb.org/conf/1990/P314.PDF

23. Papon, T., Athanassoulis, M.: A parametric I/O model for modern storage disks. In: Proceedings of the 17th International Workshop on Data Management on New Hardware (DaMoN) (2021). https://doi.org/10.1145/3465998.3466003

24. Setzer, S., Steidl, G., Teuber, T., Moerkotte, G.: Approximation related to quotient functionals. J. Approx. Theory **162**(3), 545–558 (2010). https://doi.org/10.1016/J.JAT.2009.08.009

25. Shapiro, L.D.: Join processing in database systems with large main memories. ACM Trans. Database Syst. **11**(3), 239–264 (1986). https://doi.org/10.1145/6314.6315

26. Shatdal, A., Kant, C., Naughton, J.F.: Cache conscious algorithms for relational query processing. Technical Report, University of Wisconsin-Madison Department of Computer Sciences (1994). https://minds.wisconsin.edu/bitstream/handle/1793/59896/TR1234.pdf?sequence=1

27. Siddiqui, T., Jindal, A., Qiao, S., Patel, H., Le, W.: Cost models for big data query processing: learning, retrofitting, and our findings. In: Proceedings of the ACM International Conference on Management of Data (SIGMOD), pp. 99–113 (2020). https://doi.org/10.1145/3318464.3380584

28. Sun, J. Li, G.: An end-to-end learning-based cost estimator. Proc. VLDB Endowment **13**(3), 307–319 (2019). https://doi.org/10.14778/3368289.3368296

29. Tanaka, T., Ishikawa, H.: Measurement-based cost estimation method for multi-table join operation in an in-memory database. Int. J. Adv. Softw. **10**(3&4), 459–476 (2017)

30. Tanaka, T., Ishikawa, H.: Measurement-based cost calculation method focusing on CPU architecture for database query optimization. In: Proceedings of the 11th International Conference on Management of Digital EcoSystems (MEDES), pp. 56–65 (2020). https://doi.org/10.1145/3297662.3365823

31. Wu, W., Chi, Y., Zhu, S., Tatemura, J., Hacigümüs, H., Naughton, J.F.: Predicting query execution time: are optimizer cost models really unusable? In: Proceedings of the 29th IEEE International Conference on Data Engineering (ICDE), pp. 1081–1092 (2013). https://doi.org/10.1109/ICDE.2013.6544899

32. Zhu, Q., Larson, P.A.: Building regression cost models for multidatabase systems. In: Proceedings of the 4th International Conference on Parallel and Distributed Information Systems (ICPADS), pp. 220–231 (1996). https://doi.org/10.1109/PDIS.1996.568684

DGGS-Based Continuous Trajectory Similarity Comparison

Taehoon Kim[ID], Wijae Cho, and Kyoung-Sook Kim[✉][ID]

Artificial Intelligence Research Center, National Institute of Advanced Industrial
Science and Technology (AIST), Tokyo, Japan
{kim.taehoon,cho-wijae,ks.kim}@aist.go.jp

Abstract. Moving objects like pedestrians and vehicles change their
state and location over time. Recent advances in sensors, such as Lidar
and RGB-D sensors, have made it easier to track their movement in real-
time. In the real world, multiple sensors individually monitor a specific
area and send data streams about moving objects' locations to a cloud
server. However, the areas covered by independent sensors typically overlap, resulting in duplicate location information for moving objects. To
understand the trajectory of each object, the server needs to find these
clusters of overlapping locations and incorporate them into the trajectory for the time window of the streaming process, i.e., distance-based
self-join. This paper proposes an efficient approach to integrating location stream data into trajectory data using a discrete global grid system.
Geocode-based matching can reduce the cost of distance computation
among the stream data. In experimental results, our method outperforms distance-based processing for clustering by a geocode-based index
used to remove redundant trajectory information.

Keywords: Geocode · DGGS · Stream process · Trajectory stream

1 Introduction

In the recent era of the Internet of Things (IoT), real-time sensor data continuously stream from vast IoT sources, providing valuable insights into the dynamic
state of various systems and environments. In particular, with the widespread
use of location-aware sensors, it is now possible to collect real-time trajectories
of moving objects as a geospatial data stream. The real-time trajectory stream
can be utilized for various applications, including object tracking, geofencing,
hotspot analysis, etc. However, the vast volume and velocity of the geospatial
stream data pose significant challenges to traditional spatial data processing
systems [1]:

- Existing spatial data processing frameworks are not scalable to handle large-
  scale data processing,
- Scalable spatial data processing frameworks lack support for streaming processing,

R. Chbeir et al. (Eds.): IDEAS 2024, LNCS 15511, pp. 251–264, 2025.
https://doi.org/10.1007/978-3-031-83472-1_17

- Streaming Processing Engines (SPEs) such as Apache Flink [4], Apache Spark Structured Streaming [17], and Apache Kafka Streams [14] lack capabilities for spatial data processing.
- Several spatial stream processing frameworks with SPE, like Apache Sedona [16], SPEAR [2] and GeoFlink [12], do not support trajectory stream processing.

Trajectory stream processing is different from ordinary spatial stream processing. A trajectory stream is essentially the same as a point stream, where position information is represented as points over observed time. However, since the position of a moving object is continuous in the real world and is discretized into points for representation in the digital world, it is more appropriate to represent it as a trajectory. Since (spatial) stream data is unbounded, to represent it as a trajectory, we need to 1) collect data within a given time interval using the window operation, 2) classify by moving objects, 3) sort in chronological order, and 4) represent as a trajectory collection, which is called a trajectory stream.

TStream [11] is an extension of GeoFlink and a distributed, scalable and real-time trajectory SPE. However, there are several drawbacks:

- Only support local Cartesian coordinates (x,y), not global coordinate reference systems such as WGS84 [8], which is a widely used coordinate reference system as a form of latitude and longitude.
- To make geometric operations efficient, TStream uses a local grid index, which means that users always need to provide the information needed to create the grid index (spatial data stream boundaries and grid cell sizes) before the query request.

To address these issues, we propose a geocode-based approach to facilitate trajectory stream processing on top of existing SPEs. It should be noted that geocoding is a method of compressing and efficiently representing spatial information by encoding latitude and longitude coordinates into a single string, which is referred to as a geocode. Geocode is a unique identifier that is useful for spatial indexing and data searching. Depending on the length of the string (resolution), it can represent locations with varying degrees of precision. It also has hierarchical properties based on resolution. Therefore, geocode makes it a good fit to address the shortcomings of TStream:

- Convert from global coordinates to geocode and use it.
- Utilize the geocode itself as a spatial index.

This paper presents a method for creating a geocode-based trajectory stream with a proposal for effectively solving the problem in a particular situation: In the case of multiple sensors generating duplicate location information for the same mobile object.

2   Related Works

Comparing the similarity of trajectories is a traditional problem in GIS [13]. The methods that can be used depend on the specific problem setting, including 1) the dimensionality of the trajectory coordinates, 2) differences in temporal granularity, 3) staggered sampling times, and 4) whether the direction of the trajectory is considered. In this work, we propose a new problem that differs from previous studies in that it is a comparison between trajectory streams that are updated in real-time.

One of the recent works, QuadGridSIM [9], is similar to the method proposed in this paper; It uses code divided into arbitrary grids to represent trajectories and compares the similarity of trajectories at the code level. However, Quad-GridSIM is intended for comparison between static trajectory datasets and has the following drawbacks:

- It uses global coordinates (geodetic coordinate system) such as latitude-longitude coordinates, but when generating the grid, it simply divides the coordinates into specified grid levels as if using a Cartesian coordinate system. Therefore, it has the same problem as GeoHash that the size of the grid cell changes depending on the latitude.
- It has the same disadvantage as TStream [11] in that the user must manually set the query area or size, and all the code must be regenerated and calculated each time. This means that the code cannot be reused once it is generated, and there is a problem of inefficiency in regenerating it every time the query area changes.

Salman et al. [10] propose an effective way to perform continuous real-time trajectory similarity comparisons in a distributed environment using the minimum bounding box (MBB) of sub-trajectories and an R-tree for similarity comparisons between static trajectory datasets and trajectory streams.

While performing real-time trajectory similarity comparisons is similar to our work, [10] differs in that it performs comparison queries on a static trajectory dataset, while our work performs continuous real-time trajectory similarity comparisons between real-time trajectories.

3   Problem Definition

The present study assumes that sensors capable of detecting and tracking moving objects are capable of generating location information, represented by points on the moving object, at each sampling time.

Definition 1 (Point location p). *A moving object's point location is $p = (id,$* ***lon, lat, t)***, *where*

- ***id*** *is a unique identifier of a moving object,*
- ***lon*** *and* ***lat*** *represent the longitude and latitude, relatively,*
- ***t*** *means data creation timestamp.*

Definition 2 (Trajectory T). *A trajectory T of a moving object is a sequence of discrete points $\{p_1, p_2, \ldots, p_n\}$ in ascending order by their timestamp, where $p_i.id = p_j.id$ and $p_i.t < p_{i+1}.t$ when $1 \leq i, j \leq n$.*

Let $\mathbb{T} = \{T_1, T_2, \ldots\}$ denotes an unbounded set of trajectories, then a trajectory stream can be defined as:

Definition 3 (Trajectory stream $S_\mathbb{T}$). *A trajectory stream $S_\mathbb{T}$ is a stream of points generated continuously from multiple moving objects in ascending order by their timestamp.*

$S_\mathbb{T}$ essentially contains trajectory points. Therefore, it can also be called a point stream. However, in contrast to an ordinary point stream where no share id, point in $S_\mathbb{T}$ share id and can be converted into sub-trajectories by windowing and groupBy. Note that windowing is a critical concept in stream processing, as it allows data to be processed in small, manageable chunks over a specified period w.

Definition 4. *Window-based sub-trajectory* T^w *A sub-trajectory $T^w \in S_\mathbb{T}$ consists of a sequence of discrete point locations $\{p_1, p_2, \ldots, p_n\}$ in ascending order by their timestamp, where*

- *$p_i.id = p_j.id$ when $1 \leq i, j \leq n$,*
- *$w_s \leq p_i.t < p_{i+1}.t \leq w_e$ and w is window period $[w_s, w_e]$*

Problem 1. **Find duplicate sub-trajectories:** For all sub-trajectories $\mathbb{T}^w$ within a given time interval w and a trajectory stream $S_\mathbb{T}$, find pairs of sub-trajectories (T_i^w, T_j^w) whose trajectories are nearly identical in shape, i.e., the distance between two trajectories is within a set threshold δ.

The problem situation is assumed to be as follows: As shown in Fig. 1, multiple sensors have overlapping capture ranges, which can cause them to recognize the same moving object as separate moving objects. To address this issue, we compare the similarity between the trajectories of moving objects to find pairs of trajectories with high similarity. The resulting pairs can then be used to transform the separate moving objects into the same moving object.

The Dynamic Time Warping (DTW) algorithm [3] is fundamentally used to compare the similarity of trajectories, but DTW requires computing the distance between the points that make up the trajectory, which is computationally expensive. Note that for two trajectories that have a number of points N and M, the time complexity of the DTW algorithm can be presented as $O(N \times M)$. However, these computations need to be handled more efficiently in systems that require real-time response.

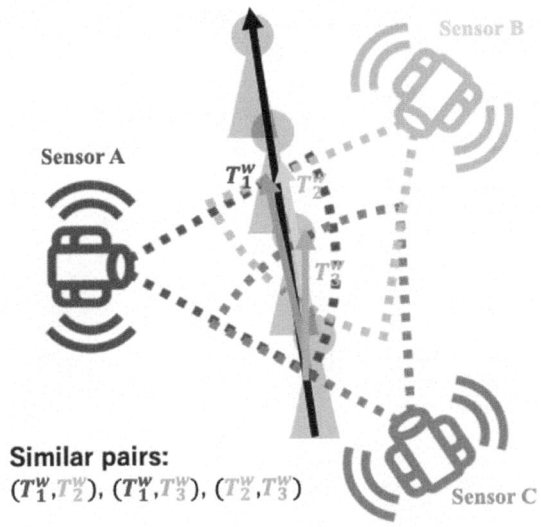

Fig. 1. Problem situation: multiple sensors have overlapping capture ranges; the black arrow line is ground truth, and the other arrow line represents captured trajectory from sensors respectively

4   Geocode-Based Sub-Trajectory Similarity Comparisons

In this paper, one important factor in the efficient generation of window-based sub-trajectories is using geocodes as keys for partitions. A messaging system publishes/subscribes data when using the Stream Processing Engine (SPE). The messaging system keeps the data in partitions to support parallel and distributed processing to ensure high scalability. Different messaging systems use different partitioning methods. This paper uses Apache Kafka [14] as a messaging system to perform partitioning with a key-based hashing method. Therefore, ensuring that data in geographically similar spaces are kept in the same partition has a performance impact. The traditional way to do this is to use a geohash, a kind of geocoding, as the key. Geocoding divides the Earth's surface into a grid of cells and converts it into a string code that represents the corresponding cell for a given coordinate. Geocodes have a hierarchical structure, with longer codes representing more precise locations. This essentially means that geocode itself can be utilized as an index.

There are many types of geocoding, such as GeoHash, Uber H3, rHEALPix, P-code, etc. GeoHash is the most widely used method for geocoding, but it has some drawbacks [2]:

– Geohash has irregular cell sizes/areas depending on latitude.
– The maximum precision of a geohash implementation is 12 (when using 64 bits).

Resolution	Length l	$2r$ = diameter
1	7,674.457 km	6,646.275 km
2	3,837.228 km	3,323.137 km
3	1,918.614 km	1,661.568 km
...		
19	14.638 m	12.677 m
20	7.319 m	6.338 m
21	3.659 m	3.169 m
22	1.830 m	1.585 m
23	91.487 cm	79.230 cm
24	45.743 cm	39.615 cm
25	22.872 cm	19.807 cm
...		
32	1.787 mm	1.547 mm

Part of the example for P-code's resolution with the grid cell's length and diameter

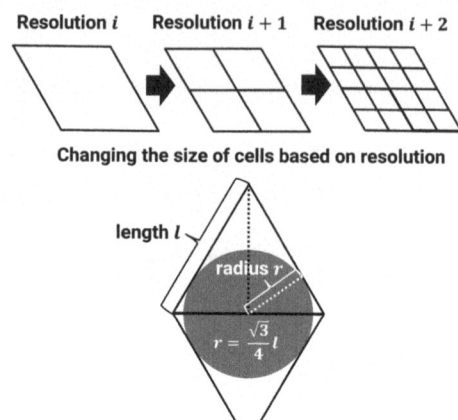

Changing the size of cells based on resolution

length l

radius r

$$r = \frac{\sqrt{3}}{4} l$$

Fig. 2. Example of P-code: shape, resolution, and cell's length l and diameter $2r$

A recently emerged solution, Discrete Global Grid Systems (DGGS), has been designed to address the aforementioned issues, particularly those pertaining to irregular cell areas. DGGS provides a unified data representation and facilitates the storage, integration, and analysis of spatio-temporal data with equal area/volume cells by partitioning the Earth's surface into a hierarchy of tessellations. There are multiple implementations of DGGS; we selected P-code [7], which encodes and decodes based on a DGGS for a point geographic location (latitude, longitude, and altitude). P-code's primary strength is its capacity to generate fine-grained resolution code (approximately $1mm^3$ scale at resolution 32), a capability that distinguishes it from other geocoding methods. The precise and selective resolution of the geocode is of significant importance, as it affects the query area and performance. Figure 2 illustrates an example of P-code. The cell of the P-code is shaped like a rhombus with two equilateral triangles connected. It is also divided into four subcells for every one-unit increase in resolution, with the child cells being completely contained by the parent cell. The length of one side of the cell and the diameter of the inscribed circle of the cell can be derived mathematically, and the values for some resolutions are shown in the table below.

The process of geocode-based trajectory similarity comparisons is conducted in the following manner. Figure 3 serves to understand the proposed methods described below in a more intuitive manner.

1. **Generate a high-resolution geocode**: This involves generating personal-level geocodes (pl-codes) for each point location. When generating pl-codes, it is necessary to record the number of times the same pl-code appears within a trajectory.
2. **Generate sub-geocodes**: Two types of sub-geocodes are generated by taking the partial value of the geocode generated in the previous step: the partition-level (pa-code) and the segment-level (sl-code). The pa-code is

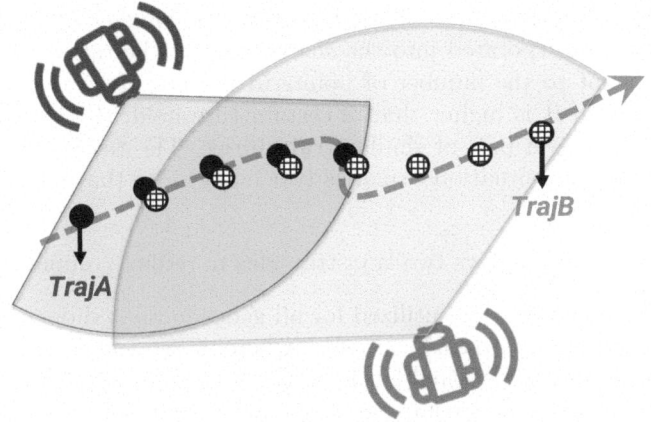

a) Problem situation

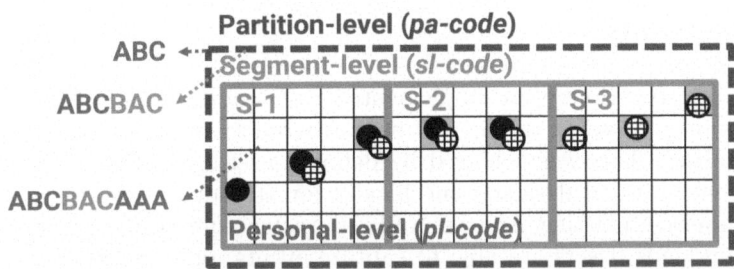

Resolution (geocode-length): *pl-code* > *sl-code* > *pa-code*

b) Example of the geocode-based grid cell index

S-1: **Shared *pl-code* count: 2**
S-2: **Shared *pl-code* count: 2**
S-3: **N/A**

Total shared *pl-code* count: 4
p_A: **Ratio with *TrajA*** → 4/5 = 0.8
p_B: **Ratio with *TrajB*** → 4/7 = 0.56
if $(p_A > \delta\,(0.5)$ *or* $p_B > \delta\,(0.5))$ *is True*:
:=> TrajA and TrajB is similar

c) Similarity comparison

Fig. 3. Geocode-based sub-trajectory similarity comparisons

employed to determine the geographical area for the generation of $\mathbb{T}^w$, while the *sl*-code is utilized to facilitate the comparison of partial trajectories (or line segments) in lieu of the entire trajectory.

3. **Compute the share count**: The share count is computed as follows: For each trajectory T^w in $\mathbb{T}^w$, the first step is to ascertain whether the same *sl*-code exists. If this is the case, the comparison is then performed at the *pl*-code

level. In the event that a *pl*-code is identical, the smaller of the calculated count values is incorporated into the shared count. The ratio of the value of the shared count to the number of points in the two trajectories should be calculated, and if it is higher than a certain threshold, the two trajectories will be returned as a pair of similar trajectories. The ratio value allows us to consider complex situations, such as two trajectories that intersect, as not similar cases.

The proposed method employs two key strategies to reduce computational cost:

- First, bitwise operations are utilized for all geocode-related operations.
- Second, a top-down approach is adopted, whereby segment-level comparisons are conducted prior to personal-level comparisons. This approach is designed to minimize the number of comparison operations required, thereby reducing the overall computational burden.

5 Implements

The proposed functions were implemented in Confluent ksqlDB [6], an open-source stream processing engine that supports functions akin to those of a relational database. This was achieved by defining user-defined functions (UDFs). ksqlDB is a database designed specifically for stream processing applications and can leverage existing Apache Kafka infrastructure to deploy stream processing workloads. Users are able to capture events, process, and serve queries using a SQL-like query language, which is both familiar and lightweight, without the necessity of any other programming languages or services. However, ksqlDB does not currently support built-in geospatial functions, such as (geo-)spatial data types and (geo-)spatial operations, except for calculating a distance between two locations. Nevertheless, in addition to the functions that are built into ksqlDB, users have the option of extending the suite of functions that are built into ksqlDB to include UDFs.

We defined four UDFs: one UDF for encoding P-code from point location $(p.lon, p.lat)$, one UDF for getting a partial code from P-code, one UDF for generating a window-based trajectory stream $S_\mathbb{T}^w$, and one UDF for getting a set of similar pairs of (T_i^w, T_j^w) by P-code-based method.

Definition 5 (TO_PCODE). *Given point p and resolution r, encoding P-code pcode from p.lon, p.lat, and r, where $1 \leq r \leq 32$*

Definition 6 (SUB_PCODE). *Given P-code pcode and resolution r, taking a sub P-code from r, where $1 \leq r \leq pcode.length$*

Definition 7 (POINTS_TO_TRAJECTORIES). *Given a stream of point p, P-code and window period w, generating window-based sub-trajectory stream $S_\mathbb{T}^w = \{T_1^w, T_2^w, \ldots, t_m^w\}$, sorted points by time p.t and then grouped by p.id*

Definition 8 (GET_SIMILAR_TRAJECTORY). *Given a window-based sub-trajectory stream* $\overline{S}_{\mathbb{T}}^{w}$, *sl-code resolution* r_s, *pl-code resolution* r_p, *and a ratio of the shared code* δ, *generating a set of sub-trajectory pairs* (T_i^w, T_j^w), *where* $T_i^w, T_j^w \in S_T^w$ *and* $|T_i^w, T_j^w| < \delta$

The implemented UDFs can operate on real-time generated point location data coming through Kafka using a simple SQL-like syntax, as shown in Fig. 4.

Fig. 4. How implemented UDFs work

There are examples of three SQL-like queries for getting similar pairs of trajectories. First, Query 1 creates a point stream from the point location stream. Query 2 creates a window-based sub-trajectory stream from the point stream with window size parameters. Lastly, Query 3 generates a set of pairs of similar trajectories. Note that *sl*-level and a ratio of the shared code δ are pre-defined in the function because of the limitation of the UDF structure. In this paper, we used 24 resolution (approximately 40 cm, see Fig. 2) as *pl*-code because the size of the intimate zone is 45 cm [5].

Query 1. Create a point stream

```
CREATE STREAM {point_stream_name}
AS SELECT id, t, lon, lat,
    TO_PCODE (lat, lon, {pl-resolution}) AS pl-code
FROM {raw_stream_name} EMIT CHANGES;
```

Query 2. Create a window-based sub-trajectory stream

```
CREATE TABLE {trajectory_table_name}
AS SELECT
    POINT_TO_TRAJECTORIES(ARRAY[id, t, lat, lon, pl-code]) AS sub-traj,
    SUB_PCODE(pl-code, {pa-resolution}) AS pa-code
FROM {point_stream_name}
WINDOW TUMBLING (SIZE {window_size} SECONDS)
GROUP BY pa-code EMIT FINAL;
```

Query 3. Get similar pairs of trajectories

```
CREATE STREAM {result_stream_name}
AS SELECT
    GET_SIMILAR_TRAJECTORY(sub-traj) AS results,
    pa-code
FROM {trajectory_table_name}
WINDOW TUMBLING (SIZE {window_size} SECONDS)
GROUP BY pa-code EMIT FINAL;
```

6   Experiments

To conduct our experiments, we generated synthetic data in the following manner:

– The following variable values should be set:
  • The number of moving objects to be generated: [10, 300],
  • The number of duplicate objects: [5, 100],
  • The duplication probability: 50%, and
  • The coordinate generation frequency: 0.1 s.
– A given number of random origin coordinates is generated for the given number of objects within a given box area (e.g., 100 m by 100 m).
– A random lifetime (e.g., [15, 100], in seconds) should be set for each object.
– Establishing the range of potential movement and subsequently generating random destination coordinates requires knowing the lifetime and average walking speed of pedestrians (1.33 m/s [15]).
– Generate coordinates that travel in a straight line from the generated start-destination coordinates at the set coordinate generation interval. At this juncture, it is necessary to incorporate a random coordinate error (averaging approximately 10 mm) that follows a Gaussian distribution to account for the sensor's inherent error.
– Duplicate objects are created for the original moving object based on the probability of creating duplicate objects during coordinate generation. These duplicate objects can possess different random lifetime values. No further duplicates should be made once the number of duplicate objects reaches the set maximum value.

An overall synthetic data generation process is shown in Fig. 5.

A comparative experiment was conducted using the prepared data to ascertain the frequency of calls to the basic operations of each method (P-code and DTW). The basic operation is as follows:

– For P-code, it is the search for the presence of a given geocode within each trajectory. When using a structure such as a HashMap, the time complexity is $O(1)$.
– For DTW, the distance between two points is calculated, and the time complexity is $O(1)$

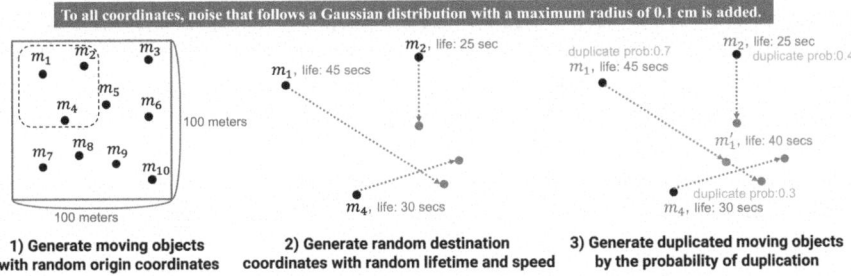

Fig. 5. Synthetic data generation for experiment

However, if we compute the time complexity of the similarity for two trajectories (of size N and M, respectively) at the trajectory level, it can be shown that P-code is $O(N \times M)$ in the worst-case. This is the same complexity as DTW, but P-code has less complexity than DTW. This is because the worst-case scenario is unlikely to occur with sl-code. Furthermore, the nature of the data assumption (pedestrian data with high-level time granularity) suggests that a significant amount of data is likely to be aggregated into the same pl-code.

The experiments were conducted by varying the number of moving objects and the window size for each method to ascertain the impact of these variables on the results. Other variables that need to be set are as below.

- Set parameters for the proposed method with P-code:
  - pa-code resolution r_{pa}: 15 (around 200 m)
  - sl-code resolution r_{sl}: 20 (around 6.3 m)
  - pl-code resolution r_{pl}: 24 (around 40 cm)
  - The ratio of the shared code δ: 0.5 (50%)
- Set parameters for DTW:
  - Distance calculation: Euclidean distance
  - The ratio of the threshold value δ: 5 m

The experiment results are presented in Table 1. Note that experiments with the same number of moving objects used the same synthetic data set for both P-code and DTW, i.e., the data set varied by the number of moving objects. The number of basic operations per object is calculated by dividing the number of basic operations by the number of objects. The ratio to another method is calculated relative to that method. For example, for P-code, the percentage is calculated by dividing the number of basic operations in P-code by the number of basic operations in DTW. In contrast, for DTW, the calculation is reversed.

A straightforward calculation of the ratio of basic operations between the two methods revealed that DTW required approximately 42 to 69 times more computational operations than P-code. This phenomenon intensifies with an increase in the number of moving objects and the window size.

The number of basic operations is correlated with the number of objects for both methods. As the number of objects increases, the number of basic operations also increases. Furthermore, the number of basic operations for window size is also found to be correlated for both methods. As the window size increases, the number of basic operations also increases. However, the proportional increase in DTW is significantly more significant than that of P-code. Specifically, P-code increases by approximately 11–22% when the window size goes from 5 to 10 s, while DTW increases by 28–58%.

The results demonstrate that the proposed geocode-based similarity search method is significantly more efficient than the existing common method (i.e., DTW).

7   Conclusion and Future Work

Processing trajectory streams is essential to support the diverse geoprocessing capabilities required by applications. This study presents methods that 1) geocode-based trajectory stream generation and 2) geocode-based (especially using P-code) efficient similarity measure with calculated shared counts. For geocode-based methods, we chose P-code. To substantiate the efficacy of the proposed method, a comparative experiment was conducted with DTW, and the complexity and computational cost of the basic operations were evaluated. The findings revealed that DTW necessitates approximately 42 to 69 times more basic operations.

However, there is still more work to be done. It is necessary to verify that the comparison is not merely in terms of the number of basic operations but also in terms of the real-time latency observed in a real system. Furthermore, it is essential to verify that the system behaves properly in real-world cases, utilizing real data.

Acknowledgments. This paper is partially based on results obtained from a project, JPNP20006, commissioned by the New Energy and Industrial Technology Development Organization (NEDO).

References

1. Almeida, A., Brás, S., Sargento, S., Pinto, F.C.: Time series big data: a survey on data stream frameworks, analysis and algorithms. J. Big Data **10**(1), 83 (2023). https://doi.org/10.1186/s40537-023-00760-1
2. Baig, F., Teng, D., Kong, J., Wang, F.: SPEAR: dynamic spatio-temporal query processing over high velocity data streams. In: Procedings of the 37th IEEE International Conference on Data Engineering (ICDE), pp. 2279–2284 (2021).https://doi.org/10.1109/ICDE51399.2021.00237
3. Berndt, D.J., Clifford, J.: Using dynamic time warping to find patterns in time series. In: Proceedings of the AAAI Workshop on Knowledge Discovery and Data Mining (KDD), pp. 359–370 (1994). https://doi.org/10.5555/3000850.3000887
4. Carbone, P., Katsifodimos, A., Ewen, S., Markl, V., Haridi, S., Tzoumas, K.: Apache Flink: stream and batch processing in a single engine. In: Bulletin of the IEEE Computer Society Technical Committee on Data Engineering, vol. 38, no. 4 (2015). http://sites.computer.org/debull/A15dec/issue1.htm

Table 1. Computational Efficiency Comparison: P-code vs. DTW

Methods	# of objects	Window size (seconds)	# of basic operations	# of basic operations (per object)	Ratio to another one
P-code	10	5	974	63.13	2%
		10	1,156	77.07	2%
	75	5	6,475	86.33	2%
		10	7,441	99.21	1%
	150	5	14,405	96.03	2%
		10	16,109	107.39	2%
	300	5	31,190	103.97	2%
		10	36,546	121.82	1%
DTW	10	5	40,421	2,694.73	4,268%
		10	56,823	3,788.20	4,915%
	75	5	378,116	5,041.55	5,840%
		10	521,010	6,946.80	7,002%
	150	5	766,908	5,112.72	5,324%
		10	983,882	6,559.21	6,108%
	300	5	1,585,972	5,286.57	5,085%
		10	2,517,963	8,393.21	6,890%

5. Hanafi, I., El Araby, M., Al Hagla, K., El Sayary, S., et al.: Human social behavior in public urban spaces: towards higher quality cities. Spaces and flows: Int. J. Urban ExtraUrban Stud. **3**(2), 23–35 (2013). https://doi.org/10.18848/2154-8676/CGP/v03i02/53690

6. Jafarpour, H., Desai, R., Guy, D.: KSQL: streaming SQL engine for Apache Kafka. In: Proceeding of 22nd International Conference on Extending Database Technology (EDBT), pp. 524–533 (2019). https://doi.org/10.5441/002/edbt.2019.48

7. Kim, T., Lee, J., Kim, K.S., Matono, A., Li, K.J.: Utilizing extended geocodes for handling massive three-dimensional point cloud data. World Wide Web **24**, 1321–1344 (2021). https://doi.org/10.1007/s11280-020-00783-1

8. Kumar, M.: World geodetic system 1984: a modern and accurate global reference frame. Mar. Geodesy **12**(2), 117–126 (1988). https://doi.org/10.1080/15210608809379580

9. Liu, J., et al.: QuadGridSIM: a quadrilateral grid-based method for high-performance and robust trajectory similarity analysis. Trans. GIS **28**(1), 83–107 (2024). https://doi.org/10.1111/tgis.13126

10. Shaikh, S.A., Kitagawa, H., Matono, A.: Distributed, continuous and real-time trajectory similarity search. In: Proceedings of the 23rd International Symposium on Parallel and Distributed Computing (ISPDC), pp. 1–8 (2024).https://doi.org/10.1109/ISPDC62236.2024.10705397

11. Shaikh, S.A., Kitagawa, H., Matono, A., Kim, K.s.: TStream: a framework for real-time and scalable trajectory stream processing and analysis. In: Proceedings of the 30th International Conference on Advances in Geographic Information Systems (SIGSPATIAL) (2022). https://doi.org/10.1145/3557915.3560964

12. Shaikh, S.A., Kitagawa, H., Matono, A., Mariam, K., Kim, K.S.: GeoFlink: an efficient and scalable spatial data stream management system. IEEE Access **10**, 24909–24935 (2022). https://doi.org/10.1109/ACCESS.2022.3154063

13. Tao, Y., et al.: A comparative analysis of trajectory similarity measures. GISci. Remote Sens. **58**(5), 643–669 (2021). https://doi.org/10.1080/15481603.2021.1908927

14. Thein, K.M.M.: Apache Kafka: next generation distributed messaging system. Int. J. Sci. Eng. Technol. Res. **3**(47), 9478–9483 (2014). https://ijsetr.com/uploads/436215IJSETR3636-621.pdf

15. Willen, C., Lehmann, K., Sunnerhagen, K.: Walking speed indoors and outdoors in healthy persons and in persons with late effects of polio. J. Neurol. Res. **3**(2), 62–67 (2013). https://doi.org/10.4021/jnr187w

16. Yu, J., Zhang, Z., Sarwat, M.: Spatial data management in Apache Spark: the GeoSpark perspective and beyond. GeoInformatica **23**, 37–78 (2019). https://doi.org/10.1007/s10707-018-0330-9

17. Zaharia, M., Chowdhury, M., Franklin, M.J., Shenker, S., Stoica, I.: Spark: cluster computing with working sets. In: Proceedings of the 2nd USENIX Conference on Hot Topics in Cloud Computing (HotCloud) (2010). https://doi.org/10.5555/1863103.1863113

Machine Learning and Rules

Predictive Modeling of Key Performance Indicators for Greenhouse Gas Emission Reduction Using Machine Learning

Claudia Diamantini[1] , Tarique Khan[1(✉)] , Alex Mircoli[2] ,
and Domenico Potena[1]

[1] Department of Information Engineering, Ancona, Italy
{c.diamantini,t.khan,d.potena}@univpm.it
[2] Department of Economic and Social Sciences, Università Politecnica delle Marche,
Ancona, Italy
a.mircoli@univpm.it

Abstract. As demand for digital technologies grows, so does the impact of greenhouse gas (GHG) emissions. This paper explores innovative strategies to reduce carbon dioxide (CO2e) emissions in the product and technology industry, focusing on sustainable solutions to minimize environmental impact during product processes and transformations. The integration of artificial intelligence and Key Performance Indicators (KPIs) is significant within industrial sectors. This study proposes a machine learning-based approach to predict KPIs aimed at minimizing GHG emissions across product processes and transformations. A dataset obtained from the Environmental Protection Agency (EPA) portal was used; feature selection is performed using the Recursive Feature Elimination (RFE) method. We extract relevant KPIs from the dataset and mathematically represent these KPIs. We train six machine learning (ML) classifiers: Random Forest (RF), Decision Tree (DT), K-nearest neighbors (K-NN), Gradient Boosting (GB), Adaboost, and Light Gradient Boosting Machine (LGBM). Grid search optimization is applied to enhance the classifier's performance. The results are evaluated using accuracy, precision, recall and the F1 score. The study achieves a maximum accuracy of 96.55% with GB, while AdaBoost attains the lowest accuracy at 93.91%.

Keywords: Key Performance Indicators (KPIs) · Machine Learning (ML) · Greenhouse Gas (GHG) · Fluorinated Gases · Sustainability

1 Introduction

According to a report by the US Environmental Protection Agency, greenhouse gas (GHG) emissions are increasing worldwide, with the transportation sector contributing approximately 27% of total emissions [1]. GHG emissions play an important role in the product and technology sector, affecting manufacturing

R. Chbeir et al. (Eds.): IDEAS 2024, LNCS 15511, pp. 267–280, 2025.
https://doi.org/10.1007/978-3-031-83472-1_18

processes, energy consumption, and product lifecycle. These emissions, measured in metric tons, have profound effects on the environment, product quality, and workers [2]. In this context, organizations rely heavily on quantitative information obtained through Key Performance Indicators (KPIs) to gauge their processes and structures [3]. KPIs, also referred to as performance measures or metrics, provide a framework to evaluate critical aspects of an organization's performance and future success [4].

In the rapidly advancing fields of product manufacturing and technology, accurately forecasting KPIs is essential to maintain operational efficiency and competitive advantage. KPIs are critical metrics that provide insight into the performance of various organizational processes, allowing companies to assess success and optimize their strategies. However, traditional forecasting models often fall short of capturing the complex, non-linear patterns inherent in large-scale industrial data. This can lead to inaccurate predictions, causing organizations to rely on reactive rather than proactive decision-making. Such an approach can result in missed opportunities, inefficient resource allocation, and a decline in market competitiveness. The integration of ML techniques into KPI forecasting presents a significant opportunity to overcome these challenges. By leveraging ML algorithms, it is possible to develop predictive models that are more accurate, adaptable, and capable of handling the intricate dynamics of industrial data. This can empower organizations to make informed, data-driven decisions, improving their operational efficiency and strategic agility. In this research, we introduce a machine learning (ML)-based model to predict Key Performance Indicators (KPIs) to minimize greenhouse gas (GHG) emissions during product manufacturing. Our approach utilizes a variety of ML classifiers and employs a data set sourced from the EPA portal. The different KPIs were extracted using statistical analysis and interpreted using mathematical expressions. To validate those KPIs, we trained six-variate ML classifiers and applied a grid search to obtain more optimized results.

The structure of the paper is outlined as follows: In Sect. 2, we review previous studies on Key Performance Indicators (KPIs) and Machine Learning (ML) models, highlighting their shortcomings. Section 3 details the proposed approach, including the techniques for feature selection and the algorithms used. Section 4 summarizes the experimental results. Section 5 delves into the discussion. Lastly, Sect. 6 draws conclusions and suggests possible directions for future research.

2   Background

Key Performance Indicators (KPIs) play a crucial role in efficiently achieving goals. In reference [5], researchers introduce a deep learning-based model to predict KPIs for electrical machines. This model incorporates various features such as rotation of the magnet, torque, minimum power, and mass of iron and copper, among others. To facilitate the training process, a data-driven deep learning meta-model is employed, utilizing both the Deep Convolutional Neural Network (DCNN) and Deep Neural Network (DNN). These networks serve to expedite the optimization process and mitigate computational expenses, which are typically high due to the time-consuming nature of optimizing KPIs. The evaluation is conducted using two distinct types of datasets. Renewable energy sources play a crucial role in reducing CO2 emissions.

In reference [6], the authors explore how technological innovation significantly contributes to enhancing energy efficiency and reducing carbon footprints across various industries. They analyzed data from 1990 to 2018 to gain deeper insights. The paper integrates concepts of renewable energy adoption, technological advancements, and export quality standards to propose a comprehensive strategy for mitigating CO2 emissions and promoting a more sustainable economy. Liu et al. [7] employ a blend of crisp and fuzzy models to produce a near-real-time dataset encompassing daily CO2 emissions from fossil fuels and cement production across 31 countries. Their objective is twofold: to mitigate carbon emissions and enhance the overall value of the supply chain system.

In reference [8], a cohesive modeling framework is outlined, enabling the projection of greenhouse gas (GHG) emissions by integrating multiple sectors like transportation, industry, and agriculture. This study offers policy-makers a valuable resource for validating the efficacy of sector-specific strategies and assessing emission reduction efforts. The work proposed in [9] advocates for implementing predictive maintenance in the automotive sector based on a systematic literature review. The study covers statistical inference methods, stochastic techniques, and AI approaches. The findings indicate substantial benefits, including reduced maintenance costs by 25–30%, increased production by 20–25%, and prevention of 70–75% of asset breakdowns.

In reference [10], various ML classifiers, including SVR, optimized genetic SVR, RandomForest, extreme gradient boosting, and artificial neural networks, were employed to forecast overall equipment effectiveness (OEE). Performance evaluation using metrics like RMSE, MAE, and MAPE showed strong performance by SVR and good results by XGB. Forecasting KPIs from historical data poses a significant challenge due to the volume and complexity of past datasets. In a study [11], researchers propose a machine learning-based regression model to predict KPIs by leveraging spatial and temporal historical data, particularly air temperature data. Various machine learning models, including SVR, RF, DT, XGB, NN, and MHQRF, were trained using three types of historical datasets. MHQRF demonstrated the highest performance across all datasets. Effective monitoring and control of production heavily rely on Key Performance Indicators (KPIs).

In reference [12], the authors introduce graph-based prediction techniques that utilize optimization methods and random forests to forecast missing KPIs. Implementing optimized methods enhances the accuracy of predicting missing KPIs and improves decision-making processes. Research [13] introduces a hybrid model combining RF and Extreme Learning Machine techniques to forecast CO2 emissions in Hebei, China. By analyzing historical CO2 emission data from 1995 to 2025, researchers developed a predictive model for future projections. To optimize RF performance, they incorporated the moth-flame optimization (MFO)

Table 1. Limitations of Past Studies

Work	Objectives	Datasets	Classifier	Limitations
[5]	To propose a deep learning-based model to predict KPIs for electrical machines.	Two types of datasets	DNN, DCNN	There is a lack of comparison with other machine learning techniques or traditional optimization methods.
[6]	The study explores how technological innovation plays a crucial role in enhancing energy efficiency and reducing carbon footprints in various industries.	Survey	-	The study did not include any practical scenarios illustrating the implications of its findings.
[7]	Utilizes a combination of crisp and fuzzy models to generate a near-real-time dataset covering 31 countries' daily emissions of CO2 from fossil fuels and cement production.	Real-time dataset	-	The study lacks detailed information on the specific methodologies used to collect and validate the near-real-time daily global CO2 emission data.
[8]	Unified framework for projecting emission of GHG across different sectors	Literature review	–	Data quality and availability are the key factors in predicting emissions, which is lacking in the study.
[9]	predictive maintenance within the automotive sector based on a systematic literature review	–	–	The study may not cover all the latest advancements in predictive maintenance in the automotive sector.
[10]	The objective of this study is to use various ML and DL classifiers to predict OEE.	Industrial case study	SVR, optimized genetic SVR, RF, XGB, and ANN	The study was limited tThe small samples dataset was collected.
[11]	Forecasting KPIs for historical data using a regression model.	Rossmann and Supermart-I and II	RF, DT, Linear Regression, SVR, NN, XGB, MHQRF	There is a lack of bias that may arise when historical data is incorporated into regression algorithms, which could affect the reliability of the predicted results.
[12]	Using optimization methods a graph-based prediction was proposed to predict missing KPIs.	Literature review	RF	The study was limited to using only one ML algorithm which is the RF.
[13]	The study proposes a hybrid model to predict the emission of CO2 in Hebei, China.	A case study covering the years 1995–2015	RF and Extreme Learning Techniques	The study was constrained by the utilization of traditional ML algorithms, lacking advanced DL models for predicting historical datasets of large size.
Proposed Work	Propose an ML-based model to predict KPIs to minimize the emission of GHG.	GHG dataset (EPA)	RF, DT, GB, LGBM, KNN, and AdaBoost	The proposed study relies solely on a single dataset for its case study.

method. These insights are instrumental for policymakers and environmental agencies, facilitating the implementation of tailored strategies to reduce CO2 emissions in Hebei, China. Table 1 outlines the limitations observed in previous research works. It can be noticed that a significant portion of prior studies used regression techniques to forecast CO2 emissions.

3   Methodology

This study aims to develop a predictive model to forecast Key Performance Indicators (KPIs) that can help minimize greenhouse gas (GHG) emissions during manufacturing and product destruction processes. The proposed model serves as a predictive tool for KPI estimation and incorporates feature selection strategies. Figure 1 illustrates the workflow of the proposed approach.

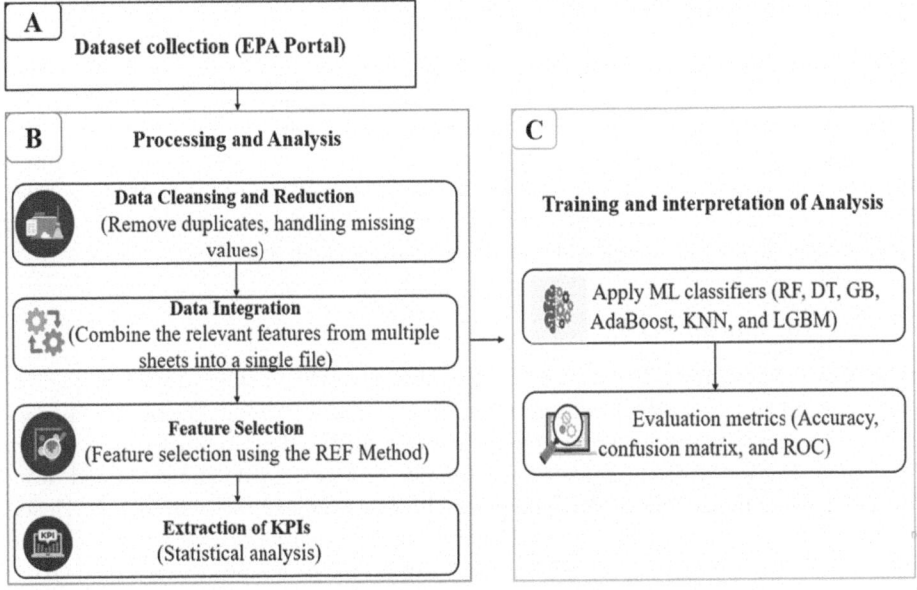

Fig. 1. Methodological Flowchart: GHG Reduction KPIs Prediction

3.1   Dataset and Preprocessing

To build predictive models, we initially acquired a dataset from a publicly accessible repository[1]. The Greenhouse Gas Reporting Program compiled data spanning from 2011 to 2022. All emissions data are expressed in metric tons of carbon dioxide equivalent, utilizing GWPs from IPCC's AR4. The dataset consists of five main categories: Emissions from P&T pro by chem file comprising (10×1527)

---

[1] https://www.epa.gov/ghgreporting/data-sets.

samples, encompassing information such as Reporting year, Facility name, Fluorinated GHG Emissions (metric tons), Fluorinated GHG Emissions (mt CO2e), Is Other GHG? and more. The emission by process file contains (43 × 5340) information such as Equipment Leak Emissions (mt CO2e), Average EF Approach, Emissions Method, Process Type, etc. Emissions Destr prev prod file contains samples of (10 × 844). The Emissions Destruction Details file contains samples of (10 × 24) and the Emissions Container Venting file contains samples of (9 × 209). The 5 categories were integrated into a single file for further analysis and feature selection. We employ pre-processing techniques to eliminate unreliable and noisy data. To handle missing values, we identified missing scores falling below 10% or exceeding 40% within each dataset. For the remaining missing values, we utilized MICE techniques alongside k-nearest neighbor algorithms.

3.2   Feature Selection Using Recursive Feature Elimination

At the beginning, our dataset included 5340 samples with 41 features. After pre-processing and feature selection, this process led us to select 19 features based on their ranking predicted by the estimator. Recursive Feature Elimination (RFE) was employed to pick the best feature subset by leveraging the learned model and classification accuracy. RFE identifies less significant features by ranking their importance, which has proven beneficial for machine learning models. Table 2 outlines the selected features along with their descriptions and ranking scores. The RFE method sets a ranking for each feature, with most features having a 0.95 ranking score and the lowest ranking score recorded as 0.80.

3.3   Extraction of KPIs

We extracted relevant KPIs that meet our objectives from the data set using statistical analysis. We calculated the mean and standard deviation of the characteristics to identify key performance indicators (KPI). The mean value portrays the central tendency of the features within the dataset, while the standard deviation signifies the extent of dispersion of the features around this mean. Based on our analysis, we identified five KPIs with high mean values and low standard deviation values, indicating consistent and impactful performance metrics within the dataset. These KPIs include features with mean values ranging from 5.333 to 7.680 and standard deviation values ranging from 0.246 to 1.434. Specifically, the KPIs are Fluorinated GHG Emissions (measured in metric tons of CO2 equivalent) with a mean of 6.880 and a standard deviation of 0.246, Equipment Leak Emissions (measured in metric tons of CO2 equivalent) with a mean of 5.333 and a standard deviation of 0.498, Emission Calculator Factor Method with a mean of 5.690 and a standard deviation of 0.345, Effective Destruction Efficiency Range with a mean of 6.717 and a standard deviation of 1.434, and Emission Factor (EF) Method with a mean of 7.680 and a standard deviation of 0.422. These KPIs are indicative of consistent and reliable performance metrics within the dataset. The highest mean value, 7.680, is observed for the Emission Factor (EF) method, indicating its prominence in the dataset. Conversely, the

Table 2. Selected Features from Dataset

No.	Name of Features	Descriptions	Ranking Score
1	Reporting Year	Year in which emission was reported	0.80
2	Facility ID	Unique identification number given to each facility	0.80
3	Facility Name	Name of reporting industry/facility	0.88
4	Process Name	Identifier of process name	0.95
5	Emissions Method	Methods to determine the mass emission of each GHG emitted from the process	0.95
6	Process Type	Indicates two types of process: either production or transformation.	0.89
7	Reaction	Characterization of Processes	0.88
8	Distillation	Characterization of Processes	0.88
9	Packaging	Characterization of Processes	0.95
10	Fluorinated GHG Emissions (metric tons)	Emissions of GHG during production and transformation processes are measured in metric tons.	0.98
11	Fluorinated GHG Emissions (mt CO2e)	Emissions of fluorinated GHG, which is equivalent to CO2 and measured in metric tons	0.97
12	Fluorinated GHG Group Name	Name of fluorinated greenhouse gases	0.89
13	Process Vent Emissions (mt CO2e)	GWP-weighted mass of all fluorinated GHG in this group and measured in metric ton CO2 equivalent	0.98
14	EF (Emission Factor) Method Used	Specific emission factor method is used to determine the mass emission of the GHG group emitted from the process vent.	0.95
15	ECF Method Used	Specific emission calculator factor method is used to determine the mass emission of GHG emitted from process vent	0.95
16	Effective Destruction Efficiency Range	Based on the CO2 equivalent, the specific efficiency range is the range for destruction	0.95
17	EPA Correlation with Site-Specific Leak Monitoring Indicator	An indicator (yes/no) to indicate whether this method was used for mass emission of GHG group emitted from equipment leaks	0.80
18	Equipment Leak Emissions (mt CO2e)	GWP-weighted mass of all fluorinated GHG in this group emitted from equipment leaks and measured in metric ton CO2 equivalent	0.95
19	Is Other GHG?	A label class or targeted class that indicates whether the GHG fluorinated was entered by a user other than selected from the list	0.95

lowest standard deviation, 0.246, is associated with fluorinated GHG emissions, indicating a relatively consistent performance or measurement precision across the dataset for this KPI.

Mathematical Representation of KPIs: After extracting KPIs from a dataset we illustrate these KPIs using Mathematical equations and examples.

1. Fluorinated GHG Emissions (mt CO2e): Under the IPCC guidelines for emissions reporting, during the production transformation process, Fluorinated GHGs are emitted, which are often measured in metric tons and are equivalent to CO_2. This can be mathematically illustrated as follows 1:

$$F_{\text{emission}} = \sum_{m=1}^{T} M_m \tag{1}$$

Here, M_m represents the emission of Fluorinated GHG from each process, with m denoting the total number of processes involved in Fluorinated GHG emission. This equation enables the determination of reduction goals to mitigate the impact of emissions on the environment and product quality. Minimizing such emissions requires a thorough assessment and implementation of mitigation strategies. The intensity of fluorinated GHG emission can be incorporated. The intensity level related to a specific process can be measured, which enables normalization and evaluation across various periods, facilities, and processes. The reduced intensity level of emissions indicates the improved efficiency of a process and can be formulated using Eq. 2.

$$\text{Intensity Level} = \frac{\sum_{i=1}^{P} D_i}{T} \tag{2}$$

Here, D_i shows the emission of Fluorinated GHG from each process while T defines the total level of a process (e.g., production transformation, destruction) linked with the emissions.

2. Equipment Leak Emissions (mt CO2e): This KPI refers to the quantity of GHG distributed into the environment due to leakage from equipment. Usually measured in metric tons, which is equivalent to (mt CO2e). With this KPI one can assess the impact on the environment as well as the operation of the industrial process. According to the Global Warming Potential (GWP) table, which is a measure of how much heat is stuck in the atmosphere over a specified period but is related to CO2, let's say methane gas has a GWP weight of 25 over a 100-year period, which means that it is 25 times more effective at catching heat compared to CO2 over that time frame. We can achieve this KPI using the following equation3:

$$EL = \sum_{r=1}^{R} (S_r \times GWP_r) \tag{3}$$

where EL denotes the equipment leak while GWP_r denotes the GWP table of each emitted gas related to CO2, S_r represents the amount of each GHG emitted equipment leaks, and R denotes the total number of GHG involved in the process.

3. Emission Calculator Factor Method: This Key Performance Indicator (KPI) is used to approximate emissions based on definite factors that are interlinked with a certain process. For example, a manufacturing plant consumed 20,345 kWh of electricity in a month. Now the emission factor for CO_2 emissions from electricity consumption is 1.3 Kg CO_2 per kWh. Using this KPI, it can be calculated as follows:

$$EFM = \text{Process} \times \text{Emission Factor} \tag{4}$$

Here, EFM defines the emission calculator factor method, while Process defines the amount of data for which the process is evaluated. The Emission Factor denotes the rate of impurity emitted per process. We can calculate the emission of CO_2 from the electricity consumption using Eq. 4. We can say that the estimated emission of CO_2 from the consumption of electricity is 26,448.5 kg.

4. Effective Destruction Efficiency Range: In this Key Performance Indicator (KPI), we measure the effectiveness of a destruction process by removing pollutants or dangerous chemicals. Usually measured in the percentage during specific treatment or disposal of a process. Mathematically, it can be used as follows 5:

$$\text{Destruction } (\%) = \left(1 - \frac{\text{outcome Absorption}}{\text{Input Absorption}}\right) \times 100 \tag{5}$$

5. Emission Factor (EF) Method: This Key Performance Indicator (KPI) provides a direct way to estimate the emissions.

Scenario: Let's assume that a travel agency wants to estimate the emission of CO_2 from their buses. The buses consumed a total amount of diesel 20,450 gallons of diesel over 24 h with an emission factor of 35 kg CO_2/gallon. We can calculate using 6:

$$Q = \text{Data of process} \times EF \tag{6}$$

Here, Q denotes the emission factor method, Data of process denotes the activity under process and EF denotes the emission factor. So, we can say that 714,400 kg of CO_2 is emitted from the buses.

3.4   Machine Learning for KPI Prediction

In this phase, we applied various machine learning models to estimate key performance indicators (KPIs). Figure 1 illustrates the working flow of the proposed study. The predictive model was designed to utilize a range of features, including process type, Fluorinated GHG Emissions (measured in metric tons), and emission methods, among others, to provide accurate predictions. The impact of features on the overall KPI outcomes was measured through the RFE technique and the less significant features were removed. Then, the dataset was split into 30% for testing and 70% for training to assess the model's performance. We compared the results based on four metrics: accuracy, precision, recall, and F1 score.

4   Experimental Results

The performance of the predictive model was evaluated using various metrics such as accuracy, precision, recall, F1 score, and confusion matrix. Table 3 presents the comparison of six machine learning (ML) models. To optimize the performance of each classifier, we applied Grid search, which is a technique used to optimize hyperparameters by exploring a specified range of parameter combinations. We trained the grid search with a minimum number of $n\_estimators$ set to 150, while the minimum depth was −1 and the maximum depths were 10 and 20. The $minimum\_samples\_leaf$ values were set to 1 and 3. For the criterion, entropy was used. Learning rates of 0.1 and 0.5 were used to train the model. Based on the nature of different classifiers, the parameters were set accordingly.

Table 3. Performance evaluation of ML classifiers

ML Classifiers	Accuracy	Precision	Recall	F1 score
GB	96.55%	95.33%	97.68%	96.49%
RF	95.91%	94.4%	97.35%	95.85%
AdaBoost	93.91%	91.66%	96.20%	93.88%
LGBM	95.67%	94.66%	96.53%	95.58%
DT	94.95%	93.86%	95.87%	94.85%
KNN	95.51%	94.78%	96.03%	95.40%

The GB classifier achieved the highest accuracy rate at 96.55%, while the AdaBoost classifier achieved an accuracy rate of 93. 91%. The GB classifier obtained the highest precision rate at 95.33%, while the AdaBoost classifier achieved a 91.66% precision rate. GB achieved the highest recall rate at 97.68%, while DT achieved a recall rate of 95. 85%. It is observed that the highest F1 score recorded by the GB classifier is 96.49%, while AdaBoost achieved the lowest F1 score at 93.88%.

Figure 2 illustrates the confusion matrix for the GB classifier and RF. It depicts the confusion matrix for the GB classifier, where the classifier impressively achieves good accuracy by accurately predicting 1206 samples from 1249 samples, while 43 samples were inaccurately predicted. We present the performance of the RF classifier, the classifier correctly predicted 1198 samples from 1249 samples, while 51 samples were incorrectly predicted. These illustrations offer valuable insight into the performance of the classifier, highlighting its ability to accurately classify the samples while providing a clear understanding of the misclassifications.

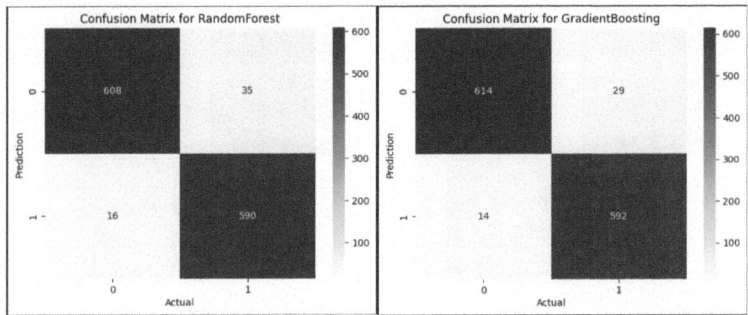

Fig. 2. Confusion Matrices: Performance Evaluation of ML Classifiers

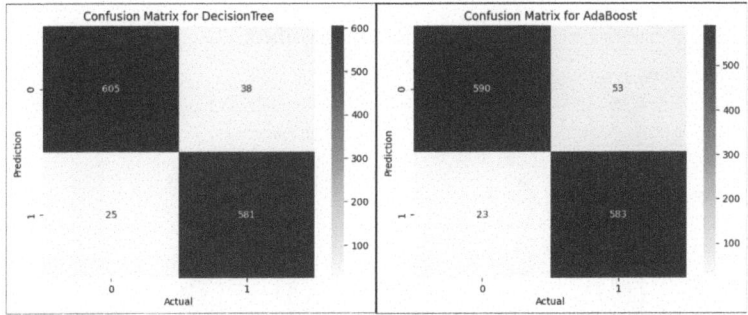

Fig. 3. Confusion Matrices: Performance Evaluation of ML Classifiers

Figure 3 depicts the confusion matrices for the AdaBoost and DT classifiers. The confusion matrix for AdaBoost shows that the classifier achieves an accuracy rate of 0.95% by correctly predicting 1173 samples from 1249 samples, while the DT obtains an accuracy equal to 0.94%. Figure 4 displays the confusion matrices for the KNN and LGBM classifiers. The first classifier correctly predicted 1193 samples from the 1249 samples, while 56 samples were incorrectly predicted. The second classifier, instead, accurately predicted 1195 samples, while 54 samples were incorrectly predicted. Figure 5 presents an overview of the performance of six ML classifiers in terms of accuracy rate. All classifiers demonstrate strong performance, with accuracy rates exceeding 90%. Notably, the GB classifier achieved the highest accuracy rate of 96.55%, while the AdaBoost classifier attained the lowest accuracy rate of 93.91%. There is minor difference among the accuracy rates of RF, LGBM, and KNN. Overall, the classifiers demonstrate strong performance across multiple metrics, indicating their effectiveness in the classification task.

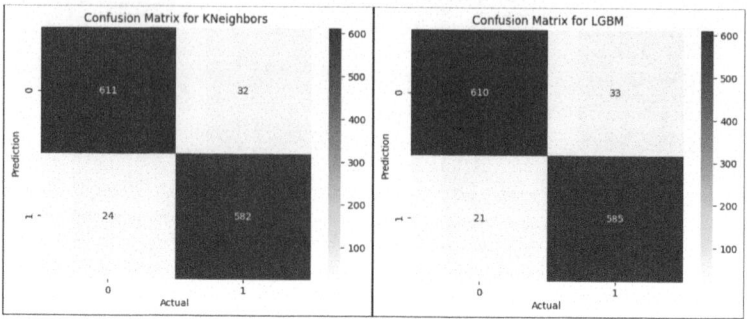

Fig. 4. Confusion Matrices: Performance Evaluation of ML Classifiers

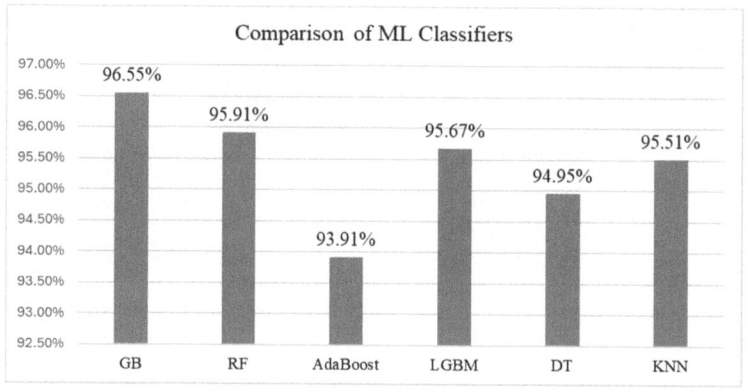

Fig. 5. Overall Performance Metrics Comparison for ML Classifiers

5   Conclusion

In this study, an ML-based model is proposed to predict Key Performance Indicators (KPIs) to minimize GHG emissions during product manufacturing and destruction utilizing various machine learning (ML) classifiers. A dataset was downloaded from the EPA portal. The data set, which contains five main categories, was integrated into a single file. The missing data were imputed using MICE techniques, and the best features were selected using the RFE method. Five different KPIs were identified from the dataset, interpreted using mathematical illustrations, and validated using ML classifiers. Six classifiers (DT, RF, GB, KNN, Adaboost, and LGBM) were trained to optimize the results and a grid search was applied to each classifier. Performance was evaluated using accuracy, precision, recall, F1 score, and confusion matrix. The classifiers GB, RF, LGBM, DT, and KNN achieved strong overall performance, while AdaBoost performed a slightly lower. The highest accuracy rate was recorded at 96.55%, achieved by GB. TThe study could be further improved by evaluating other datasets related to GHG emissions. We also intend to extend the experimentation by using

10-fold cross-validation in order to better evaluate the performance of the models. In future work, we plan to utilize advanced deep-learning models, such as the Transformer model, which has demonstrated significant success in KPI forecasting in other domains [20], and apply ensemble techniques to various datasets in order to further enhance prediction accuracy and identify more KPIs and their implications in different sectors.

Funding

This research has been funded by the Vitality, Project Code ECS00000041, CUP I33C22001330007, funded under the National Recovery and Resilience Plan (NRRP), Mission 4 Component 2 Investment 1.5, 'Creation and strengthening of innovation ecosystems,' construction of 'territorial leaders in R&D' - Innovation Ecosystems, Project 'Innovation, digitalization, and sustainability for the diffused economy in Central Italy, VITALITY' Call for tender No. 3277 of December 30, 2021, and Concession Decree No. 0001057.23-06-2022 of the Italian Ministry of University, funded by the European Union, NextGenerationEU.

References

1. Hertwich, E.G.: Increased carbon footprint of materials production driven by a rise in investments. Nat. Geosci. **14**(3), 151–155 (2021). https://doi.org/10.1038/s41561-021-00690-8
2. Brandenstein, N., Ackermann, K., Aeschbach, N., Rummel, J.: The key determinants of individual greenhouse gas emissions in Germany are mostly domain-specific. Commun. Earth Environ. **4**(1), (2023) https://doi.org/10.1038/s43247-023-01092-x
3. Gilsing, R., et al.: Defining business model key performance indicators using intentional linguistic summaries. Softw. Syst. Model. **20**(4), 965–996 (2021). https://doi.org/10.1007/s10270-021-00894-x
4. van de Ven, M., Lara Machado, P., Athanasopoulou, A., Aysolmaz, B., Turetken, O.: Key performance indicators for business models: a systematic review and catalog. Inf. Syst. e-Bus. Manage. **21**(3), 753–794 (2023). https://doi.org/10.1007/s10257-023-00650-2
5. Parekh, V., Flore, D., Schoeps, S.: Deep learning-based prediction of key performance indicators for electrical machines. IEEE Access **9**, 21786–21797 (2021). https://doi.org/10.1109/ACCESS.2021.3053856
6. Rahman, M. M., Alam, K., Velayutham, E.: Reduction of CO2 emissions: the role of renewable energy, technological innovation, and export quality. Energy Rep. **8**, 2793–2805 (2022). https://doi.org/10.1016/j.egyr.2022.01.200
7. Liu, Z. et al.: Carbon Monitor, a near-real-time daily dataset of global CO2 emission from fossil fuel and cement production. Sci. Data **7**(1), (2020). https://doi.org/10.1038/s41597-020-00708-7
8. Vashold, L., Crespo Cuaresma, J.: A unified modeling framework for projecting sectoral greenhouse gas emissions. Commun. Earth Environ. **5**(1), (2024). https://doi.org/10.1038/s43247-024-01288-9

9. Arena, F., Collotta, M., Luca, L., Ruggieri, M., Termine, F.G.: Predictive maintenance in the automotive sector: a literature review. Math. Comput. Appl. **27**(1),
2 (2021). https://doi.org/10.3390/mca27010002

10. El Mazgualdi, C., Masrour, T., El Hassani, I., Khdoudi, A.: Machine learning for
KPIs prediction: a case study of the overall equipment effectiveness within the
automotive industry. Soft. Comput. **25**(4), 2891–2909 (2021). https://doi.org/10.
1007/s00500-020-05348-y

11. Diamantini, C., Khan, T., Mircoli, A., Potena, D.: Enhancing KPI forecasting
through regression algorithms using historical data. In: Proceedings of the 9th
International Congress on Information & Communication Technology (ICICT),
439–452 (2024). https://doi.org/10.1007/978-981-97-3559-4_36

12. May, M.C., et al.: Graph-based prediction of missing KPIs through optimization
and random forests for KPI systems. Prod. Eng. **17**(2), 211–222 (2023). https://
doi.org/10.1007/s11740-022-01179-y

13. Wei, S., Yuwei, W., Chongchong, Z.: Forecasting CO2 emissions in Hebei, China,
through moth-flame optimization based on the random forest and extreme learning
machine. Environ. Sci. Poll. Res. **25**(29), 28985–28997 (2018). https://doi.org/10.
1007/s11356-018-2738-z

14. Saabith, A.L.S., Fareez, M., Vinothraj, T.: Python current trend applications - an
overview. **6**(10), (2019). https://www.researchgate.net/publication/344569950

15. Luan, J., Zhang, C., Xu, B., Xue, Y., Ren, Y.: The predictive performances of
random forest models with limited sample size and different species traits. Fisheries
Res. **227**, 105534 (2020). https://doi.org/10.1016/j.fishres.2020.105534

16. Sharma, T., Shah, M.: A comprehensive review of machine learning techniques on
diabetes detection. Vis. Comput. Ind. Biomed. Art **4**(1), 30 (2021). https://doi.
org/10.1186/s42492-021-00097-7

17. Uddin, S., Haque, I., Lu, H., Moni, M.A., Gide, E.: Comparative performance analysis of K-nearest neighbor (KNN) algorithm and its different variants for disease
prediction. Sci. Rep. **12**(1), (2022). https://doi.org/10.1038/s41598-022-10358-x

18. Hao, L., Huang, G.: An improved AdaBoost algorithm for identification of lung
cancer based on electronic nose. Heliyon **9**(3), e13633 (2023). https://doi.org/10.
1016/j.heliyon.2023.e13633

19. Mishra, D., Naik, B., Nayak, J., Souri, A., Dash, P.B., Vimal, S.: Light gradient
boosting machine with optimized hyperparameters for identification of malicious
access in IoT network. Digit. Commun. Netw. **9**(1), 125–137 (2023). https://doi.
org/10.1016/j.dcan.2022.10.004

20. Diamantini, C., Khan, T., Mircoli, A., Potena, D.: Forecasting of key performance
indicators based on transformer model. In: Proceedings of the 26th International
Conference on Enterprise Information Systems (ICEIS), 2, 280–287 (2024). https://
doi.org/10.5220/0012726500003690

Validation Without Rules: A Data Integration Case Study

Stefan Dzalev and Goran Velinov(✉)

Faculty of Computer Science and Engineering, Ss. Cyril and Methodius University,
1000 Skopje, North Macedonia
goran.velinov@finki.ukim.mk

Abstract. Data quality assessment is one of the most fundamental operations executed during data integration. Data validity is a collection of validation rules applied to the dataset's attributes. The validation rules provided by domain experts must be known during the data validation checks. In practice domain experts may not be available, or their number is insufficient, and the project timeline may be strict, resulting in unknown data validity rules. In this paper, we present a low code framework for outlier and anomaly detection as an alternative to traditional data validation rules. The framework comprises statistical and machine learning methods that recognize outliers and label them as invalid data. We evaluated the accuracy and scalability of four methods using the TPC-DI benchmark tests, and the results indicate a high level of correctness when the appropriate method is employed for a specific distribution. The framework is exceptionally effective for data with skewed and normal distributions for local and global outliers. Additionally, for different scaling factors, we observed that the ratio of validation time to data integration execution time is low and stable as datasets increase.

Keywords: Data validation · Data Integration · Outlier detection · Low code

1 Introduction

Since the emergence of data integration as a means for heterogeneous data consolidation in order to gain valuable insight from the unified data, data quality has been a crucial step in the integration process. Over time, with the popularity of the data warehouse, later the data lake and the rise in usage of machine learning by organisations, the importance of data quality has become even more apparent [1,2]. Data quality is a set of dimensions, namely accuracy, completeness, consistency, timeliness, relevance, validity, duplication, and integrity, which are the primary factors used to decide whether a data object or data point is acceptable for the system in question or not.

In recent years, as data quality needs and data itself evolved, research [3–5] has been conducted about refining and extending data quality dimensions and the importance of the validity dimension. In parallel, many advancements in

R. Chbeir et al. (Eds.): IDEAS 2024, LNCS 15511, pp. 281–294, 2025.
https://doi.org/10.1007/978-3-031-83472-1_19

data validity assessment tools and platforms have been made. In [6] the authors have introduced a data quality verification platform that supports data quality assessment for completeness, consistency, and accuracy. Their platform assesses the data validity as a subset of the consistency metric upon which predicates, intervals, and allowed values are set up. The Katara system [7] is a knowledge repository and a collaborative data cleansing platform. By comparing datasets with existing knowledge bases, it detects accurate and erroneous data entries, offering a selection of top-k potential corrections for inaccurate data. The Nadeef system [8] adopts a holistic approach to managing data quality rules. It offers users an interface to implement denial constraints and custom functions, ensuring comprehensive treatment of data validity concerns. All those tools reduce the time needed for human input in their calculation and correction. Other numerous frameworks [9–11] have showcased advancements in this area.

However, the current solutions do not eliminate the participation of domain experts in the data validation process. This means that the experts' input is still crucial, namely, the validity dimension requires very intensive involvement of the domain experts. They provide and document the validation rules needed for the tools to validate the data. But, in practice, the number of domain experts is often insufficient for the data integration project, or they may not even be available. Data integration projects may have strict deadlines and many data objects to be integrated. If validation rules are unavailable, the data validation phase could be a risk to the success of the overall data integration.

This paper presents an outlier and anomaly detection framework that can be used in the data integration process as a substitute for traditional data validation rules provided by domain experts. The framework [12] provides an automated, standardized approach to treating outliers and anomalies as non-valid data. Our goal is not to declare expert domain involvement obsolete but to examine whether the framework will result in acceptable rates of correctly identified non-valid data values in cases when validation rules are not available. To prove the concept, we have used the well-established industry-standard TPC-DI benchmark test [13]. In its specification, there are five validation rules provided by the experts. We use these rules to verify the framework's accuracy and determine the optimal thresholds per data distribution type and outlier type for all four framework methods. Besides accuracy, scalability is the second important parameter for the practical application of the framework. We have tested it for both parameters and the experimental results demonstrate its excellent performance:

- Accuracy. The accuracy depends on two metrics - precision and sensitivity. The tests show that in four out of five cases, the framework can very precisely ($\sim 99\%$) detect non-valid data points with a sensitivity rate from 80% to 99%.
- Scalability. The relative overhead ratio of using the framework in the data integration process is low, $\sim 10\%$ for the best methods, and stable for different scaling factors.

Moreover, the framework is suitable for any data integration architecture using Apache Spark, with easy installation and low-code configuration. It is also extensible by adding more validation methods.

The next sections of the paper are organized as follows: Sect. 2 presents our proposed framework. Section 3 outlines the experimental evaluation of the proposed framework. In Sect. 4, we discuss our experiment results. Section 5 presents the conclusion and further work.

2   Proposed Framework

In this section, we provide an overview of the basic technical features of the framework and explain the validation process as an additional step of the data integration process. We also cover the effectiveness and methods used in the initial implementation of the framework.

2.1   Practical Features

The processing engine of the framework is Apache Spark because of its wide recognition as the de-facto standard in distributed data processing. Furthermore, one of its core elements is an industry-proven machine learning API, which contains numerous machine learning algorithms and feature engineering methods utilized in the framework [14,15].

Packaged as a Python wheel, the framework can be installed and used as any Python library within a project. To be used, a YAML definitions file needs to be created locally within the project using the framework, and the framework's set of operations is executed based on those definitions [16]. Each operation has predefined values of the parameters also stored in the YAML definitions file. It is important to note that the framework is extensible by using the factory pattern and new operations can be easily added by introducing a new class to the code.

This low-code approach offers a simple and standardized method for implementing a validation process for any data object. To initiate the validation process, only one method needs to be called, and it can be invoked from any step of the data integration process. It only requires an Apache Spark DataFrame of the data to be validated and a path to the local YAML definitions file. Using YAML to configure the process allows for standardized, low-code, and setup-oriented usage. Also, when Jinja is employed in the YAML files, the configurations can be utilized by multiple different processes, promoting reusability.

2.2   Validation Process

Upon the validation process is invoked, using the input Apache Spark DataFrame and the path to the local YAML definitions file as parameters, the YAML file is parsed (see Fig. 1). The framework implies three categories of operations: utility, statistical method, and machine learning method. We define utility operations as logic that aids the statistical or machine learning method operations. They can range from simple operations like filter and select to more complex feature engineering ones. For the initial implementation of the framework, statistical method operations are the Inter-Quartile Range and Z-Score methods [17–19],

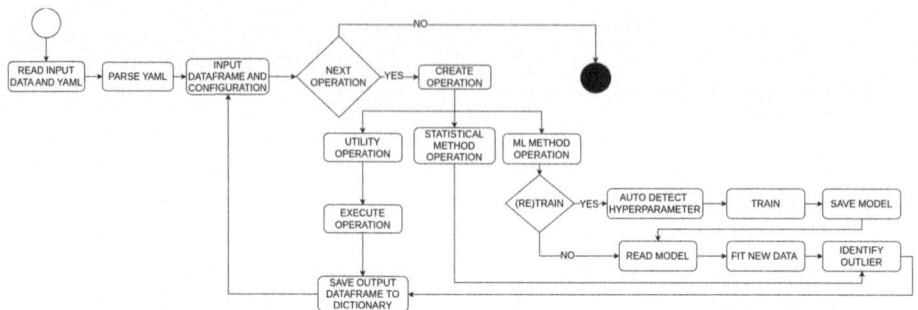

Fig. 1. Framework for outlier detection and validation process

while machine learning operations include the K-means [20, 21] and the Isolation Forest [22, 23] algorithms.

The code in the framework uses the factory pattern, so an instance of an operation class is created based on the definition in the YAML (see Fig. 2). As attributes for the class, the values in the YAML file are used. Each operation has only one method, which contains the logic for the particular operation. An Apache Spark DataFrame is the input to this method. After reading and creating an instance of the class for the first operation the method is invoked and the input data is transformed using the logic of the operation. After the Operation is finished, its output is an Apache Spark DataFrame with the transformed data. This data is saved in a dictionary data type, where the key is the name of the operation and the value is the resulting transformed data.

```
name: validation_example
operations:
 - name: isolation_forest_operation
   category: anomaly_detection
   operation_name: isolation_forest
   model_path: "/models/isolation_forest_validation"
   retrain: False
   params:
     features_column: Tier
     contamination: 0.0001

 - name: non_valid_data
   category: utils
   operation_name: filter
   params:
     conditions:
       - "outlier=1"

 - name: valid_data
   category: utils
   operation_name: filter
   data_from_operation: isolation_forest_operation
   params:
     conditions:
       - "outlier!=1"
```

Fig. 2. Configuration for Isolation forest validation

These steps are repeated until all defined operations in the YAML file are iterated. The last operation filters out the outliers as non-valid data. The data of the last operation can be read from the resulting dictionary and handled separately from the valid data.

The operations containing machine learning algorithms and statistical techniques require a threshold value for outliers. Moreover, the operations containing machine learning algorithms require a path as an input parameter to save the resulting model. These operations automatically use a model if it has been created in the past and save it if not. The models require hyperparameter setup, which is automated in the respective operations.

2.3   Effectiveness

To measure the effectiveness of the framework, we use two metrics, Sensitivity and Precision, which are important for increasing data accuracy. Sensitivity, also known as Recall, measures the number of correctly identified non-valid data values out of all true non-valid data values. Precision measures the number of correctly identified non-valid data values out of all data values declared non-valid by the method (TP - True Positives, FP - False Positives, and FN - False Negatives):

$$Sensitivity = \frac{TP}{TP + FN} \tag{1}$$

$$Precision = \frac{TP}{TP + FP} \tag{2}$$

We have selected four appropriate methods to demonstrate the effectiveness of the framework. Nevertheless, the framework can be extended by including any additional method suitable for this purpose.

InterQuartile Range. We have chosen to incorporate this method because its wide recognition, straightforward implementation, relatively fast execution time and easily interpretable results are advantageous when compared with more advanced machine learning models for this use case. We have parameterized the multiplier of the InterQuartile Range formula, as a threshold value, bearing in mind the varied data distributions.

Z-score. Similarly to the InterQuartile Range method, the Z-score's universal acknowledgment as a statistical technique for outlier detection, its understandability, and commonality determined our decision to implement it in the framework to be unquestionable. We have parameterized the threshold value to allow flexibility for various data distributions.

K-means. The first step in the K-means outlier detection procedure is the automated detection of the optimal number of clusters hyper-parameter - K before the outlier detection is initiated. We iterate using $K=2..20$, and for each K, we train the model and calculate the Silhouette Score. The procedure chooses the K for which the trained model with K has the highest Silhouette Score. The next important hyper-parameter is the number of iterations for the clusters' convergence. We have set it to have a default value of 50. The final step is calculating the distance from each data point to its centroid. The distances are then divided by the maximum distance between a data point and its centroid. The result is an outlyingness factor, a value between 0 and 1. The closer the value is to 1, the more likely it is an outlier. We consider this distance to be a threshold value and use it as a parameter in our framework.

Isolation Forest. Because this algorithm is not part of the Spark ML library, we use an open-source PySpark implementation [24]. Regarding the hyper-parameter tuning, for the number of trees hyper-parameter, we have added a step before finalizing the model to automatically find the optimal number of trees, based on evaluation results where the number of trees can be 50, 100, 150, 200, 250, 300, 350, 400, 450 and 500. The number of trees is decided based on the ROC value, i.e. the model with the highest ROC value is chosen. We consider the contamination factor to be a threshold value in the outlier detection, and it is used as a threshold parameter when the Isolation Forest method is used.

3   Experimental Evaluation

To experimentally prove the accuracy and scalability of the framework, we have conducted a series of experiments based on all four methods and diverse data volumes.

3.1   Experiment Specification

The synthetic data generated by the data generator very closely resembles the data from a real-world brokerage firm; the transformation and data quality rules are domain-oriented and reflect requirements that would be set by domain experts and business analysts in the real world. Moreover, the data quality rules have a subset of data validity checks, which is an important aspect we considered in our decision to use the TPC-DI benchmark to evaluate the framework.

The data can be created using various scaling factors, where each increment of the scaling factor adds 100 MB of data. Our experiments use data generated using scaling factors 3, 6, 12, 24, and 48; their sizes are 0.3 GB, 0.6 GB, 1.2 GB, 2.4 GB, and 4.8 GB, respectively. We use the dataset generated with a scaling factor of 3 to evaluate optimal threshold values for each method and the various data distributions and outlier types. The datasets generated using scaling factors from 6 to 48 are used in experiments regarding scalability, and the datasets with a scaling factor of 48 are used to evaluate the accuracy of the framework.

The generated data is divided into separate directories named BatchN, where N ranges from 1 to 3. Batch1 is the initial load of historical data and has the largest volume of data of all batches. Batch2 and Batch3 are delta increments. In our experiment we use all three batches. For the machine learning methods, the models are trained using Batch1. The saved models are then used for Batch2 and Batch3.

We focus on the data quality checks specified in the TPC-DI specification, which check the validity. Three data objects in total have validation rules for at least one of their attributes. These objects are DimCustomer, DimTrade, and DimCompany. We have differentiated them in 5 test cases:

- **Test Case 1**: The DOB (Date of Birth) attribute of the DimCustomer where DOB is invalid if DOB < Batch Date - 100 years or DOB > Batch Date i.e. a customer cannot be older than 100 years and cannot be born in the future. This column is not nullable, as per the specification.
- **Test Case 2**: The Tier attribute of the DimCustomer where, if the value is not null, the only valid values are 1, 2 and 3, other values are considered invalid.
- **Test Case 3**: The Fee attribute of DimTrade where, if the value is not null, it cannot exceed TradePrice * Quantity.
- **Test Case 4**: The Commission attribute of DimTrade where, if the value is not null, it cannot exceed TradePrice * Quantity.
- **Test Case 5**: The SPrating attribute of DimCompany where, if the value is not null, the only valid values are AAA, AA[+/-], A[+/-], BBB[+/-], BB[+/-], B[+/-], CCC[+/-], CC, C and D.

3.2 Experiment Implementation

In our study, we conducted a series of experiments to evaluate our proposed framework carried out using a laptop computer equipped with an AMD Ryzen 5 5600h processor running at 3.3 GHz, 16 GB of RAM, and L1 Instruction Cache: 6×32 KB, L1 Data Cache: 6×32 KB, L2 Cache: 6×512 KB and L3 Cache: 16 MB. The laptop was running the Ubuntu 20.04 LTS operating system.

For all test cases, we use data generated using different scaling factors and we use all three batches in every dataset generated for every specified scaling factor. It is important to note that DimCompany, as per the specification, has only historical data, i.e., only Batch1 data is used to load DimCompany. For every framework method, test case, and scaling factor combination, we execute a full run of the TPC-DI integration process with all objects and their batches included. We can determine the TP, FP, and FN parameters for all experiments and use them to calculate the Sensitivity and Precision performance metrics.

Since the outlier detection methods require a numeric attribute as input, during the outlier detection process, the non-numeric attributes of test case 1 and test case 5 are converted to numeric types using the utility functions within the framework. Following our experiment preparation, we have conducted a statistical analysis of the attributes included in the test cases to observe the distribution of the data and the extremity of the outlier values (see Fig. 3). The outlier values (true non-valid data values) are colored in red.

The summary of our findings for each test case is presented as follows:

- **Finding 1**: The attribute in **Test Case 1** has a normal distribution, and the outliers' values range from being local to isolated and global. The proportion of local and global outliers depends on the randomness of the data generation.

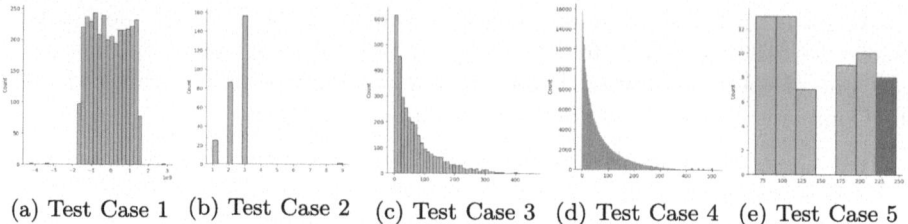

(a) Test Case 1 (b) Test Case 2   (c) Test Case 3 (d) Test Case 4 (e) Test Case 5

Fig. 3. Histograms of all test cases

- **Finding 2**: The attribute in **Test Case 2** has a skewed distribution, and the outliers are local.
- **Finding 3**: The attribute in **Test Case 3** has a skewed distribution, and the outliers are global; they deviate significantly and are isolated from the rest of the data points.
- **Finding 4**: The attribute in **Test Case 4** has a skewed distribution, and the outliers are global; they deviate significantly and are isolated from the rest of the data points.
- **Finding 5**: The attribute in **Test Case 5** has a uniform distribution, and it has no outliers. In fact, the non-valid data points have the same frequency as valid data points, making this attribute's distribution unbalanced in an outlier detection context. Therefore we exclude **Test Case 5** in our experiments.

3.3 Determining Optimal Thresholds

The primary prerequisite in our framework is to input threshold parameter values for outlyingness assertion for all methods used, for normal and skewed data distributions, and for local and global outliers. We used the Batch1 data from the dataset generated with scaling factor 3 to ascertain these parameters and to evaluate the F1 Score, a harmonic mean of Sensitivity and Precision metrics.

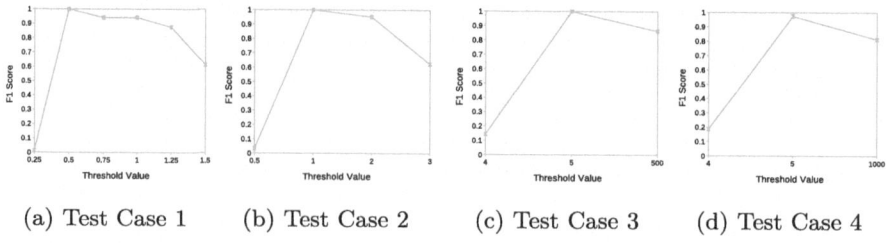

(a) Test Case 1   (b) Test Case 2   (c) Test Case 3   (d) Test Case 4

Fig. 4. InterQuartile Range optimal threshold value

InterQuartile Range performs well when the threshold value is between 0.5 and 1 and is used for attributes with normal data distribution (see Fig. 4(a)). If the attribute has a skewed data distribution but with local outliers the threshold value is optimal between 1 and 2. When the outliers are global and the data distribution is skewed, higher values should be used (see Fig. 4(c) and Fig. 4(d)).

Z-score is effective when the threshold value between 2 and 4 is used for attributes with normal data distribution (see Fig. 5(a)). Attributes with skewed data distribution but with local outliers require a threshold value higher between 3 and 5 for the method to be effective (see Fig. 5(b)). Skewed distributions do not affect the performance of Z-score, and this is why the threshold value is less sensitive (see Fig. 5(c) and Fig. 5(d)).

When using a threshold value between 0.4 and 0.6 for attributes with normal distribution, the K-means method performs well, as shown in (see Fig. 6(a)). A clear limitation of this method is when the data is skewed and the outliers are local, resulting in poor performance (see Fig. 6(b)). This is because the locality of the outliers negatively affects the model, i.e. they add bias during the training step. A threshold value of 0.1 is best when the attribute has skewed distribution and global outliers (see Fig. 6(c) and Fig. 6(d)).

The threshold value for the Isolation Forest method is the contamination parameter, so the expected proportion of non-valid data points has the most crucial role when deciding which threshold value should be used (see Fig. 7). A contamination parameter with a value close to the proportion of non-valid data points will provide best results. Test Case 1 has 0.13% non-valid data, Test Case 2 has 0.17%, Test Case 3 has 0.007%, and Test Case 4 has 0.01%.

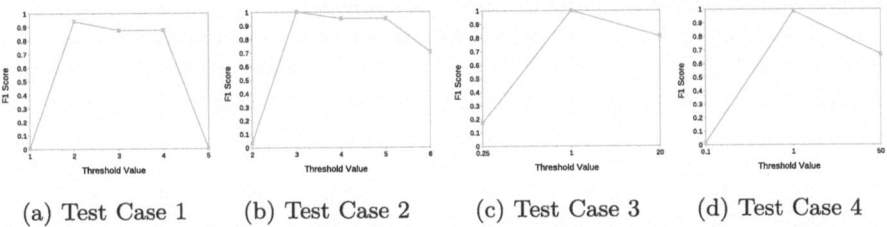

(a) Test Case 1   (b) Test Case 2   (c) Test Case 3   (d) Test Case 4

Fig. 5. Z-score Optimal Threshold Value

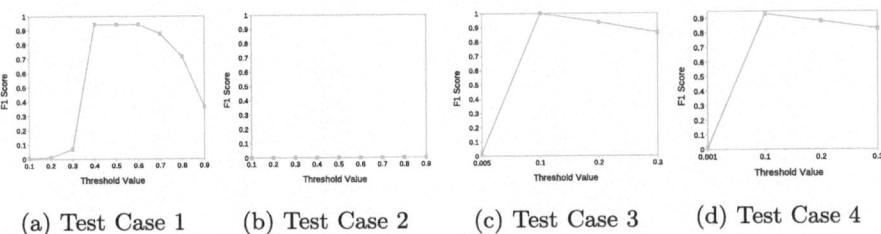

(a) Test Case 1   (b) Test Case 2   (c) Test Case 3   (d) Test Case 4

Fig. 6. K-means optimal threshold value

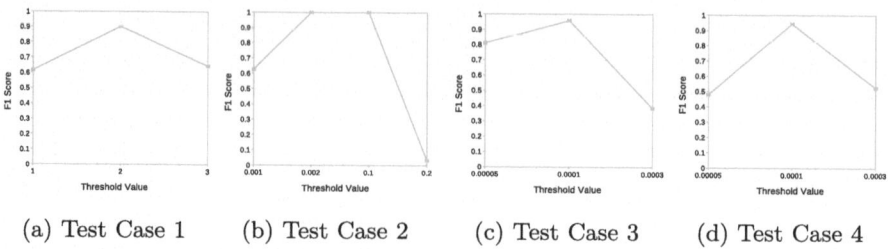

(a) Test Case 1   (b) Test Case 2   (c) Test Case 3   (d) Test Case 4

Fig. 7. Isolation Forest optimal threshold value

3.4   Accuracy Assessment

The accuracy assessment was conducted using datasets with a scaling factor of 48, as it closely resembles a real-world scenario due to the volume of data. We executed the experiments using all four methods for each test case and every batch of the dataset. We used the precision and sensitivity metrics to evaluate the quality of non-valid data points determination.

For Test Case 1, we observe that InterQuartile Range, Z-score and K-means methods have similar accuracy in non-valid data point identification (see Fig. 8(a)). They can very precisely detect the non-valid data points but with a lower sensitivity rate (not detect every non-valid data point). Due to the mixed global and local outliers in Test Case 1, Isolation Forest can detect most non-valid data points, but it results in many false positives.

For Test Case 2 (see Fig. 8(b)), we observe that InterQuartile Range and Z-score methods have similar accuracy. They can very precisely detect the non-valid data points but with a lower sensitivity rate. The Isolation Forest method

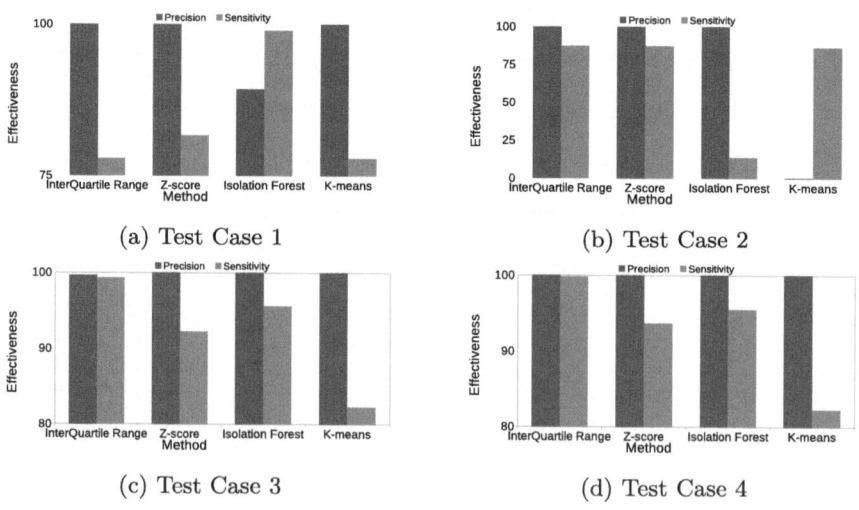

(a) Test Case 1   (b) Test Case 2

(c) Test Case 3   (d) Test Case 4

Fig. 8. Validation accuracy of each method

is not able to identify all non-valid data points, due to all of the outliers in this test case are local. The K-means method exhibits a limitation to precisely identify non-valid data points in this test case.

For Test Cases 3 and 4, we observe (see Fig. 8(c) and Fig. 8(d)) that all of the methods have similar accuracy, which is high. The InterQuartile Range method, not being affected by skewed distributions is able to identify non-valid data points best. Second is the Isolation Forest method, a method specialized in the detection of global outliers as in the scenario for these test cases. The Z-score method is not able to correctly identify all non-valid data points because it is affected by the skewed distribution, but like the K-means method, it exhibits acceptable accuracy.

3.5   Addressing Scalability

To evaluate the scalability of our framework, we executed the benchmark for each scaling factor and assessed the overhead when a method was used. During each run, we used one of our methods for all test cases and summed the overhead times. We observe the historical load of the first batch separately from the incremental loads of the subsequent batches because, at that point, an automated main hyper-parameter detection is performed for the machine learning methods. This step is time-consuming and only occurs during the historical load.

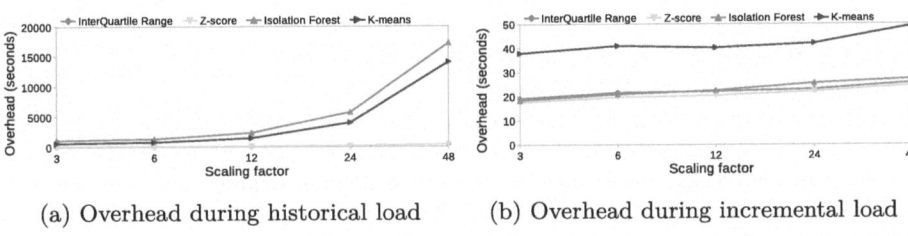

(a) Overhead during historical load   (b) Overhead during incremental load

Fig. 9. Overhead of the validation process

When K-means and Isolation Forest are utilized, the automated main hyper-parameter value detection drastically increases the overhead compared to the statistical methods (See Fig. 9(a)). This is the trade-off of utilizing machine learning methods. We also evaluate the average overhead time in incremental loads (See Fig. 9(b)). The K-means has the highest overhead, while the other methods have similar overheads. The linear increase in overhead, both in the historical and incremental executions, indicates that our framework can scale.

Another proof of the framework's scalability is the stable relative overhead ratio between the total validation time and total execution time (see Fig. 10.). We show the ratio per method for the particular run across all our experimentation scaling factors. Even though the ratio for a historical (one-time) load is higher, the ratio for the incremental loads is stabilized to 10% for the best methods like InterQuartile Range and Z-score and between 25%-35% for the Isolation Forest and K-means methods.

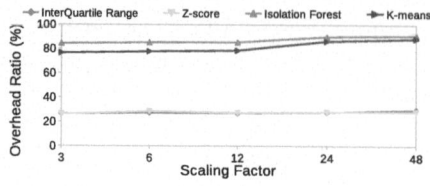
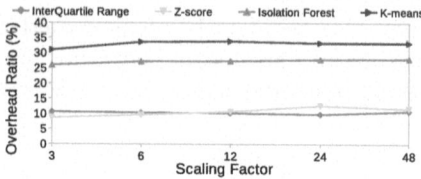

(a) Overhead ratio of historical load   (b) Overhead ratio of incremental load

Fig. 10. Overhead ratio between total validation and total execution time

4   Discussion

Before using the framework, it is crucial to conduct a visual and statistical data exploration to determine the most suitable method and threshold values. In our experiment setup, we explored and tested the framework's methods using various threshold values to identify the optimal value for the specific dataset characteristics.

When the optimal threshold value is used, the InterQuartile Range and Z-score provide the best results, while the Isolation Forest and K-means give results with similar quality if their limitation of inability to detect local outliers is taken into perspective. From an accuracy perspective, the choice of which method to utilize upon using the framework when the outliers are global can be based on personal preference. However, InterQuartile Range and Z-score would be the most appropriate choice when the outliers are local.

All data elements identified by the framework as non-valid will not be considered in the next steps of data integration and analysis. So, minimizing incorrect identifications is important for further implementations of the framework in real-world data integration endeavors. The results of our accuracy evaluation indicate that, depending on the data distribution and outlier type, the framework is very effective when the appropriate method is used. It can accurately identify the majority of non-valid values and with a very small number of valid data elements labeled as non-valid.

The execution time for Isolation Forest and K-means includes an additional overhead for their main hyper-parameter automated detection. This characteristic does not impose a significant drawback in the case of Isolation Forest because the overhead is drastic only on the first historical load. In the case of K-means, however, there is a more significant overhead for incremental loads when compared with other methods. But the most vital point, in the context of execution time, is that the framework can scale to the volume of data.

A key insight from our results is that the proposed framework is effective in the precise and correct identification of non-valid data points when the number of non-valid data points is small enough for the non-valid data points to be considered outliers or anomalies. Hence, the main limitation of our framework is the ability to detect non-valid data points when the number of distinct true non-valid data points is of a similar frequency to distinct valid data points.

5   Conclusion and Future Work

Over the past decades, there has been much research into the results of data validation for machine learning, more so than research on the results of machine learning for data validation. We have showcased and evaluated our framework, which is an attempt to achieve a low code, parameterized outlier and anomaly detection process, and proved that it can be used as a substitute for traditional data validation rules provided by domain experts. Our framework, as a collection of statistical and machine learning methods, is able to detect non-valid data with a high degree of accuracy when the appropriate method is used after data exploration is conducted using descriptive statistics and visual inspection. We demonstrated that the framework can be used for attributes with local or global non-valid data points in normal and skewed distributions if the frequency of non-valid data points is not similar to that of valid data points. We also proved the scalability of our framework, which is imperative to its usability in real-world use cases.

In our future work, we focus on automating the data exploration process needed for choosing the appropriate method. We aim to have automated data distribution assertion, potential outlier frequency, and locality detection. This will aid in decreasing the preparation time and eliminate human error in choosing the appropriate method. Furthermore, in pursuing state-of-the-art performance, we plan to expand our framework to be self-supervised, following the work of [25].

References

1. Aldoseri, A., Al-Khalifa, K.N., Hamouda, A.M.: Re-thinking data strategy and integration for artificial intelligence: concepts, opportunities, and challenges. Appl. Sci. **13**(12), 7082 (2023). https://doi.org/10.3390/app13127082
2. Ridzuan, F., Zainon, W.M.N.W.: A review on data quality dimensions for big data. In: Proceedings of the 7th Information Systems International Conference (ISICO), pp. 341–348 (2024). https://doi.org/10.1016/j.procs.2024.03.008
3. Gabr, M.I., Helmy, Y.M., Elzanfaly, D.S.: Data quality dimensions, metrics, and improvement techniques. Future Comput. Inf. J. **6**(1), 3 (2021). https://doi.org/10.54623/fue.fcij.6.1.3
4. Pandey, R.D., Snigdh, I.: Validity as a measure of data quality in internet of things systems. Wireless Pers. Commun. **126**, 933–948 (2022). https://doi.org/10.1007/s11277-022-09777-w
5. Martín, L., Sánchez, L., Lanza, J., Sotres, P.: Development and evaluation of artificial intelligence techniques for IoT data quality assessment and curation. Internet Things **22**, 100779 (2023). https://doi.org/10.1016/j.iot.2023.100779
6. Schelter, S., et al.: Automating large-scale data quality verification. Proc. VLDB Endowment **11**(12), 1781–1794 (2018). https://doi.org/10.14778/3229863.3229867
7. Chu, X., et al.: KATARA: a data cleaning system powered by knowledge bases and crowdsourcing. In: Proceedings of the ACM International Conference on Management of Data Pages (SIGMOD), 1247–1261 (2015). https://doi.org/10.1145/2723372.2749431

8. Dallachiesa, M., et al.: NADEEF: a commodity data cleaning system. In: Proceedings of the ACM International Conference on Management of Data (SIGMOD), pp. 541–552 (2013). https://doi.org/10.1145/2463676.2465327

9. Peng, J., et al.: Self-supervised and interpretable data cleaning with sequence generative adversarial networks. Proc. VLDB Endowment **6**(3) 433–446 (2022). https://doi.org/10.14778/3570690.3570694

10. Rekatsinas, T., Chu, X., Ilyas, I.F., Ré, C.: HoloClean: holistic data repairs with probabilistic inference, Proc. VLDB Endowment **10**(11), 1190–1201 (2017). https://doi.org/10.14778/3137628.3137631

11. Neutatz, F., Chen, B., Abedjan, Z., Wu, E.: From cleaning before ML to cleaning for ML. IEEE Data Eng. Bull. **44**(1), 24–41 (2021). http://sites.computer.org/debull/A21mar/p24.pdf

12. https://github.com/stefandzalev/validation-framework , Accessed 10 Oct 2024

13. Poess, M., Rabl, T., Jacobsen, H.A., Caufield, B.: TPC-DI: the first industry benchmark for data integration. Proc. VLDB Endowment **7**(13), 1367–1378 (2014). https://doi.org/10.14778/2733004.2733009

14. Hasan, Z., Jie Xing, H., Idrees Magray, M.: Big data machine learning using Apache Spark Mllib. Mesopotamian J. Big Data **2022**, 1–11 (2022). https://doi.org/10.58496/mjbd/2022/001

15. Polak, A.: Scaling Machine Learning with Spark, O'Reilly Media, Inc., (2023). https://www.oreilly.com/library/view/scaling-machine-learning/9781098106812

16. Wang, B.: Programming for qualitative data analysis: towards a YAML workflow. In: Proceedings of the Australasian Conference on Information System (ACIS) (2022). https://aisel.aisnet.org/acis2022/17

17. Frery, A.C.: Interquartile range. In: Encyclopedia of Mathematical Geosciences, pp. 1–3. Springer (2022). https://doi.org/10.1007/978-3-030-26050-7_165-1

18. Dash, C.S.K., Behera, A.K., Dehuri, S., Ghosh, A.: An outliers detection and elimination framework in classification task of data mining. Decis. Anal. J. **6**, 100164 (2023). https://doi.org/10.1016/j.dajour.2023.100164

19. Gong, X., Zhang, F., Lu, T., You, W.: Comparative analysis of three outlier detection methods in univariate data sets. In: Proceedings of the 3rd International Conference on Electronic Communication & Artificial Intelligence (IWECAI), pp. 209–213 (2022). https://ieeexplore.ieee.org/document/9750786

20. Sinaga, K.P., Yang, M.-S.: Unsupervised K-means clustering algorithm. IEEE Access **8**, 80716–80727 (2020). https://doi.org/10.1109/access.2020.2988796

21. Huang, Z., Zheng, H., Li, C., Che, C.: Application of machine learning-based K-means clustering for financial fraud detection, Acad. J. Sci. Technol. **10**(1), 33–39 (2024). https://doi.org/10.54097/74414c90

22. Liu, F.T., Ting, K.M., Zhou, Z.-H.: Isolation forest. In: Proceedings of the 8th IEEE International Conference on Data Mining (ICDM), pp. 413–422 (2008). https://doi.org/10.1109/icdm.2008.17

23. Barbariol, T., Dalla Chiara, F., Marcato, D., Susto, G.A.: A review of tree-based approaches for anomaly detection. Chapter in: Control Charts and Machine Learning for Anomaly Detection in Manufacturing, pp. 149–185. Springer (2022). https://doi.org/10.1007/978-3-030-83819-5_7

24. Isolation Forest PySpark Implementation. https://github.com/titicaca/spark-iforest, Accessed 10 Oct 2024

25. Sehwag, V., Chiang, M., Mittal, P.: SSD: a unified framework for self-supervised outlier detection. In: Proceedings of the 9th International Conference on Learning Representations (ICLR) (2021). https://openreview.net/pdf?id=v5gjXpmR8J

Recommendation

Fairness in Group Recommender Systems Using Variational Autoencoders

Muhammad Shahzaib Ali and Kostas Stefanidis$^{(\boxtimes)}$ (iD)

Tampere University, Tampere, Finland
konstantinos.stefanidis@tuni.fi

Abstract. Recommender systems are integral to enhancing user experiences on platforms, like Amazon and Netflix, by providing personalized suggestions. However, these systems often face significant fairness challenges, particularly in group settings where diverse preferences must be aggregated. In this paper, we explore the use of Variational Autoencoders (VAEs) to improve fairness in group recommendations. By introducing stochastic elements into the VAE framework, we aim to generate diverse and equitable recommendations. Extensive evaluations using the MovieLens 20M dataset demonstrate that incorporating noise during the recommendation process significantly enhances fairness with a minimal impact on ranking quality. The study identifies the Hybrid aggregation method paired with uniform noise as the optimal tradeoff, balancing group satisfaction, dissatisfaction, and ranking quality.

Keywords: Fairness · Group Recommendations · Variational Autoencoders

1 Introduction

Recommender systems assist users in navigating extensive information spaces by providing personalized suggestions for products, movies, music, and various other data items. Fairness in recommender systems has emerged as a critical area of concern [16]. Fairness pertains to the equitable treatment of all items and users within the recommendation process, ensuring no particular group is disadvantaged. The complexity of fairness issues becomes more pronounced in the context of group recommendations, where the preferences of multiple individuals must be aggregated into a single recommendation list. Group recommender systems face the intricate task of balancing diverse preferences to propose options satisfactory for the entire group. Existing methods, such as averaging individual preferences or using voting mechanisms, often fail to address the nuanced fairness concerns within groups. These approaches can overlook minority preferences and fail to ensure that recommendations are fair to all group members, leading to dissatisfaction and perceived unfairness.

Variational Autoencoders (VAEs) present a promising avenue for addressing several challenges. Specifically, VAEs are a type of neural network that can

R. Chbeir et al. (Eds.): IDEAS 2024, LNCS 15511, pp. 297–311, 2025.
https://doi.org/10.1007/978-3-031-83472-1_20

effectively handle data sparsity and enhance recommendation quality by learning efficient latent representations of user preferences and item characteristics. By introducing stochastic elements into the recommendation process, VAEs can generate more diverse and potentially fairer recommendations. This capability is particularly valuable in mitigating biases and ensuring a more balanced distribution of recommendations across different items and users.

In this paper, we investigate how VAEs can be leveraged to improve fairness in group recommendations. We study the impact of introducing stochasticity into the model, enhancing the variability and fairness of the recommendations generated. We evaluate different methods for aggregating group preferences to identify the most effective approaches for ensuring fair treatment of all group members. Overall, the goal is to develop a recommendation system that not only provides high-quality suggestions but also ensures that these recommendations are fair and equitable for all users within a group. Our experimental results suggest that moderate levels of noise can enhance fairness metrics, such as reduced disparity in exposure across items, while maintaining a high Normalized Discounted Cumulative Gain (NDCG), a measure of ranking quality. Additional experiments indicate that a uniform noise level strikes an optimal balance, effectively improving fairness without degrading recommendation accuracy.

2   Related Work

Variational Autoencoders (VAEs): Unlike traditional autoencoders that map input data to a latent vector, VAEs convert the input into parameters of a probability distribution, typically the mean and variance of a Gaussian. This creates a continuous and structured latent space, advantageous for generating new data instances resembling the input data. The VAE architecture consists of two main components. The **Encoder** translates the input into mean and logarithm of variance parameters of a Gaussian distribution within a latent space, from which the latent vector is sampled. The **Decoder** reconstructs the input data from the sampled latent vector, minimizing reconstruction error and balancing it with regularization from the latent space distribution. The entire model is trained by optimizing the Evidence Lower Bound (ELBO), balancing the reconstruction loss with the Kullback-Leibler divergence between the learned and prior distributions, thus regularizing the model.

Integrating VAEs into recommender systems addresses challenges, such as managing large datasets and providing personalized recommendations from incomplete user data. **Matrix Factorization and VAEs:** While traditional recommender systems use matrix factorization to reveal latent factors, VAEs provide a probabilistic approach, embedding user and item features into a latent space for more accurate user preference and item characteristic representation [5]. **Handling Sparse Data:** VAEs are robust to data sparsity in collaborative filtering systems, effectively learning to impute missing values, providing a denser latent space representation [10]. **Improving Quality:** VAEs can incorporate additional information, such as content features and demographic data, enhancing recommendation quality by understanding user behavior and content properties better [5]. **Fairness and Diversity:** Recent studies explore how VAEs can address fairness and diversity in recommendations. For instance, F2VAE introduces terms in VAE's loss function for fair representation and recommendation, increasing precision and reducing unfairness [4].

Group Recommendations: Group Recommender Systems (GRS) aid decision-making for groups by merging diverse preferences into a group consensus. Unlike individual recommender systems, GRS must aggregate preferences of all group members [9]. Traditional methods, such as averaging or plurality voting, may neglect minority interests. Advanced methods like the "least misery" method ensure no member is heavily displeased, while the "maximum satisfaction" strategy maximizes overall happiness [12]. Emerging approaches include interactive systems allowing iterative feedback, such as the Interactive Multi-Party Critiquing system [6] and iterative voting under uncertainty [11]. Fair sequential recommendation methods dynamically adjust recommendations to balance cumulative satisfaction [13]. A novel framework using reinforcement learning models group recommendations as a sequential decision-making process, optimizing recommendations based on long-term user satisfaction [14].

Amortized Fairness in Recommender Systems: Amortized fairness ensures fair treatment in ranking systems over time, addressing cumulative attention distribution. This is crucial in scenarios where items with similar relevance scores must be ranked. Amortized fairness ensures that the cumulative attention each item receives is proportional to its relevance [1]. The concept involves defining subjects $u_n...u_n$ and rankings $p_1...p_n$. For each ranking p_j, subjects u_i are assigned relevance r_{ji} and attention scores a_{ji}. The goal is to ensure proportional cumulative attention and relevance, expressed as: $\frac{\sum_{l=1}^{m} a_{li1}}{\sum_{l=1}^{m} r_{li1}} = \frac{\sum_{l=1}^{m} a_{li2}}{\sum_{l=1}^{m} r_{li2}}$, $\forall u_{i1}, u_{i2}$. Unfairness is the distance between attention and relevance distributions, quantified using metrics like KL-divergence or L1-norm [1]. To balance fairness and quality, the Discounted Cumulative Gain (DCG) metric measures ranking quality, emphasizing higher relevance scores at top positions. [2] suggests using DCG to measure quality: $DCG@k(r) = \sum_{i=1}^{k} \frac{2^{r(i)}-1}{\log_2(i+1)}$. Normalized DCG (NDCG) provides a measure for new item positioning: $NDCG(\rho\rho*) = \frac{DCG(Q_k(\rho*))}{DCG@k(\rho)}$. [3] shows that VAEs can mitigate popularity bias, providing a balanced representation of items and ensuring a fairer recommendation process.

3   Methodology

3.1   Variational Autoencoder Architecture

We will be using Variational Autoencoders (VAEs) to produce single user recommendations. VAEs represent a significant advancement in the modeling of user preferences within collaborative filtering systems. These models excel in encoding high-dimensional data into a lower-dimensional interpretable latent space, offering a robust way to handle sparsity and scalability in recommender systems.

Encoder: The encoder in a VAE for recommender systems transforms user-item interaction data into a compressed latent space representation characterized typically by its mean (μ) and variance (σ^2). These parameters define a Gaussian distribution from which latent variables are sampled, capturing the underlying user preferences and item characteristics. The encoder approximates the intractable true posterior distribution of the latent variables given the input (user-item ratings) by a tractable distribution: $q_\phi(z_u \mid x_u) = N(\mu_\phi(x_u), diag(\sigma_\phi^2(x_u)))$, where ϕ denotes the parameters of the encoder, learned to minimize the Kullback-Leibler divergence from the approximate to the true posterior, effectively capturing the user's latent preferences. This minimization is crucial for enhancing the model's ability to generalize and personalize recommendations.

Reparameterization Trick: To facilitate gradient-based optimization via stochastic nodes, VAEs employ the reparameterization trick: $Z_u = \mu_\phi(x_u) + \epsilon * \sigma_\phi(x_u)$, where $\epsilon \sim N(0, I_k)$. This trick allows for the backpropagation of gradients directly through the stochastic latent variables, maintaining the stochastic nature of the model while enabling efficient training.

Decoder: In the decoding phase, the latent representations are transformed back into the original space, predicting the likelihood of unrated items. The decoder models the probability of user u interacting with items I as: $p_\theta(x_u \mid z_u) = Mult(N_u\pi(z_u))$, where $\pi(z_u)$ is a probability distribution over items computed using a softmax function applied to a transformation of the latent representation: $\pi(z_u) \propto exp(f_\theta(z_u))$. This setup allows the system to estimate the user's preferences for items they haven't interacted with based on their latent interests.

3.2   Amortized Fairness in Variational Autoencoders

Amortized fairness, as conceptualized in recommender systems, seeks to ensure that items of similar relevance receive equitable attention over multiple rounds of recommendations, rather than in a single instance. This is particularly crucial in systems where the visibility of recommendations can significantly impact user choice and item success. The objective is to distribute the exposure of items in a manner that aligns with their relevance, reducing the risk of bias towards certain items only due to their positioning in a recommendation list [8].

In the context of VAEs used for collaborative filtering, implementing amortized fairness involves modifying the encoder and decoder functions to incorporate stochasticity in both the training and the evaluation phase. This modification aims to produce varied outputs for similar inputs across different recommendation sessions, addressing the inherent unfairness in static ranking systems.

The encoder in a fairness-aware VAE remains a function that maps input data (user-item interactions) to a latent space, but with an additional stochastic component. This component introduces noise into the encoding process, allowing the latent representation to vary slightly each time depending on the noise realization. Mathematically, the encoder can be expressed as: $q_\phi(z_u \mid x_u) = N(\mu_\phi(x_u), diag(\sigma_\phi^2(x_u))) + \epsilon$, where ϵ is a noise vector sampled from a predefined distribution, enhancing the diversity of the latent representations.

The decoder, which reconstructs the input data from the latent representations, also incorporates randomness in its process. This randomness ensures that the reconstruction and hence the recommendation output varies with each iteration, even under the same input conditions. The modified decoder function can be modeled as: $p_\theta(x_u \mid z_u) = softmax(f_\theta(z_u + \eta))$, where η represents a noise vector introduced in the decoding phase, further promoting variability in the output recommendations. The effectiveness of this amortized fairness in VAE's is measured using metrics such as NDCG to measure ranking quality.

3.3 Fairness Enhancement

By incorporating random noise into the latent variables during the evaluation or deployment phase, the model can reduce bias and variance in item recommendations, leading to a more balanced exposure of items across different user groups. We will be using four different noise distributions and expect this to have a direct impact on the ranking of group recommendations.

Gaussian Noise ($\sigma^2 = 0.5$): Introducing a moderate level of Gaussian noise with a variance of 0.5 marginally enhances fairness. This level of noise is expected to lead to subtle changes in the unfairness metrics, reflecting a slight improvement. The impact on the ranking quality measured by Normalized Discounted Cumulative Gain (NDCG) should be minimal, and the item positions in the final ranking should also be close to their original placements.

Gaussian Noise ($\sigma^2 = 1$): Implementing Gaussian noise with a variance of 1.0, identical to that used during the training phase, is expected to reduce the measures of unfairness. This adjustment should also affect the NDCG similarly, a notable shift is expected in both fairness and ranking accuracy.

Gaussian Noise ($\sigma^2 = 2$): Doubling the variance of the noise to 2.0 in the test phase compared to what was applied during training is expected to further decrease the unfairness metrics. This increase in noise variance would lead to more frequent changes in the ranking positions of items, with an expected reduction in NDCG due to these adjustments.

Uniform Noise: The application of uniform noise serves to investigate the effects of variance independent of the mean. This approach allows for an examination of how varying levels of uniformity impact the distribution and fairness of item rankings across different scenarios.

3.4 Evaluation Metrics

We use the following evaluation metrics as defined by [15];
1. Satisfaction: User satisfaction is a critical measure that gauges how well the recommendations align with the user's preferences. Satisfaction can be measured using the equation: $sat(u_i, Gr) = \frac{\sum_{d \in Gr} P(u_i, d)}{\sum_{d \in A(u_i)} P(u_i, d)}$, where Gr is the group recommendation list, $P_{ui}(d)$ is the preference score of user ui for item d, and $A(ui)$

is the individual recommendation list for user ui. This measure normalizes user satisfaction by comparing the quality of group recommendations against the user's ideal individual recommendations. For group recommendations, overall group satisfaction must consider the aggregated preferences of all members. The group satisfaction is calculated as: $groupSat(G, Gr) = \frac{\sum_{u_i \in G} sat(u_i, Gr)}{|G|}$. This measure averages the satisfaction scores of all group members, ensuring that the recommendations provided are collectively acceptable [13].

2. Dissatisfaction: Dissatisfaction in recommender systems arises when the recommended items do not align well with the user's preferences, which can be expressed indirectly through user disagreement scores. For a single user ui, the user disagreement score is given by: $userDis(u_i, G, Gr) = 1 - sat(u_i, Gr)$. Dissatisfaction within groups can be measured by examining the variance in satisfaction scores among group members, termed as group disagreement. Group disagreement is: $GroupDis(G, Gr) = \max_{u_i \in G} sat(u_i, Gr) - \min_{u_i \in G} sat(u_i, Gr)$. This formula captures the difference in satisfaction between the most and least satisfied members within a group, reflecting the extent to which a particular user's preferences are not met by the group recommendations [13].

3. Discounted Cumulative Gain (DCG): DCG is a metric used to evaluate the ranking quality of recommendations. It measures the usefulness of a recommendation list based on the positions of relevant items. Higher-ranked items are given more importance using a logarithmic discount factor, reflecting the diminishing returns of user satisfaction as they scroll down the list. DCG is computed using the formula: $DCG@k(r) = \sum_{i=1}^{k} \frac{2^{r(i)} - 1}{\log_2(i+1)}$, where $r(i)$ denotes the relevance score of the item at position i. This approach ensures that items at higher ranks contribute more to the overall gain than those at lower ranks, aligning with user behavior where top recommendations are more likely to be viewed and interacted with.

4. Normalized Discounted Cumulative Gain (NDCG): NDCG extends DCG by providing a normalized score that allows for comparison across different recommendation lists. NDCG is obtained by dividing the DCG by the Ideal DCG (IDCG), which represents the highest possible DCG for a perfectly ordered list of recommendations. It is calculated as: $NDCG@k = \frac{DCG@k}{IDCG@k}$. The normalization process ensures that the evaluation metric is scale-independent, facilitating fair comparisons between different algorithms or systems. By employing NDCG, we can assess how closely the recommender system's output approximates an ideal ranking, thereby providing a robust measure of its effectiveness in placing the most relevant items at the top of the list. This is crucial for improving user satisfaction and engagement, as users are more likely to interact with highly relevant recommendations presented earlier in the list.

5. Discounted Fairness (DFH): DFH evaluates the fairness of recommendations by considering both the relevance and the position of items in the recommendation list. It ensures that items with similar relevance are treated equitably over multiple rounds of recommendations. DFH can be formalized as:

$DFH = \frac{1}{|G|} \sum_{u \in G} |\sum_{k=1}^{n} a_{uk} - \sum_{k=1}^{n} r_{uk}|$, where a_{uk} is the attention received by the item k for user u and r_{uk} is the relevance score. This metric helps identify the disparity in the exposure of relevant items across different users.

6. F-Score: F-Score combines precision and recall into a single metric, providing a balanced measure of a system's accuracy, useful for binary classification tasks. Precision (P) and Recall (R) are defined as $P = \frac{TP}{TP+FP}$ and $R = \frac{TP}{TP+FN}$, where TP is true positives, FP is false positives, and FN is false negatives. The F-Score is then calculated as: $F = \frac{2 \cdot P \cdot R}{P+R}$.

By utilizing these metrics, the evaluation provides a holistic assessment of the recommendation system's performance, focusing on both the accuracy of recommendations and the fairness of item exposure and user satisfaction.

4 Experimental Evaluation

We aim to improve fairness in recommendations by integrating noise distributions into the test phase of VAE, inspired by [2]. We form three distinct group types, each with 100 instances with eight members, and utilize four different aggregation methods to generate group recommendations. We assess the performance of group recommendations using Satisfaction, Dissatisfaction, NDCG, DFH, and F-Score. For NDCG and DFH, we compute them for each group member and average the results. Additionally, we average the metrics across the 100 instances to obtain an overall evaluation for each group type. We use the Movie-Lens dataset with 20M ratings for 27,300 movies contributed by 138,500 users [7]. The data was transformed into a binary format to reflect implicit feedback. Users who interacted with fewer than five movies were excluded (see Table 1).

Group Formation. We construct three group types, each with eight members, reflecting varying levels of similarity among members. The similarity between any two users is quantified using the Pearson Correlation.

Mixed Similarity Group: This group consists of four similar and four dissimilar users. Similarity is quantitatively defined using a threshold model; members with a Pearson correlation coefficient above 0.5 are considered similar. This setup allows the exploration of how a recommendation system manages diversity within a group that contains equally divided opinion clusters.

Homogeneous Group: Comprising members who all exhibit a similarity score above 0.5 with each other, this group's dynamic is expected to have a high degree

Table 1. Data description after filtering.

Dataset	Ratings	Users	Items	Heldout Users
MovieLens	9,990,682	136,677	20,720	10,000

of agreement on recommendations, serving as a test bed for the system's ability to maintain engagement and satisfaction in a uniformly agreeable environment.

Heterogeneous Group: All members in this group have similarity scores below 0.5 with respect to each other, representing a challenging scenario for the recommendation system in terms of maximizing satisfaction and ensuring fairness across highly diverse preferences.

Aggregation Methods for Group Recommendations. We will use the following aggregations to generate recommendation outcomes for each group, each designed to capture different dimensions of group satisfaction:

Average Method: The average score of recommendations across all group members is calculated, providing a straightforward measure that considers all opinions equally but may overlook individual dissatisfaction.

Least Misery Method: This method focuses on the least satisfied member of the group, aiming to ensure that the recommendation is at least tolerable for everyone, which can prevent any member from being extremely dissatisfied.

Most Pleasure Method: In contrast to least misery, this method seeks to maximize the satisfaction of the happiest group member, potentially favoring items that are highly rated by at least one person.

Hybrid Method: Combining the average and least misery methods with an α_j value of 0.5, this method strives to balance general satisfaction with the need to avoid extreme discontent, providing a compromise between group fairness and individual satisfaction: $score(G, dz, j) = (1 - \alpha_j) \cdot avgScore(G, dz, j) + \alpha_j \cdot leastScore(G, dz, j)$.

5   Experimental Results

The predicted ratings obtained by using VAEs in the single-user recommendation system were used as an input to the group recommendation system. We evaluated the impact of different noise levels (0, 0.5, 1, 2, and Uniform) on the performance of our group recommender system using various aggregation methods.

5.1   Impact of Noise Levels on Group Recommendations

Homogeneous Groups: Table 2 shows that the Avg method achieved the highest satisfaction score (0.974) and NDCG (0.834), indicating strong agreement among users with similar preferences. HB followed closely with a slightly lower satisfaction score (0.971) and NDCG (0.810). Both methods exhibited low dissatisfaction and DFH metrics, suggesting minimal unfairness. These results indicate that in the absence of noise, homogeneous groups benefit from consistent and equitable recommendations, as the preferences are highly aligned among users. When a moderate level of noise (0.5) was introduced, a slight reduction in satisfaction (0.968 for Avg) and NDCG (0.798) was observed. This suggests

Table 2. Homogenous Group Data.

Aggregation Methods	Satisfaction	Dissatisfaction	DCG	NDCG	DFH	F-Score
Noise Level 0						
Avg	0.974046	0.056707	5.874677	0.834439	0.086123	0.958098
LM	0.96132	0.038674	5.447713	0.773793	0.070124	0.961237
MP	0.944823	0.078806	5.306357	0.753715	0.069952	0.932527
HB	0.970747	0.043077	5.70562	0.810427	0.086124	0.963679
Noise Level 0.5						
Avg	0.967731	0.063467	5.619626	0.798212	0.089497	0.951578
LM	0.956021	0.050591	5.218595	0.741249	0.070182	0.95255
MP	0.940887	0.084091	5.158706	0.732743	0.069953	0.927809
HB	0.964185	0.053586	5.452808	0.774517	0.089497	0.955067
Noise Level 1						
Avg	0.95133	0.086796	5.117147	0.72684	0.087508	0.931066
LM	0.935593	0.072125	4.667618	0.662989	0.070176	0.93154
MP	0.916218	0.123764	4.593431	0.652451	0.069953	0.89486
HB	0.946427	0.073862	4.937306	0.701295	0.087509	0.935966
Noise Level 2						
Avg	0.878097	0.271894	3.833712	0.544541	0.081386	0.785494
LM	0.830451	0.156315	2.872237	0.407973	0.070744	0.835224
MP	0.822651	0.318537	3.384876	0.480788	0.069958	0.731081
HB	0.86466	0.192625	3.423653	0.486296	0.081528	0.831895
Uniform Noise Level						
Avg	0.975493	0.057552	5.901089	0.838191	0.084449	0.958365
LM	0.964838	0.042643	5.540445	0.786965	0.07012	0.960991
MP	0.948091	0.078939	5.343702	0.75902	0.069952	0.934031
HB	0.972904	0.047045	5.763103	0.818591	0.08445	0.962695

that even a small amount of noise can impact the accuracy of recommendations. However, HB maintained its relative performance, showing a slight resilience to the introduction of minor noise. This robustness is attributed to HB's ability to balance overall satisfaction and mitigate dissatisfaction, ensuring that even with noise, the recommendations remain fairly accurate and satisfactory. When the noise level increased to 1, noticeable drops in satisfaction (0.951 for Avg) and NDCG (0.727) were observed. This highlights the sensitivity of homogeneous groups to higher levels of noise. Despite this, HB continued to perform relatively well compared to other methods, indicating its capacity to handle noise better. The significant reduction in satisfaction and NDCG explains the challenges that noise introduces, particularly in maintaining the precision of recommendations. With the highest noise level (2), there were significant declines in satisfaction

Table 3. Heterogeneous Group Data.

Aggregation Methods	Satisfaction	Dissatisfaction	DCG	NDCG	DFH	F-Score
Noise Level 0						
Avg	0.843899	0.204308	2.424769	0.344414	0.089826	0.817851
LM	0.822256	0.171953	1.988137	0.282395	0.070908	0.824639
MP	0.686066	0.315421	1.833002	0.26036	0.069962	0.680089
HB	0.838045	0.182596	2.226325	0.316227	0.089851	0.826918
Noise Level 0.5						
Avg	0.841392	0.207662	2.444075	0.347156	0.09453	0.814854
LM	0.821099	0.164429	2.042484	0.290114	0.070934	0.827841
MP	0.745413	0.321997	2.637756	0.374667	0.069957	0.70326
HB	0.834962	0.177366	2.247982	0.319303	0.094555	0.828137
Noise Level 1						
Avg	0.837992	0.23032	2.376873	0.337611	0.09401	0.797518
LM	0.81081	0.171658	1.841741	0.261601	0.070958	0.818721
MP	0.660717	0.391212	1.740307	0.247193	0.069963	0.618309
HB	0.829211	0.191696	2.114046	0.300279	0.094036	0.817783
Noise Level 2						
Avg	0.786713	0.362919	2.03816	0.2895	0.085744	0.688059
LM	0.735394	0.211473	1.330639	0.189004	0.071394	0.759575
MP	0.640243	0.478197	1.64885	0.234203	0.069963	0.558172
HB	0.768677	0.255092	1.654434	0.234996	0.085828	0.755059
Uniform Noise Level						
Avg	0.840183	0.214399	2.466898	0.350398	0.087812	0.810318
LM	0.816268	0.165511	2.015902	0.286339	0.07083	0.824647
MP	0.665965	0.341078	1.807378	0.25672	0.069963	0.654299
HB	0.832524	0.180226	2.243656	0.318689	0.087835	0.8253

(0.878 for Avg) and NDCG (0.545 for Avg), indicating that substantial noise disrupts the consistency of recommendations. This level of noise makes it difficult for the recommendation system to accurately predict user preferences, resulting in lower satisfaction and greater variability in the quality of recommendations. When uniform noise was applied, the performance metrics were comparable to those observed with 0 Noise. Avg showed the highest satisfaction (0.975) and NDCG (0.838), suggesting that uniform noise maintains stability like no-noise conditions. This indicates that uniform noise can be a viable strategy to introduce diversity without compromising the accuracy and fairness of recommendations. This consistent performance reflects its potential to balance randomness in recommendations while preserving the integrity of user satisfaction.

Heterogeneous Groups: In Table 3, for heterogeneous groups with no noise, satisfaction was notably lower (0.844 for Avg) compared to homogeneous groups with higher dissatisfaction and lower NDCG (0.344). This outcome reflects the inherent challenge of catering to diverse user preferences. The variability in user tastes makes it difficult for the recommendation system to achieve high levels of satisfaction and agreement, leading to more pronounced dissatisfaction and lower ranking quality. The introduction of 0.5 noise, led to minor improvements in DCG and NDCG, but overall satisfaction slightly decreased (0.841 for Avg). This mixed impact suggests that while some diversity can enhance ranking quality, it may not significantly improve overall user satisfaction in highly diverse groups. The subtle shifts in performance metrics indicate that minor noise does not drastically alter the dynamics of heterogeneous group recommendations. At the 1 noise level, satisfaction (0.838 for Avg) and NDCG (0.338) showed slight improvements. This indicates that moderate noise might help balance recommendations by introducing variability that can potentially align better with diverse user preferences. The modest gains suggest that a moderate level of noise can contribute positively by preventing overfitting specific user preferences and promoting broader appeal in recommendations. A noticeable decrease in satisfaction (0.787 for Avg) and NDCG (0.290) was observed at noise level 2. This highlights the difficulty in maintaining fairness and satisfaction with high noise in diverse groups. The significant drop in metrics suggests that excessive noise exacerbates the challenge of delivering accurate and fair recommendations, further complicating the task of meeting varied user expectations. Performance remained relatively stable with uniform noise, with Avg showing the highest satisfaction (0.840) and NDCG (0.350). This stability indicates that uniform noise can be managed effectively even in diverse groups, providing a balanced approach to introducing variability without severely impacting satisfaction and ranking quality. The results suggest that uniform noise can offer a compromise, maintaining reasonable performance while enhancing fairness and diversity.

Mixed Similarity Groups: In Table 4, at noise level 0, satisfaction (0.880 for Avg) and NDCG (0.483) were intermediate between homogeneous and heterogeneous groups, reflecting the balanced nature of these groups. The results indicate that mixed groups benefit from moderate alignment in preferences, leading to relatively high satisfaction and decent ranking quality. The introduction of a slight noise-like noise level 0.5 decreases satisfaction (0.877 for Avg) and NDCG (0.470), showing a minor impact. This suggests that mixed similarity groups can tolerate minor noise without significant detriment to recommendation quality. Increasing the noise level to 1 further decreases the satisfaction (0.861 for Avg) and NDCG (0.451), indicating sensitivity to moderate noise. The results highlight that mixed similarity groups, while more resilient than homogeneous or heterogeneous groups, still experience a decline in performance with increasing noise levels. The moderate noise impacts the alignment of recommendations with user preferences, reducing overall satisfaction and ranking quality. Higher noise levels like noise level 2 lead to significant drops in satisfaction (0.812 for Avg) and NDCG (0.366), suggesting that high noise disrupts balanced group recommendations. The results indicate that excessive noise can destabilize the balance in mixed similarity groups, leading to lower satisfaction and poorer ranking quality. The significant declines emphasize the challenge of maintaining effective recommendations with high levels of noise. Avg maintained the highest satisfaction (0.880) and NDCG (0.502) under uniform noise, showing that this type of noise can maintain stability in mixed groups. The results suggest that uniform noise is effective in preserving the balance and diversity of recommendations without significantly compromising performance. The consistent metrics indicate that uniform noise can provide a reliable approach to managing variability in recommendations for mixed similarity groups. Uniform Noise emerges as the optimal choice for enhancing fairness and satisfaction across all group types. Uniform Noise consistently maintained high satisfaction and NDCG scores comparable to the 0 noise condition, indicating its ability to introduce diversity without significantly disrupting the accuracy and fairness of recommendations.

5.2   Impact of Aggregation Methods on Group Recommendations

Satisfaction and Dissatisfaction: Across all noise levels, Avg consistently achieved the highest satisfaction in homogeneous, heterogeneous, and mixed similarity groups. For instance, with no noise, Avg in homogeneous groups achieved a satisfaction score of 0.974, while the MP method showed the lowest satisfaction at 0.945. This trend remained consistent across noise levels, indicating that Avg provides the most balanced recommendations across diverse user groups.

For dissatisfaction, LM generally resulted in lower dissatisfaction scores compared to other methods. This method aims to prevent extreme dissatisfaction by focusing on the least satisfied group member. For example, with 0.5 noise, LM in homogeneous groups had a dissatisfaction score of 0.051, which is lower compared to the Avg, MP, and Hybrid aggregation methods. This trend suggests that LM is effective in minimizing dissatisfaction among group members.

NDCG: NDCG is a critical metric for evaluating the ranking quality of recommendations. The results indicate that Avg consistently yielded the highest NDCG scores across all noise levels and group types. For instance, with uniform noise, Avg in mixed similarity groups achieved an NDCG of 0.502. In contrast, MP generally had the lowest NDCG scores, reflecting its focus on the preferences of the most satisfied group member rather than overall ranking quality.

DFH: DFH evaluates the fairness of the recommendations by considering both the relevance and position of items. The results show that DFH scores were highest with the Avg and HB methods across all noise levels. For example, with 1 noise, HB in homogeneous groups had a DFH of 0.088, indicating a more balanced exposure of relevant items across different users. This suggests that Avg and HB are more effective in promoting fairness in recommendations.

F-Score: F-Score showed that the LM and HB methods generally outperformed the Avg and MP methods. This trend was consistent across all noise levels and group types. For instance, with noise level 2, LM and HB in homogeneous groups

Table 4. Mixed Similarity Group Data.

Aggregation Methods	Satisfaction	Dissatisfaction	DCG	NDCG	DFH	F-Score
Noise Level 0						
Avg	0.880495	0.20513	3.39986	0.482916	0.069997	0.834111
LM	0.848745	0.128293	2.539375	0.360693	0.070437	0.859542
MP	0.756013	0.289878	2.659737	0.377789	0.069957	0.725553
HB	0.869162	0.143684	2.945936	0.418441	0.070016	0.861985
Noise Level 0.5						
Avg	0.876706	0.202606	3.311273	0.470333	0.069993	0.834023
LM	0.845147	0.131166	2.483763	0.352794	0.07053	0.856309
MP	0.74391	0.292459	2.568146	0.36478	0.069957	0.720402
HB	0.865499	0.144086	2.874143	0.408243	0.070015	0.860019
Noise Level 1						
Avg	0.860858	0.235052	3.174154	0.450857	0.069992	0.806159
LM	0.827804	0.151473	2.376574	0.337569	0.070493	0.837355
MP	0.724167	0.324307	2.425481	0.344515	0.069959	0.693514
HB	0.849497	0.175426	2.751251	0.390788	0.070013	0.835896
Noise Level 2						
Avg	0.812139	0.346677	2.575832	0.365871	0.070013	0.709917
LM	0.758951	0.207645	1.678201	0.238372	0.070942	0.773213
MP	0.665062	0.426709	1.955654	0.277781	0.069962	0.595689
HB	0.791997	0.24938	2.09719	0.297885	0.070032	0.768789
Uniform Noise Level						
Avg	0.879807	0.216471	3.534236	0.502003	0.06999	0.827057
LM	0.847213	0.127661	2.685867	0.381501	0.070486	0.859059
MP	0.759654	0.281608	2.75755	0.391683	0.069957	0.732726
HB	0.867165	0.152728	3.066453	0.435559	0.070009	0.856393

had an F-Score of 0.835 and 0.831, respectively, while Avg and MP had an F-Score of 0.785 and 0.731. This indicates that Avg and HB provide a more balanced approach to accuracy in recommendations.

Avg consistently performed the best in terms of satisfaction, DCG, NDCG, DFH, and F-Score across all noise levels and group types. It ensures that all opinions are considered equally, resulting in balanced and fair recommendations. LM excelled in minimizing dissatisfaction, making it suitable for scenarios where it is crucial to avoid extreme discontent among group members. However, it often had lower DCG and NDCG scores, indicating potential trade-offs in ranking quality. MP focused on maximizing the satisfaction of the happiest group member, which often led to lower overall satisfaction and fairness scores. This method may be suitable for groups with highly similar preferences but less effective for diverse groups. HB combined the strengths of Avg and LM, resulting in balanced performance across all metrics. It effectively minimized dissatisfaction while maintaining high satisfaction, DCG, NDCG, and DFH scores. According to the evaluation metrics, HB ensures that the recommendations are generally satisfactory for the entire group while avoiding extreme dissatisfaction for any individual member, thus offering a perfect trade-off for group recommendations.

Overall, the evaluation of various noise levels and aggregation methods indicates that HB, when paired with uniform noise, provides the best trade-off for different group types. By merging the strengths of Avg and LM, HB delivers balanced performance across all metrics. It minimizes dissatisfaction, while maintaining high levels of satisfaction, DCG, NDCG, and DFH scores. Therefore, HB emerges as the optimal strategy, ensuring generally satisfactory recommendations for the entire group and avoiding extreme dissatisfaction for any individual member, thus offering the best compromise for group recommendations.

6 Summary

This paper develops a fairness-aware group recommendation model that integrates stochastic components within the VAE framework. This model introduces randomness into the encoding and decoding processes, enhancing the variability and fairness of recommendations across different items and user groups. Our experimental evaluation provided significant insights into the trade-offs between satisfaction, dissatisfaction, and ranking quality. By systematically varying noise distributions and comparing various aggregation methods, we demonstrated how different levels and types of noise influence the performance of the recommendation model. The findings indicate that moderate levels of noise, particularly uniform noise levels, effectively balance satisfaction, dissatisfaction, and ranking quality across all types of groups. HB emerged as a robust approach, delivering balanced performance across satisfaction, DCG, NDCG, DFH, and F-Score.

References

1. Biega, A.J., Gummadi, K.P., Weikum, G.: Equity of attention: amortizing individual fairness in rankings. In: Proceedings of the 41st International ACM Conference on Research & Development in Information Retrieval (SIGIR), pp. 405–414 (2018)
2. Borges, R., Stefanidis, K.: Enhancing long term fairness in recommendations with variational autoencoders. In: Proceedings of the 11th International Conference on Management of Digital EcoSystems (MEDES), pp. 95–102 (2019)
3. Borges, R., Stefanidis, K.: On mitigating popularity bias in recommendations via variational autoencoders. In: Proceedings of the 36th ACM/SIGAPP Symposium on Applied Computing (SAC), pp. 1383–1389 (2021)
4. Borges, R., Stefanidis, K.: Feature-blind fairness in collaborative filtering recommender systems. Knowl. Inf. Syst. 64(4), 943–962 (2022)
5. Gupta, K., Raghuprasad, M. Y., Kumar, P.: A hybrid variational autoencoder for collaborative filtering. arXiv:1707.00093 (2018)

6. Guzzi, F., Ricci, F., Burke, R.: Interactive multi-party critiquing for group recommendation. In: Proceedings of the ACM Conference on Recommender Systems (RecSys), pp. 265–268 (2011)
7. Harper, F., Konstan, J.: The MovieLens datasets: history and context. ACM Trans. Interact. Intell. Syst. **5**(4), 19:1-19:19 (2016)
8. Imtiaz, W.: Fairness in variational autoencoders recommenders. MSc thesis, Tampere University, Finlald (2020)
9. Jameson, A., Smyth, B.: Recommendation to groups. Chapter in: The Adaptive Web, pp. 596–627. Springer (2007)
10. Liang, D., Krishnan, R.G., Hoffman, M.D., Jebara, T.: Variational autoencoders for collaborative filtering. In: Proceedings of the World Wide Web Conference on World Wide Web (WWW), pp. 689–698 (2018)
11. Naamani Dery, L., Kalech, M., Rokach, L., Shapira, B.: Iterative voting under uncertainty for group recommender systems. In: Proceedings of the ACM Conference on Recommender Systems (RecSys), pp. 265–268 (2010)
12. Ricci, F., Rokach, L., Shapira, B., Kantor, P.B.: Introduction to recommender systems handbook. In: Recommender Systems Handbook, Springer (2010)
13. Stratigi, M., Nummenmaa, J., Pitoura, E., Stefanidis, K.: Fair sequential group recommendations. In: Proceedings of the 35th ACM/SIGAPP Symposium on Applied Computing (SAC), pp. 1443–1452 (2020)
14. Stratigi, M., Pitoura, E., Stefanidis, K.: SQUIRREL: a framework for sequential group recommendations through reinforcement learning. Inf. Syst. **112**, 102128 (2023)
15. Stratigi, M., Pitoura, E., Nummenmaa, J., Stefanidis, K.: Sequential group recommendations based on satisfaction and disagreement scores. J. Intell. Inf. Syst. **52**(2), 227–254 (2022)
16. Pitoura, E., Stefanidis, K., Koutrika, G.: Fairness in rankings and recommendations: an overview. VLDB J. **31**(3), 431–458 (2022)

A Supervised Contrastive Learning Framework for Aspect-Based Recommendations

Padipat Sitkrongwong[1]([✉]) [iD] and Atsuhiro Takasu[1,2] [iD]

[1] National Institute of Informatics, Tokyo, Japan
{padipat,takasu}@nii.ac.jp
[2] The Graduate University for Advanced Studies, SOKENDAI, Tokyo, Japan

Abstract. Aspect-based recommender systems (RSs) aim to provide accurate and explainable recommendations by leveraging fine-grained features from user reviews. However, existing methods do not utilize the accompanying sentiment information to enhance the discriminability of positive and negative preferences. To determine whether a user likes or dislikes an item, it is intuitive to assess the alignment or contradiction of item positive and negative aspects with the user's preferred and rejected aspects. To realize this intuition, we propose a novel supervised contrastive learning approach that models relationships between ratings and aspect preferences for making recommendations. Our aspect representations are explicitly enriched with sentiments, capturing both semantic and sentimental aspects of user preferences. Additionally, we introduce a constraint to model the semantic relationship between observed and unobserved aspect preferences, enhancing recommendation accuracy. Extensive experiments demonstrate that our proposed framework consistently outperforms state-of-the-art RS methods not only in terms of accuracy but also in robustness to negative items.

Keywords: contrastive learning · negative preferences · review-based recommendations · aspect features · sentiment analysis

1 Introduction

Learning user and item representations that adeptly capture user preferences and item characteristics has emerged as a primary objective in recommender systems (RSs). Aspect-based RSs enhance the quality of such representations by exploring the fine-grained product features, i.e., *aspects*, mentioned in reviews. For example, [1,7,24] rely on aspect-level sentiment analysis tools [14,23] to extract aspect-sentiment lexicons from reviews. These lexicons consist of aspect words (or phrases) and, in some cases, their associated sentiment polarities ("pos": positive or "neg": negative), as shown in Fig. 1. Leveraging aspects is valuable for recommendations in multiple ways. Firstly, when users and items share common aspects in their reviews, it suggests potential recommendation suitability. For example,

R. Chbeir et al. (Eds.): IDEAS 2024, LNCS 15511, pp. 312–328, 2025.
https://doi.org/10.1007/978-3-031-83472-1_21

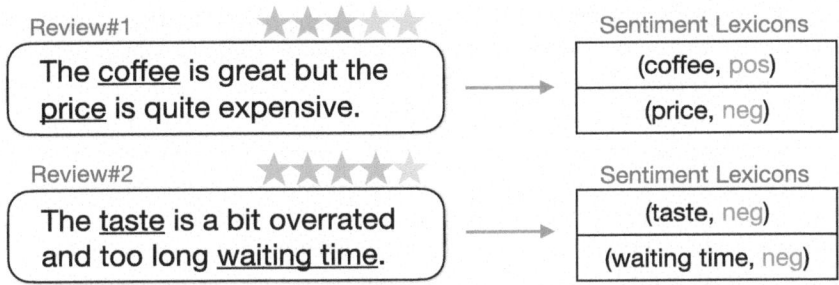

Fig. 1. Example of aspect-sentiment lexicons extracted from reviews.

if a user often mentions "free Wi-Fi," recommending certain places offering this service is logical. Additionally, aspects provide intuitive explanations for recommendations. In the previous example, it is understandable that items are being recommended because they are frequently mentioned with free Wi-Fi.

Nevertheless, several aspect-based RSs still encounter common challenges. Firstly, many aspect-based RSs prioritize on recommending items that shared common aspects with users, often overlooking sentiment information. However, aspects are rarely mentioned without accompanying sentiment polarities [14,21], which are crucial for accurately modeling user preferences and preventing misguided recommendations. For instance, upon analyzing the sentiment polarities associated with "coffee" and "taste" in Fig. 2, it becomes evident that the restaurant has received many negative opinions and should not be recommended.

Secondly, although some aspect-based RSs attempt to incorporate sentiment information [1,9,10,24], none of them explicitly consider a fundamental intuition regarding the relationship between user and item sentimental aspects. Specifically, when a user "likes" an item, it is expected that the user's preferred aspects align with the item's positive aspects, while the item should lack negative aspects that the user rejects. Conversely, when a user "dislikes" an item, it is likely that the item contains negative aspects of concern to the user, but rarely includes the positive aspects desired by the user. We refer to this concept as *contrastive properties* between positive and negative aspects of users and items, and argue that considering these properties is essential for producing more accurate and intuitive recommendations.

Aspect Term	Aspect Segment	Sentiment	Aspect Term	Aspect Segment	Sentiment
coffee	strong **coffee** and no bitter taste	pos	coffee	bitter **coffee** than you will ever like	neg
coffee	this is a really good cup of **coffee**	pos	coffee	the **coffee** turned out rather weak	neg
taste	this **tastes** real butter	pos	taste	it **tastes** like burnt tires	neg

John ⟵ recommend? ⟶ ABC restaurant

Fig. 2. Example of aspect terms, segments and their associated sentiments.

Thirdly, aspect-based RSs often incorporate aspects into the model as isolated words or phrases, referred to as *aspect terms*, while completely ignoring the surrounding contexts. In fact, such contexts provide more accurate insights into users' intentions towards those aspects, which is crucial to capture both semantic meaning and sentiment polarities of aspects. Thus, incorporating both aspect terms and their contexts, referred to as *aspect segments* (as illustrated by the examples in Fig. 2), could enhance the preference modeling.

Moreover, since the positive and negative representations of users and items are derived from aspects and their sentiments, it is crucial to ensure that the representations of these aspects, i.e., aspect embeddings, not only capture their semantic meanings but also incorporate their corresponding sentiment polarities. While many pre-trained language models can effectively produce aspect embeddings that preserve semantic regularities, they may fail to naturally integrate sentiment polarities into these embeddings [19,22]. Consequently, this could lead to aspect embeddings that are unable to discern sentiment, as depicted in Fig. 4(a). Utilizing such embeddings in making recommendations would be equivalent to disregarding sentiment information.

Finally, relying solely on aspects mentioned in reviews may inadequately model user preferences. Considering the second review in Fig. 1, a high-rated item review scarcely highlights positive aspects, implying that there are other hidden factors influencing user preferences for items. To address this issue, many aspect-based RSs have combined the representations learned from aspects with latent factors associated with specific users and items to refine recommendations [7,24]. However, these latent factors are often defined independently without explicit relationships with the observed aspects, potentially impacting the interpretability of the representations.

To address these limitations, we propose a novel framework called PACLR (**P**reference-**A**ware **C**ontrastive **L**earning for **R**ecommendations) for leveraging sentiments in aspect-based RSs. Centering on the intuition of contrastive properties, we design a novel supervised contrastive learning that models relationships between ratings and aspect preferences to learn positive and negative representations. We introduce an effective technique to encode sentiment polarities into aspect segment representations, resulting in both semantically and sentimentally discriminable aspect embeddings. Finally, we apply contrastive properties as a constraint to incorporate hidden factors as unobserved aspects. Extensive experiments demonstrate that PACLR outperforms existing state-of-the-art RS methods. The primary contributions of this work can be summarized as:

- We propose a novel framework for sentiment-enriched aspect based recommendations. Our supervised contrastive learning integrates two preference types: aspect preferences from reviews, generating fine-grained positive and negative representations, and rating preferences, supervising their distances and recommendation scores.
- Differing from the existing methods, our aspect embeddings are generated from aspect segments and explicitly enriched with sentiment information, enable both semantic and sentiment based discrimination.

- Lastly, we introduce a distance-based constraint to model observed and unobserved aspect relationships, enhancing recommendation accuracy.

2   Related Work

2.1   Aspect-Based Recommendations

The aspect-based recommendations can be classified into two main approaches based on how they represent aspect information: latent and explicit approaches. The latent approach aims to infer aspect information from reviews as latent variables and utilize them for making recommendations [3–5,12]. For example, A3NCF [3] defines aspects as topic distributions over a set of reviews associated with users and items. However, associating the latent variables with specific product features is challenging for explainability. On the other hand, the explicit approach relies on techniques to detect aspects from reviews before using them in their recommendation algorithms. These techniques vary from aspect-level sentiment analysis [14,21,23] to more advanced deep learning methods [17,20]. Once the aspects are extracted, they are utilized by various aspect-based RS methods. For example, EFM [24] represents the importance of each aspect to each user/item based on its frequency of occurrence in reviews. AARM [7] models interactions between synonymous and similar aspects while considering the varying aspect interests of users on different items. Benefiting from fine-grained aspect information, the explicit approach could provide recommendations with intuitive explainability. Nonetheless, their performance relies on aspect extraction quality. Extracted aspects are often isolated aspect terms, lacking contextual information. Many methods also try to address feature sparsity issue by incorporating the model with latent features, but they are defined independently from the observed aspects, limiting the interpretability.

Finally, it is important to note that many aspect-based RSs tend to underutilize or completely neglect the sentiment polarities of aspects when modeling user preferences. However, it is essential to recognize that not all aspects have positive influences, as negative preferences may have a stronger impact on user satisfaction with the systems. For example, in music recommendations, users tend to feel more upset when being recommended songs they disliked than when not being recommended songs they liked [13].

2.2   Negative Preference Modeling

Some works in aspect-based RSs also attempt to incorporate sentiment information to model both user positive and negative preferences [1,9,10,24]. For example, CARP [9] proposes a latent approach to model aspects and sentiments and learns separate positive and negative representations using a capsule network. However, their aspect embeddings are generated by applying convolutional and self-attention techniques on reviews, which limits the explainability of the model. RPR [10] treats the sentiment polarity of each aspect as identical to the sentiments (i.e., rating scores) of its review, which are then used to compute user-preferred and rejected aspect scores for rating prediction. However, we disagree with the notion that the polarity of each aspect necessarily aligns with its review's rating, as depicted in Fig. 1. Additionally, in both CARP and RPR, positive and negative scores are computed separately without considering any relationships between positive and negative aspect preferences. This limitation restricts the models' capability to distinguish the characteristics of positive items from negative ones.

Additionally, self-supervised learning offers another approach to utilize negative preferences. For instance, RGCL [18] proposes self-supervised contrastive learning to enhance node and edge discriminations on the user-item bipartite graph, considering non-interacted items and their reviews as negative samples. However, non-interacted items may not necessarily align with items actually disliked by users. Furthermore, [13] introduce supervised contrastive learning based on explicit user preferences (like or dislike), optimizing item representation distances tailored to individual users. Nonetheless, they overlook the distances between user and item representations, which is more relevant for RSs.

3   Proposed Framework

In this section, we provide a detailed explanation of PACLR. The architecture of our framework is presented in Fig. 3, which consists of five main modules: *Aspect-Sentiment Extractor*, *Aspect Encoder*, *Sentiment Projector*, *Aspect Aggregator*, and *Contrastive Recommender*. Firstly, the Aspect-Sentiment Extractor extracts the aspects and their associated sentiments from reviews of a target user and a target item. The Aspect Encoder then maps those aspects to their dense representations using a pre-trained language model. Next, the Sentiment Projector encodes the associated sentiment polarities to create aspect embeddings that are both semantically and sentimentally discriminable. The Aspect Aggregator then models these embeddings to generated positive and negative representations of target user and item. Finally, the Contrastive Recommender models both observed and unobserved aspects and generate recommendation scores based on a contrastive properties between positive and negative preferences.

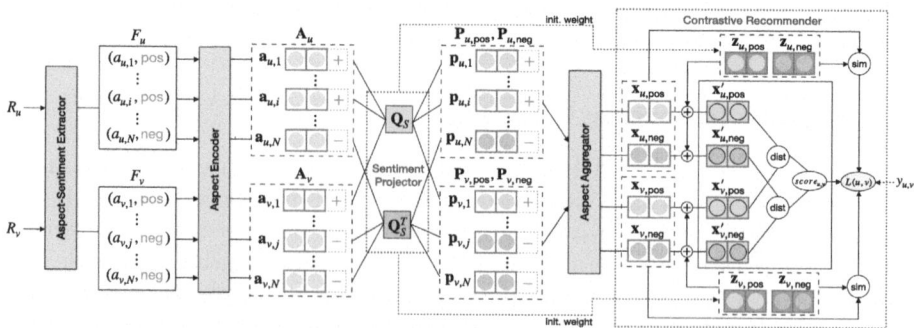

Fig. 3. An overview of PACLR's architecture.

3.1 Aspect-Sentiment Extractor

The objective of this module is to extract aspects together with their associated sentiment polarities from reviews, which will be used as input for our framework. To achieve this, any of the aspect extraction and sentiment analysis methods such as [14,17,20,21,23] can be applied. Specifically, let R_u and R_v represent sets of reviews of user u and item v, respectively. The task is to extract the sets of user aspects A_u and item aspects A_v along with their associated sentiment polarities S_u and S_v from R_u and R_v. Since the sizes of A_u and A_v are of variable length, we follow the approach in [7] to select the top N most relevant aspects for particular users and items based on the TF-IDF scores. The output of this module consists of sets of aspect-sentiment lexicons $F_u = \{(a_{u,1}, s_{u,1}), \dots, (a_{u,N}, s_{u,N})\}$ and $F_v = \{(a_{v,1}, s_{v,1}), \dots, (a_{v,N}, s_{v,N})\}$, where $a_{u,i}, a_{v,j} \in A_u, A_v$ and $s_{u,i}, s_{v,j} \in S_u, S_v$ represent the aspects and their sentiment polarities of user u and item v, respectively. Note that in our work, the values of $s_{u,i}$ and $s_{v,j}$ can be either "pos" (positive) or "neg" (negative), while the values of $a_{u,i}$ and $a_{v,j}$ can support either aspect terms or segments.

3.2 Aspect Encoder

The purpose of this module is to utilize a text encoder function f_{enc} to map each aspect $a_{u,i}$ in F_u (also $a_{v,j}$ in F_v) to its dense representation $\mathbf{a}_{u,i} \in \mathbb{R}^d$, where d is the embedding dimension. The aspect embeddings of semantically similar aspects are expected to be embedded closely together, while dissimilar ones should be further away. To achieve this, the choice of f_{enc} depends on the format of the extracted aspects. Since we want to explore whether the contexts of aspect terms can provide more accurate semantic meaning of aspects, we decide to represent $a_{u,i}$ as an aspect segment and select Sentence-BERT [15] as the encoder for our work. Sentence-BERT is an extension of BERT [6], fine-tuned to generate sentence embeddings by utilizing a Siamese network [16].

The output of this module consists of the sets of aspect embeddings $\mathbf{A}_u$ for user u and $\mathbf{A}_v$ for item v. In Fig. 4(a), we present the t-SNE [11] visualization of aspect embeddings to demonstrate the quality of the representations. The aspect embeddings are generated from 1000 randomly sampled aspect segments, with 100 segments from each of the 10 aspect terms from the Amazon Grocery and Gourmet Food dataset. This visualization clearly demonstrates that the aspect embeddings are semantically discriminable based on their associated aspect terms. For instance, aspect segments such as "taste" and "flavor" are embedded close together, indicating their strong semantic relationship. To preserve these relationships, the embedding weights are frozen throughout the training process.

3.3 Sentiment Projector

Although our Aspect Encoder can generate aspect embeddings with semantic regularities, those aspect embeddings are not discriminable based on their associated sentiment polarities. Considering the Fig. 4(a), for example, there is no

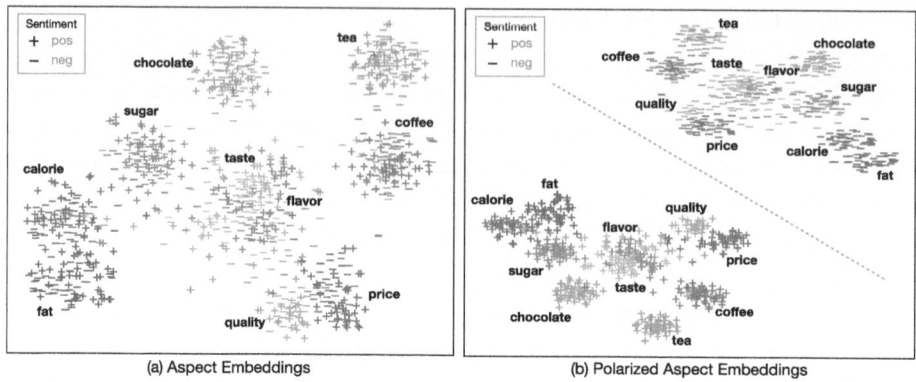

| (a) Aspect Embeddings | (b) Polarized Aspect Embeddings |

Fig. 4. Example of aspect embeddings: (a) before, and (b) after sentiment projection.

obvious clusters among aspects with the same sentiment polarity, and most aspects with different sentiment polarities are mixed together. Utilizing those aspect embeddings directly would fail to discriminate the user's opinion toward such aspects (i.e., whether a user actually likes or dislikes such aspects) and could mislead the recommendations.

To address this issue, we introduce the Sentiment Projector with the goal of explicitly encoding the accompanying sentiment polarities of aspects into their embeddings, making them both semantically and sentimentally discriminable. Specifically, given a pair of $(a_{u,i}, s_{u,i}) \in F_u$, an aspect embedding $\mathbf{a}_{u,i}$ of $a_{u,i}$ is projected into a *polarized* aspect embedding $\mathbf{p}_{u,i} \in \mathbb{R}^d$ with respect to its accompanying sentiment polarity $s_{u,i}$ by:

$$\mathbf{p}_{u,i} = \begin{cases} \mathbf{Q}_S\, \mathbf{a}_{u,i}, & \text{if } s_{u,i} = \text{pos}, \\ \mathbf{Q}_S^T\, \mathbf{a}_{u,i}, & \text{otherwise}, \end{cases} \tag{1}$$

where $\mathbf{Q}_S \in \mathbb{R}^{d \times d}$ is a *polarization matrix* that enforces the aspect embeddings with different sentiment polarities to be projected further away in the embedding space. To ensure the discriminability of the $\mathbf{p}_{u,i}$, $\mathbf{Q}_S$ is initialized as an orthonormal matrix, which can be seen as a simple rotation matrix that rotates vectors to a specific location in the embedding space. By multiplying with the transpose of $\mathbf{Q}_S$, the negative aspect embeddings are rotated in the opposite direction from the positive aspect embeddings, while both the length and orientation of such embeddings are preserved. This indicates that even though the positive and negative aspect embeddings are further away from each other due to the projection, their semantic regularities among aspects are not affected, as visualized by Fig. 4(b). Note that the polarized aspect embedding of item v, $\mathbf{p}_{v,j}$, is computed similarly. The output of this module consists of sets of positive aspect embeddings $\mathbf{P}_{u,\text{pos}}$ and $\mathbf{P}_{v,\text{pos}}$, as well as negative aspect embeddings $\mathbf{P}_{u,\text{neg}}$ and $\mathbf{P}_{v,\text{neg}}$ for both users and items, respectively.

3.4   Aspect Aggregator

The purpose of this module is to utilize the sentiment-enriched aspect embeddings to model the positive and negative user preferences and item characteristics. Specifically, the task is to apply the aggregation function f_{agg} to aggregate all polarized aspect embeddings in $\mathbf{P}_{u,s}$ and $\mathbf{P}_{v,s}$ to create user and item aspect representations $\mathbf{x}_{u,s}$ and $\mathbf{x}_{v,s}$ for sentiment s, where $s \in \{\text{pos}, \text{neg}\}$. Similar to the Aspect Encoder, the choice of f_{agg} also varies from simple aggregation operations such as vector summation or averaging, to more advanced techniques such as self-attention used in [9]. In this paper, we adopt a state-of-the-art *aspect-interaction* technique proposed by [7] as f_{agg}. The key idea is to generate a user representation considering not only the user aspects but also how well each aspect of this user is matched with aspects of a target item. Specifically, we first modify $\mathbf{p}_{u,i}$ to capture how suitable its interactions are with every aspect of item v with the same sentiment s by:

$$
\mathbf{p}'_{u,i} = \sum_{\mathbf{p}_{v,j} \in \mathbf{P}_{v,s}} \beta_{i,j}(\mathbf{p}_{u,i} \odot \mathbf{p}_{v,j}),
$$

$$
\beta_{i,j} = \sigma_1(\mathbf{w}^T_{\text{att}_1}(\mathbf{p}_{u,i} \odot \mathbf{p}_{v,j})),
$$

(2)

where $\sigma_1(\cdot)$ is a softmax function over interactions between $\mathbf{p}_{u,i}$ and all $\mathbf{p}_{v,j} \in \mathbf{P}_{v,s}$, and $\mathbf{w}^T_{\text{att}_1} \in \mathbb{R}^d$ is a model parameter. The aspect-level attention score $\beta_{i,j}$ allows the interactions of semantically similar aspect pairs (e.g., flavor and taste) to have more contributions than those unrelated pairs (e.g., coffee and parking lot). The user aspect representation on sentiment s is then generated by a weighted sum over all $\mathbf{p}'_{u,i} \in \mathbf{P}'_{u,s}$ as follows:

$$
\mathbf{x}_{u,s} = \sum_{\mathbf{p}'_{u,i} \in \mathbf{P}'_{u,s}} \alpha_{u,v,i}\, \mathbf{p}'_{u,i},
$$

$$
\alpha_{u,v,i} = \sigma_2(\mathbf{w}^T_{\text{att}_2}(\mathbf{p}_{u,i} \sum_{\mathbf{P}_{v,s}} \mathbf{p}_{v,j}),
$$

(3)

where $\sigma_2(\cdot)$ is a softmax function over all $\mathbf{p}_{u,i} \in \mathbf{P}_{u,s}$, and $\mathbf{w}^T_{\text{att}_2} \in \mathbb{R}^d$ is a model parameter. The user-level attention score $\alpha_{u,v,i}$ indicates how important aspect $a_{u,i}$ is for item v from the perspective of user u. For instance, if an item is frequently mentioned positively with the "coffee" aspect, a user who also shows a strong interest in coffee in his past reviews should pay more attention to this aspect when considering items. Regardless of the choices for f_{agg}, the output of this module consists of: 1) user positive and negative aspect representations $\mathbf{x}_{u,\text{pos}}$ and $\mathbf{x}_{u,\text{neg}}$, capturing preferred and rejected aspects of user u, and 2) item positive and negative aspect representations $\mathbf{x}_{v,\text{pos}}$ and $\mathbf{x}_{v,\text{neg}}$, capturing positive and negative aspects of item v.

3.5   Contrastive Recommender

Preference-Aware Contrastive Learning. Having generated positive and negative aspect representations, our subsequent objective is to effectively utilize

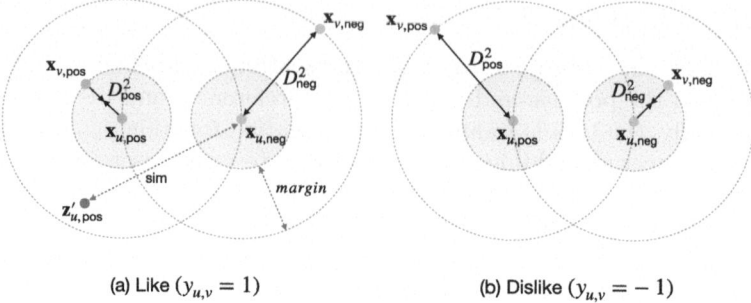

(a) Like $(y_{u,v} = 1)$ (b) Dislike $(y_{u,v} = -1)$

Fig. 5. Illustration of the contrastive properties between positive and negative representations: (a) when users like items, and (b) when users dislike items.

them for recommendations. To formalize the intuition of contrastive properties, we have derived a supervised contrastive learning approach that incorporates two kinds of preferences: the *aspect preferences* from positive and negative aspect representations of users and items, and *rating preferences*, which act as labels to supervise their distances and generate recommendation scores. Given a user-item pair (u, v), the recommendation score and the preference-aware contrastive loss $\mathcal{L}_{PACL}$ are computed as Eq. 4 and 5, respectively.

$$
\begin{aligned}
score_{u,v} &= D_{\text{neg}}^2(u, v) - D_{\text{pos}}^2(u, v) \\
&= \text{dist}^2(\mathbf{x}_{u,\text{neg}}, \mathbf{x}_{v,\text{neg}}) - \text{dist}^2(\mathbf{x}_{u,\text{pos}}, \mathbf{x}_{v,\text{pos}}),
\end{aligned}
\tag{4}
$$

$$
\mathcal{L}_{PACL}(u, v) = \max(0, margin - y_{u,v}\, score_{u,v}),
\tag{5}
$$

where $D_{\text{pos}}(u, v)$ and $D_{\text{neg}}(u, v)$ respectively denotes distances between positive and negative sentiment pairs of user u and item v, $\text{dist}(\cdot)$ is a distance function, $y_{u,v} \in \{-1, 1\}$ is a preference label denoting whether the user u likes (1) or dislikes (-1) an item v, and *margin* is a hyperparameter for controlling the separation between positive and negative representations.

The $\mathcal{L}_{PACL}$ draws inspiration from the triplet loss commonly used in Siamese networks [16] and adapts it for the context of recommender systems, to guide the learning process of user and item representations based on their contrastive properties. Figure 5 demonstrates how PACL is computed based on two scenarios. When a user likes an item $(y_{u,v} = 1)$, the distance between positive pairs (i.e., $\mathbf{x}_{u,\text{pos}}$ and $\mathbf{x}_{v,\text{pos}}$) should be smaller than the distance between negative pairs. In this case, positive scores are produced. If the computed score exceeds the specified margin, there will be no penalty. In contrast, when a user dislikes an item $(y_{u,v} = -1)$, the negative pairs should have smaller distances than the positive pairs, resulting in negative scores. No penalty is incurred if the score multiplication with $y_{u,v}$ exceeds the margin.

Modeling Unobserved Preferences. Apart from the observed aspect information in reviews, there may be unobserved features that influence each individual user's preferences on items. In this work, these features are referred to as latent positive and negative aspects, which are not explicitly mentioned in the reviews. Specifically, we define $\mathbf{z}_{u,\text{pos}}, \mathbf{z}_{u,\text{neg}} \in \mathbb{R}^d$ as the latent positive and negative aspect representations of user u, respectively. To incorporate these latent aspects into the model, we project them using the user positive and negative polarization matrices $\mathbf{Q}_{U,\text{pos}}, \mathbf{Q}_{U,\text{neg}} \in \mathbb{R}^{d \times d}$ which shared among all users as: $\mathbf{z}'_{u,\text{pos}} = \mathbf{Q}_{U,\text{pos}} \mathbf{z}_{u,\text{pos}}$ and $\mathbf{z}'_{u,\text{neg}} = \mathbf{Q}_{U,\text{neg}} \mathbf{z}_{u,\text{neg}}$. To ensure that $\mathbf{z}'_{u,\text{pos}}$ and $\mathbf{z}'_{u,\text{neg}}$ are projected into the same space as the explicit aspect representations $\mathbf{x}_{u,\text{pos}}$ and $\mathbf{x}_{u,\text{neg}}$, we initialize $\mathbf{Q}_{U,\text{pos}}$ and $\mathbf{Q}_{U,\text{neg}}$ with the weights of $\mathbf{Q}_S$ and its transpose (from Eq. 1), respectively. The final preference representation of user u for sentiment s is then generated by combining both explicit and latent aspect representations as: $\mathbf{x}'_{u,s} = (\mathbf{W}_{\text{agg}} \mathbf{x}_{u,s} + \mathbf{b}_{\text{agg}}) + \mathbf{z}'_{u,s}$, where $\mathbf{W}_{\text{agg}} \in \mathbb{R}^{d \times d}$ and $\mathbf{b}_{\text{agg}} \in \mathbb{R}^d$ are model parameters shared by all explicit aspect representations. Similarly, the final representation of item v with sentiment s, $\mathbf{x}'_{v,s}$, is computed in a similar manner. The parameters $\mathbf{x}_{u,s}$ and $\mathbf{x}_{v,s}$ used to compute recommendation scores in Eq. 4 can now be replacedwith $\mathbf{x}'_{u,s}$ and $\mathbf{x}'_{v,s}$ for all $s \in \{\text{pos}, \text{neg}\}$. To provide a better interpretation of latent aspect preferences, we further realize the intuition of contrastive properties to model relationships between latent and explicit aspect representations with a *distant-based contrastive constraint*:

$$\mathcal{L}_{Dist}(u,v) = \sum_{t \in \{u,v\}} \sum_{s_1 \neq s_2} \max(0, \text{sim}(\mathbf{x}_{t,s_1}, \mathbf{z}'_{t,s_2})), \tag{6}$$

where $\text{sim}(\cdot)$ denotes a similarity function. This constraint encourages the explicit and latent aspect representations with different sentiment polarities to be further away from each other. For instance, if a user expresses a dislike for expensive food, it is intuitive to imply that their unobserved positive aspects should be distant from expensive restaurants.

The overall loss for the user-item pair (u,v) is computed by combining $\mathcal{L}_{Dist}(u,v)$ with $\mathcal{L}_{PACL}(u,v)$ defined as:

$$\mathcal{L}(u,v) = \mathcal{L}_{PACL}(u,v) + \lambda \mathcal{L}_{Dist}(u,v), \tag{7}$$

where λ is a hyperparameter determining the constraint's impact relative to the loss. Adam [8] is utilized as our optimizer.

4 Experimental Evaluation

Our experiments are designed to address these research questions: (**RQ1**) How does the integration of aspects enhance recommendations? (**RQ2**) What are the advantages of representing aspects as segments rather than isolated terms? (**RQ3**) Why is it imperative to incorporate sentiment information when modeling aspects? (**RQ4**) How do sentiment-enriched aspects and the proposed supervised contrastive learning contribute to the performance of PACLR? (**RQ5**) How

Table 1. Statistics for Amazon review datasets

	H. Care	G. Food	V. Games
Interactions	8,163	15,721	15,748
Users	211	502	433
Items	334	415	1,573
Density	0.1149	0.0755	0.0231
Pos/Neg Ratio	1.0719	1.3003	1.5081
Aspects/User	57.76	65.44	141.79
Aspects/Item	36.55	79.16	39.03

does modeling the relationships between observed and unobserved preferences enhances the performance of PACLR? (**RQ6**) How can recommendations provided by PACLR be explained by aspect preferences? The subsequent sections provide detailed insight into our experimental setup. We proceed to present the results and discussions.

4.1   Experimental Settings

Data Preparation. We employed the Amazon product review dataset [12] spanning three categories: Health and Personal Care, Grocery and Gourmet Food, and Video Games, for our comprehensive evaluations. We selected Sentires [23] as a tool to extract aspect terms and their corresponding aspect segments from reviews within each dataset, and utilized Happy Transformer[1] to classify the sentiments of the identified aspect segments. Subsequently, users and items devoid of both positive and negative aspects were systematically excluded from the dataset. To ensure an ample dataset for effective model training and evaluation, only users with a minimum of 20 interactions were retained. Furthermore, items with a minimum of 10 interactions were selected for the first two datasets, while a minimum of 5 interactions were chosen for the video games dataset. Table 1 provides an overview of the basic statistics of these datasets after preprocessing. Finally, we employ a stratified sampling technique to construct the training and test datasets. For each user, a subset of 10 interactions is randomly chosen as test data, while the remaining interactions serve as training data.

Baselines and Parameter Settings. We compared the performance of our PACLR framework against several state-of-the-art RS baselines. **DeepCoNN** [25] and **NARRE** [2] are deep learning-based approaches for modeling reviews without leveraging fine-grained aspect information. **RGCL** [18] proposes a self-supervised contrastive learning to model a user-item graph where ratings and reviews are treated as edges. The remaining methods are all aspect-based RS approaches. Both **EFM** [24] and **AARM** [7] utilize Sentires to extract aspect terms and incorporate them into their recommendations. **A3NCF** [3], **CARP**

---

[1] https://happytransformer.com/.

Table 2. Performance comparison of PACLR and the baseline RS methods

	Health and Personal Care						Grocery and Gourmet Food						Video Games					
	HR		NDCG		AUC	FPR(↓)	HR		NDCG		AUC	FPR(↓)	HR		NDCG		AUC	FPR(↓)
	@3	@5	@3	@5			@3	@5	@3	@5			@3	@5	@3	@5		
DeepCoNN	0.8664	0.8274	0.8931	0.9103	0.5604	0.8064	0.7789	0.7622	0.8605	0.8814	0.5833	0.7068	0.8037	0.7912	0.8641	0.8812	0.5784	0.6686
NARRE	0.8695	0.8585	0.8961	0.9135	0.5732	0.772	0.8061	0.7777	0.8809	0.8942	0.5652	0.8075	0.8299	0.8134	**0.8855**	**0.8987**	0.5791	0.7948
RGCL	0.8616	0.8491	0.8929	0.91	0.5007	0.9665	0.7749	0.7748	0.8605	0.8767	0.5019	0.9332	0.8122	0.8051	0.8791	0.8931	0.5272	0.9378
EFM	0.8239	0.7991	0.8691	0.8809	0.538	0.6883	0.7023	0.6996	0.8198	0.8458	0.5174	0.7256	0.8129	0.8028	0.8705	0.8878	0.5818	**0.5092**
A3NCF	0.8585	0.8453	0.8882	0.9054	0.5719	0.7824	0.7849	0.7665	0.871	0.8878	0.5644	0.7347	0.7798	0.7718	0.8499	0.8689	0.523	0.8921
CARP	0.8318	0.8302	0.8769	0.8993	0.5372	0.8745	0.7716	0.7606	0.8589	0.8805	0.5056	0.6936	0.7906	0.7889	0.8564	0.8774	0.5226	0.8183
RPR	0.8711	0.8462	0.894	0.9076	0.5648	0.6987	0.7829	0.7518	0.8689	0.8808	0.5818	0.6515	0.8329	0.8143	0.8717	0.8914	0.5673	0.516
AARM	0.8443	0.8198	0.8781	0.8923	0.5007	0.9979	0.7729	0.7542	0.8615	0.8804	0.5032	0.9915	0.7906	0.7889	0.8578	0.8761	0.5133	0.9514
AARM-S	0.8789	0.8453	0.8994	0.9087	0.578	0.7741	0.8035	0.7759	0.882	**0.8945**	0.5709	0.7685	0.8206	0.8079	0.8762	0.8929	0.5677	0.8027
PACLR	**0.8789**	**0.8651**	**0.8969**	**0.9143**	**0.6267**	**0.6004**	**0.8207**	**0.7793**	**0.8832**	**0.8945**	**0.6367**	**0.5468**	**0.8368**	**0.8185**	**0.8855**	**0.8975**	**0.6354**	**0.5831**

[9], and **RPR** [10] adopt a latent approach for representing aspects. Note that among these methods, only EFM and our proposed PACLR leverage explicit sentiment paired with each aspect, while RPR determines each aspect's sentiment based on a review's rating. For CARP, sentiment information is inferred through the training process. We fine-tune all hyperparameters according to their respective original works to achieve the optimal performance.

For PACLR, we set the maximum length of aspect segments (i.e., the number of word tokens) to 16. These aspect segments are transformed into embeddings using Sentence-BERT with its default configuration. The embedding dimension d is set to match the dimension of the aspect embeddings, which is 384. The maximum number of aspects N is fine-tuned from {16, 32, 64}. We obtain the preference labels $(y_{u,v})$ by binarizing the ratings using a threshold of 4. The Euclidean distance is selected as a distance function dist($\cdot$) with $margin = 1$. For the distance-based constraint, we opt for the cosine similarity as sim($\cdot$), and λ is chosen from {0.1, 1, 10}. The batch size and learning rate of Adam [8] are optimized from {32, 64, 128} and {0.001, 0.1, 1}, respectively. Additionally, we apply L2 regularization with a coefficient chosen from {0.001, 0.01, 0.1}.

4.2   Results and Discussion

Comparison with Baselines. We evaluate the performances of the models using a combination of ranking and classification metrics. Among them, the false positive rate (FPR) provides insight into how frequently the system recommends items that users dislike. The evaluation results are summarized in Table 2. Clearly, PACLR consistently outperforms all baselines across all datasets. These findings underline the effectiveness of PACLR in achieving state-of-the-art performance from multiple angles: aspect integration, aspect representation, and sentiment incorporation.

Aspect Integration (RQ1): Through the integration of aspects, PACLR excels in generating user and item representations that encapsulate fine-grained preferences at the feature-level. The utilization of our contrastive framework to model these representations results in more precise recommendations compared to a review-based approach that overlooks aspect information (i.e., DeepCoNN, NARRE, and RGCL).

Table 3. Performance comparison of PACLR variants

	HR		NDCG		AUC	FPR($\downarrow$)
	@3	@5	@3	@5		
PACLR	**0.8207**	**0.7793**	**0.8832**	**0.8945**	**0.6367**	**0.5468**
PACLR-W	0.7749	0.759	0.8593	0.8815	0.6041	0.6385
PACLR-NP	0.7444	0.7402	0.8411	0.8684	0.5489	0.855
PACLR-NL	0.7703	0.7398	0.8557	0.8726	0.5758	0.6456
PACLR-NC	0.7829	0.7594	0.8574	0.878	0.5874	0.6216

Aspect Representation (RQ2): We compare the performance of PACLR with other aspect-based RS methods. By representing aspects as segments, PACLR clearly outperforms other aspect-based RS methods that represent aspects as isolated terms (EFM and AARM) or latent variables (A3NCF). To further highlight the value of aspect segments, we also introduce AARM-S, which utilizes the same aspect-interaction function as PACLR but represents aspects by segments instead of terms. As expected, AARM-S surpasses the original AARM across all datasets. This indicates that, even without access to sentiment information, the utilization of aspect segments enhances user intention understanding, leading to more meaningful representations and better recommendation accuracy.

Sentiment Incorporation (RQ3 and RQ4): PACLR, leveraging sentiments, consistently outperforms aspect-based RS baselines that disregard sentiments. For example, while AARM's aspect-interaction function is advantageous for suggesting synonymous aspects, overlooking sentiments can result in misguided recommendations, as indicated by its high FPR across datasets. Conversely, PACLR effectively utilizes sentiment information to distinguish between positive and negative preferences, facilitating accurate recommendations through contrastive properties, which consequently yields higher AUC and lower FPR. Comparing with sentiment-aware RS methods, EFM shows variable FPR performance due to feature sparsity, while CARP and RPR, respectively relying on latent approach and reviews' ratings, risk inadequate sentiment capture at the aspect-level. Additionally, self-supervised methods like RGCL consistently show low AUC and high FPR. In contrast, PACLR's sentiment-enriched embeddings and novel supervised contrastive learning enhance both recommendation reliability and interpretability. Thus, PACLR demonstrates more dependable recommendations, prioritizing user-preferred items (high HR and NDCG), avoiding disliked items (low FPR), and improving discriminability (high AUC).

Ablation Studies. In this section, we introduce various variants of PACLR to further investigate our research questions. Their performances on the Amazon Grocery and Gourmet Food dataset are presented in Table 3.

Rank	ItemID	Score	protein	taste	package	Latent Sim.	sugar	chocolate	sauce	Latent Sim.
1	351	2.28	0.23	0.12	0	0.29	0	0	0	-0.59
3	242	1.91	0	0.24	0	0.55	0.09	0	0	-0.33
5	270	0.47	0	0.08	0.07	0.26	0.14	0.41	0	-0.1
9	59	-0.44	0	0.19	0	-0.76	0.38	0	0.25	0.89
10	136	-1.11	0	0	0	-0.14	0.17	0.33	0	0.56

Fig. 6. PACLR's ranked list example showing common aspects of items with a user.

Aspect Segment Modeling (RQ2): To emphasize the advantages of representing aspects as segments, we introduce PACLR-W, a variant in which aspect terms and Word2Vec are used instead of aspect segments and Sentence-BERT. PACLR consistently outperforms PACLR-W across all evaluation metrics, reaffirming that the surrounding contexts of aspect terms provide valuable information that enhances the accuracy of aspect-based RSs.

Sentiment-Enriched Embeddings (RQ4): We introduce PACLR-NP by excluding the Sentiment Projector and all polarization matrices from PACLR. This variant represents a model that takes sentiment information into account but refrains from explicitly encoding it into aspect embeddings. The outcomes show that PACLR-NP yields the lowest AUC and the highest FPR among all variants. This suggests that without the sentiment enrichment, the model lacks the ability to effectively distinguish between liked and disliked items. This highlights the importance of including sentiment information to accurately distinguish user preferences for items through positive and negative representations.

Distance-Based Constraint (RQ5): We also examine PACLR-NC, a variant that integrates latent features into the model without establishing a meaningful connection with observed aspect preferences. However, PACLR-NC does not show substantial improvement compared to PACLR-NL, which disregards latent information entirely. In contrast, introducing a distance-based constraint significantly enhances the performance of PACLR on both accuracy and discriminability. This suggests that modeling the connections between observed and unobserved preferences leads to improved interpretability and accuracy.

Explainability (RQ6). Finally, we show how aspect preferences explain PACLR's recommendation results. Figure 6 displays items ranked by PACLR scores for user #125 on the Amazon Grocery and Gourmet Food dataset. For each item, we compute TF-IDF scores for user #125's top 3 positive aspects ("protein," "taste," "package") and top 3 negative aspects ("sugar," "chocolate," "sauce"), and calculate cosine similarities between the item's latent aspect embeddings and those of user #125. Items frequently mentioned with positive aspects and fewer negative aspects, like item #351, obtained higher PACLR scores compared to those with high negative aspects, like item #136. This illustrates that considering contrastive properties between positive and negative aspects enhances the ranking accuracy. The ranking can be explained through the relationship between user and item latent aspects. For instance, item #242, sharing only one positive aspect (taste) with user #125, still ranks high due to similar latent positive aspect embeddings and distant negative aspect embeddings. This demonstrates that detailed explanations require access to both negative and latent aspects.

5   Conclusion

We propose a novel framework for aspect-based recommender systems. Through comprehensive evaluations, we have demonstrated the significant impact of modeling aspect segments, incorporating sentiments with contrastive properties, and establishing relationships between observed and unobserved preferences. There are two potential research directions for our work. Firstly, we intend to investigate the potential of generative models in aiding the suggestion of aspects, especially in scenarios with limited labeled data. Additionally, we aim to study further into the impact of unobserved preferences on enhancing the explainability of recommendations.

Acknowledgments. This research was supported by MEXT under the program "Developing a Research Data Ecosystem for the Promotion of Data-Driven Science".

Disclosure of Interests. There is no competing interests to declare that are relevant to the content of this article.

References

1. Bauman, K., Liu, B., Tuzhilin, A.: Aspect based recommendations: recommending items with the most valuable aspects based on user reviews. In: Proceedings of the 23rd ACM SIGKDD International Conference on Knowledge Discovery & Data Mining (KDD), pp. 717–725 (2017)
2. Chen, C., Zhang, M., Liu, Y., Ma, S.: Neural attentional rating regression with review-level explanations. In: Proceedings of the 2018 World Wide Web Conference on World Wide Web (WWW), pp. 1583–1592 (2018)
3. Cheng, Z., Ding, Y., Zhu, L., Kankanhalli, M.: Aspect-aware latent factor model: rating prediction with ratings and reviews. In: Proceedings of the 2018 World Wide Web Conference on World Wide Web (WWW), pp. 639–648 (2018)
4. Cheng, Z., et al.: A3NCF: an adaptive aspect attention model for rating prediction. In: Proceedings of the 27th International Joint Conference on Artificial Intelligence (IJCAI), pp. 3748–3754 (2018)
5. Chin, J.Y., Zhao, K., Joty, S., Cong, G.: ANR: aspect-based neural recommender. In: Proceedings of the 27th ACM International Conference on Information & Knowledge Management (CIKM), pp. 147–156 (2018)
6. Devlin, J., Chang, M.W., Lee, K., Toutanova, K.: BERT: pre-training of deep bidirectional transformers for language understanding. In: Proceedings of the 2019 Conference of the North American Chapter of the Association for Computational Linguistics (ACL), vol. 1 (Long & Short Papers), pp. 4171–4186 (2019)

7. Guan, X., et al.: Attentive aspect modeling for review-aware recommendation. ACM Trans. Inf. Syst. **37**(3), 1–27 (2019)
8. Kingma, D., Ba, J.: Adam: a method for stochastic optimization. In: Proceedings of the International Conference on Learning Representations (ICLR) (2015)
9. Li, C., et al.: A capsule network for recommendation and explaining what you like and dislike. In: Proceedings of the 42nd ACM International Conference on Research & Development in Information Retrieval (SIGIR), pp. 275–284 (2019)
10. Liu, H., et al.: Review polarity-wise recommender. IEEE Trans. Neural Netw. Learn. Syst. **34**(12), 10039–10050 (2023)
11. van der Maaten, L., Hinton, G.: Viualizing data using t-SNE. J. Mach. Learn. Res. **9**, 2579–2605 (2008)
12. Ni, J., Li, J., McAuley, J.: Justifying recommendations using distantly-labeled reviews and fine-grained aspects. In: Proceedings of the 2019 Conference on Empirical Methods in Natural Language Processing and the 9th International Joint Conference on Natural Language Processing (EMNLP-IJCNLP), pp. 188–197 (2019)
13. Park, M., Lee, K.: Exploiting negative preference in content-based music recommendation with contrastive learning. In: Proceedings of the 16th ACM Conference on Recommender Systems (RecSys), pp. 229–236 (2022)
14. Qiu, G., Liu, B., Bu, J., Chen, C.: Opinion word expansion and target extraction through double propagation. Comput. Linguist. **37**(1), 9–27 (2011)
15. Reimers, N., Gurevych, I.: Sentence-BERT: sentence embeddings using Siamese BERT-Networks. In: Proceedings of the 2019 Conference on Empirical Methods in Natural Language Processing and the 9th International Joint Conference on Natural Language Processing (EMNLP-IJCNLP), pp. 3980–3990 (2019)
16. Schroff, F., Kalenichenko, D., Philbin, J.: FaceNet: a unified embedding for face recognition and clustering. In: Proceedings of the IEEE Conference on Computer Vision and Pattern Recognition (CVPR), pp. 815–823 (2015)
17. Shi, T., Li, L., Wang, P., Reddy, C.K.: A simple and effective self-supervised contrastive learning framework for aspect detection. In: Proceedings of the 35th AAAI Conference on Artificial Intelligence (AAAI), 33rd Conference on Innovative Applications of Artificial Intelligence (IAAI) 2021, 11th Symposium on Educational Advances in Artificial Intelligence (EAAI), pp. 13815–13824 (2021)
18. Shuai, J., et al.: A review-aware graph contrastive learning framework for recommendation. In: Proceedings of the 45th ACM International Conference on Research & Development in Information Retrieval (SIGIR), pp. 1283–1293 (2022)
19. Tang, D., et al.: Sentiment embeddings with applications to sentiment analysis. IEEE Trans. Knowl. Data Eng. **28**(2), 496–509 (2016)
20. Tulkens, S., van Cranenburgh, A.: Embarrassingly simple unsupervised aspect extraction. In: Proceedings of the 58th Annual Meeting of the Association for Computational Linguistics (ACL), pp. 3182–3187 (2020)
21. Xu, L., et al.: Mining opinion words and opinion targets in a two-stage framework. In: Proceedings of the 51st Annual Meeting of the Association for Computational Linguistics (ACL), vol. 1, pp. 1764–1773 (2013)
22. Yu, L.C., Wang, J., Lai, K.R., Zhang, X.: Refining word embeddings for sentiment analysis. In: Proceedings of the 2017 Conference on Empirical Methods in Natural Language Processing (EMNLP), pp. 534–539 (2017)
23. Zhang, Y., et al.: Do users rate or review? Boost phrase-level sentiment labeling with review-level sentiment classification. In: Proceedings of the 37th ACM International Conference on Research & Development in Information Retrieval (SIGIR), pp. 1027–1030 (2014)

24. Zhang, Y., et al.: Explicit factor models for explainable recommendation based on phrase-level sentiment analysis. In: Proceedings of the 37th ACM International Conference on Research & Development in Information Retrieval (SIGIR), pp. 83–92 (2014)
25. Zheng, L., Noroozi, V., Yu, P.S.: Joint deep modeling of users and items using reviews for recommendation. In: Proceedings of the 10th ACM International Conference on Web Search & Data Mining (WSDM), pp. 425–434 (2017)

Multi-attribute Bias Mitigation in Recommender Systems

Uzair Ahmed and Kostas Stefanidis[✉][ORCID]

Tampere University, Tampere, Finland
konstantinos.stefanidis@tuni.fi

Abstract. Variational Autoencoder (VAE) based recommender systems have successfully matched users with potentially relevant items. VAEs work on the assumption that similar user profiles have similar likenesses and behavior and can be suggested items by finding a pattern in their item relevancy. User profiles can be grouped up based on various factors, the most important one being their history of item rankings, and other personal attributes such as age, country, and sex. An optimal output from a VAE should take into account the most relevant items from the user groups but can also develop a bias towards their attributes which has to be mitigated or it can propagate as the data increases, i.e. learning a pattern that deduces that a certain nationality finds a certain item relevant, which can add unfairness and bias in the results. In this work, we propose a VAE-based framework to minimize bias in recommendations. We take into account multiple sensitive attributes at a time and target to minimize user unfairness as much as possible and improve precision. We compare our results with a state-of-the-art method and document our findings through multiple metrics like precision, unfairness, normalized discounted cumulative gain, and recall.

Keywords: Bias · Unfairness · Recommender Systems

1 Introduction

Variational Autoencoders have emerged as a powerful tool for collaborative filtering tasks by enabling the modeling of intricate patterns in user-item interactions [9]. It provides a multinomial generative model which is useful to find latent-variable space that is non-linear. Advancements in this domain saw a rise in many VAE-based neural networks for recommender systems. For example, in Collaborative VAEs (CVAE) [10], the rating matrix's construction involves utilizing user-item embeddings that represent items in a latent space, while in contrast, Hybrid VAEs (HVAE) [8] rely solely on variational autoencoders to reconstruct the complete user history.

Recommender systems, integral to shaping user experiences, are susceptible to bias and unfairness due to their reliance on extensive data. Fairness is paramount in their development and deployment, as biased recommendations

© The Author(s), under exclusive license to Springer Nature Switzerland AG 2025
R. Chbeir et al. (Eds.): IDEAS 2024, LNCS 15511, pp. 329–342, 2025.
https://doi.org/10.1007/978-3-031-83472-1_22

can perpetuate inequalities and limit exposure to diverse perspectives. Biases may favor specific demographics or content, exacerbating systemic inequalities and reinforcing stereotypes. Addressing fairness concerns is crucial for ensuring equitable recommendations and fostering inclusivity online [2,3].

Our primary focus in this work is to refine and enhance the existing F2VAE [4] recommender system model, with a particular emphasis on improving sensitivity to user diversity, personalization, and addressing biases related to sensitive attributes. The core problem we are tackling involves the limitations of the original model which relies on a binary classifier to predict a single sensitive attribute. The binary classifier might oversimplify user preferences. A user behavior pattern might significantly be influenced by a group of sensitive attributes and the attributes might also be correlated. We investigate whether replacing this binary classifier with a more sophisticated and deeper multi-label classifier can effectively capture a broader range of user characteristics, providing a more nuanced representation of diverse attributes. The investigation into the efficacy of a multi-label classifier aligns with the broader goal of enhancing sensitivity to user diversity, improving personalization, and promoting fairness within recommender systems. Through this research question, the study seeks to contribute to the advancement of recommender system architectures, ultimately striving for more accurate, personalized, and equitable user experiences.

Our method is aimed at maximizing the classification loss associated with multiple sensitive attributes, to strengthen the adversarial training mechanism and advance the recommender system's capabilities in terms of relevancy, personalization, and fairness. The enhanced depth of the classifier enables a more intricate analysis of user profiles by considering a broader spectrum of attributes simultaneously. This depth is instrumental in capturing the complexities inherent in user characteristics, as users often exhibit a diverse array of sensitive attributes. Consequently, the increased granularity facilitates the amplification of classification loss for multiple sensitive attributes, fostering a more robust adversarial training mechanism within the Variational Autoencoder. This advancement aligns with the overarching objective of minimizing the influence of sensitive attributes on the latent space, reinforcing user privacy, and promoting fairness in the recommender system.

We train our implementation on the same parameters as the existing F2VAE [4] implementation to measure the direct impact of our changes on the results. We use the LFM-2B [11] dataset with more than 2 billion listening events from users. The dataset is instrumental in our research as it is the only one in our knowledge with user demographic information present. In our evaluation, we use precision@k (PREC@k), recall@k, unfairness@k (UFAIR@k) and NDCG@k. The precision@k, recall@k, NDCG@k, and UFAIR@k metrics collectively provide a thorough evaluation framework for the recommender system. Precision@k gauges recommendation accuracy in the top-k list, reflecting the proportion of correct suggestions. Higher precision@k values indicate more accurate recommendations aligned with user preferences. Recall@k assesses the system's ability to retrieve relevant items within the top-k recommendations, offering insight into the coverage of user preferences. NDCG@k considers both relevance and ranking, emphasizing the importance of well-ordered recommendations. Elevated precision@k, recall@k, and ndcg@k scores collectively signify accurate, diverse, and well-ordered suggestions. Additionally, the UFAIR@k metric introduces a focus on fairness, specifically capturing unfairness in the top-k recommendations. A lower UFAIR@k score suggests reduced bias and improved fairness in the recommender system's outputs, contributing to a more equitable user experience. In essence, these metrics collectively provide a comprehensive assessment of the recommender system's effectiveness, accuracy, fairness, and overall optimization of the user experience.

2   Related Work

2.1   Fairness in Recommender Systems

A variety of data-driven systems exist nowadays and as the quantity of data increases, a system can unintentionally incorporate human biases or even introduce new ones into its results, and recommendation systems are not invulnerable to these defects [6]. For example, assume an image search with the keyword 'engineers'. The percentage of male representation in the images is an example of stereotype exaggeration which is due to human bias. A recommender should maintain a balance between representing different user groups and still producing novel results to expand a user's profile, it is unfair to ignore the wishes of a certain user group but the lack of newness is equally unfair in recommendations. Most approaches to algorithmic fairness wish to negate any discrimination in the output due to the attributes that are not relevant to the task at hand, the attributes are usually referred to as sensitive attributes [12]. [5] proposes a way of mitigating bias and unfairness without any sensitive attribute information present in the data. [15] proposes a tensor factorization method that isolates the effect of sensitive attributes during the recommendation process. Our solution proposes to remove the bias of sensitive attributes which can assume several values and also several sensitive attributes. Many different combinations of sensitive attributes can be tried and tested to find the best trade-off between training resources and results.

[14] summarizes several definitions for fairness depending on their focus, process fairness, and outcome fairness. Process fairness focuses on ensuring that the recommendation model is fair and does not incorporate any unfair features during learning, such as race or sex. Outcome fairness, on the other hand, concerns the fairness of the recommendations themselves. This can involve ensuring that recommendations are based solely on user preferences rather than stereotypes. Outcome fairness can be categorized into group fairness, which aims to ensure equitable outcomes among diverse groups, and individual fairness, which emphasizes treating similar individuals similarly.

2.2   Autoencoders

Autoencoders are a type of feed-forward artificial neural network (ANN) that excels in learning an identity function in an unsupervised way that can compress and reconstruct the input. Doing this, it learns more efficiently and compresses the representation of the original input [13]. An autoencoder consists of three significant blocks. The encoder, the bottleneck, and the decoder. The encoder block takes in the input vector and passes it through its hidden layers, compressing the input representation to compute z. It is a simple feed-forward ANN that acts as a dimensionality reduction technique that learns a compressed representation of the input.

$$z = g_\phi(x) = W_x + b_z \tag{1}$$

where g_ϕ is the activation function for the encoder output, x is the input vector, W is the weight vector for all the connections and b represents their respective biases. The bottleneck represented by z is the compressed representation learned by the encoder block. The decoder block takes in the compressed representation computed by the encoder and tries to reconstruct the original input from it. The decoder ANN tries to learn the parameters required to reconstruct the compressed representation.

$$\hat{x} = f_\theta(g_\phi(x)) = W'_z + b'_z \tag{2}$$

The reconstruction is represented by $\hat{x}$ and the reconstruction is compared by the original input to calculate the loss which can then be minimized by back-propagation. The autoencoder training step finds optimal values for the set of parameters W, W', b_z and b_x. The loss is calculated using the mean squared error function $L(x, \hat{x})$.

$$L(x, \hat{x}) = 1/n \sum_{i=1}^{n} (x^i - f_\theta(g_\phi(x^i)))^2 \tag{3}$$

3   Methodology

This section focuses on modern implementations in the domain of recommender systems. New research and improvements have paved the way for better recommenders, which are more accurate, fair, and better generalized over a large group of subjects. Autoencoders have been part of many such enhancements. A new version of Autoencoder known as the Variational Autoencoder (VAE) has spearheaded the domain of recommender systems and collaborative filtering ever since. Data-driven domains are always susceptible to unfairness issues and the system must be fair in its dealings and recommender systems are no exception. A recommender system must minimize user unfairness and bias while also giving relevant results, i.e., fair to all groups of users and negating the effects of their attributes on the recommendations. A user can have several sensitive attributes related to them like age, sex, country. A framework must be developed that gives us the control to manage sensitive attributes actively while training regardless of how many they are if they exist in the training data. To address this we used the existing F2VAE model which mitigates user bias by adding two new loss terms and by taking into account a single sensitive attribute while training [4].

3.1 Variational AutoEncoders

Variational Autoencoders (VAEs) blend autoencoders and probabilistic graphical models. VAEs learn a probabilistic mapping from input data to a latent space, treating latent variables as sampled from a Gaussian distribution [9]. This probabilistic approach allows VAEs to generate diverse and realistic samples. Training involves maximizing a lower bound on data likelihood, encouraging the model to learn meaningful representations and a structured latent space.

VAEs are responsible for mapping the input x to a probability distribution which can be represented as p_θ. θ represents the parameters for the distribution. Generating a new sample that is close to the original input x can be labeled as a two-step process: (i) A latent variable z is generated from the prior distribution $p_\theta(z)$, and (ii) a sample x' is generated from a conditional distribution $p_\theta(x'|z)$. The algorithm is aware of the input x but the characteristics of the variable z are unknown which has to be inferred:

$$p_\theta(z|x) = \frac{p_\theta(x|z)p_\theta(z)}{p_\theta(x)} \tag{4}$$

where $p_\theta(x) = \int p_\theta(x|z)p_\theta(z)dz$. This computation is intractable and to overcome, an approximation distribution $q_\phi(z|x)$ is introduced to approximate $p_\theta(z|x)$ [1]. While approximating, we must ensure that the two distributions $p_\theta(z|x)$ are as similar as possible. Kullback-Leibler divergence (KL divergence) metric is introduced here which is a measure of similarity between two probability distributions, thus to ensure similarity, we could minimize the KL divergence between the two distributions. Minimizing the aforementioned metric results in the loss function for a typical VAE, also known as Evidence Lower Bound (ELBO).

$$Loss(\theta, \phi) = -E_{z \sim q_\phi(z|x)} log p_\theta(x|z) + D_{KL}(q_\phi(z|x)||p_\theta(z)) \tag{5}$$

Just like an autoencoder, a variational autoencoder also consists of an Encoder and a Decoder (see Fig. 1). The Encoder encodes an input x into two variables; one for the mean and the other for the standard deviation of the multivariate normal distribution of the latent space. The Decoder takes the mean and standard deviation of the distribution and is tasked with sampling the output with it according to $p_\theta(x|z)$ and the loss is calculated using Eq. 5. The sampling step blocks the gradients from flowing back into the encoder while backpropagating, hence the model will not train. To solve this issue, [9] proposed a re-parameterization trick.

Rather than sampling z directly, we introduce a noise variable ϵ typically from a standard normal distribution and use it to scale our predicted mean and variance which is then used to generate the latent variable z, which gives us:

$$z = \mu + \sigma \odot \epsilon \tag{6}$$

where $\epsilon \sim N(0, 1)$.

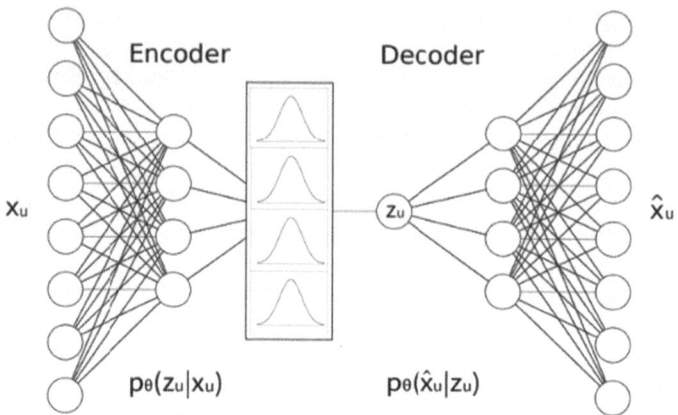

Fig. 1. Variational Autoencoder.

3.2   F2VAE

F2VAE is a framework presented in [4] dedicated to fair representation and fair recommendation. The framework builds upon the ALFR-VAE [7] and FaiRVAE implementations. ALFR-VAE was originally proposed to reduce the influence of a binary user attribute in the context of fair classification. The idea was adapted to categorical sensitive attributes in the context of recommender systems in F2VAE. FaiRVAE introduces a new metric for measuring bias in the results and is focused on mitigating systematic differences between results offered to different groups. F2VAE combines these implementations to reduce unfairness according to a single loss function.

ALFR-VAE implements a binary log-loss for measuring the influence of a binary sensitive attribute in the latent vector z. F2VAE implements a categorical log-loss of a classifier network trained to predict the sensitive attribute s from z. The classifier is trained alongside the recommender and the accuracy is calculated with:

$$C_\theta(z, s) = -E[\sum_{i=1}^{N} s \cdot log(Pred_i(z))] \tag{7}$$

where N is the number of sensitive attribute classes and $Pred_i$ the Softmax probability for the i-th class. The cumulative loss function is now given by:

$$max_\theta min_\phi L_{\theta,\phi} = E_\theta + D_\phi + C_\theta \tag{8}$$

The R.H.S. side of the equation partially comes from Eq. 5. The idea here is to maximize the classification loss to reduce sensitive information in the latent vector z.

FaiRVAE introduces a new metric for mitigating user bias:

$$F_\phi(u) = |\frac{1}{|U_s|} \sum_{c \in U_s} E[logp_\theta(x_c|z_c)] - |\frac{1}{|U|} \sum_{u \in U} E[logp_\theta(x_u|z_u)]|| \qquad (9)$$

where s represents the sensitive attribute related to the user u. The final loss function then can be written as:

$$max_\theta min_\phi L_{\theta,\phi} = E_\theta + D_\phi + C_\theta + F_\phi \qquad (10)$$

Then, the hyperparameters β, γ and τ are added to the KL Divergence metric, classification loss, and the user unfairness metric to control their impact on the loss:

$$max_\theta min_\phi L_{\theta,\phi} = E_\theta + \beta \cdot D_\phi + \gamma \cdot C_\theta + \tau \cdot F_\phi \qquad (11)$$

In Eq. 10, E_θ represents the reconstruction error and D_ϕ represents the divergence between inner representations and the approximated distribution. These loss errors need to be minimized. C_θ stands for the prediction error for the classifier which tries to predict the class of sensitive attributes from the latent vector z which needs to be maximized. F_ϕ stands for user unfairness that needs to be minimized. Gathering all these loss terms together would increase the chances for model weights to converge toward mitigating unfairness associated with a user-sensitive attribute. For a general architecture see Fig. 2.

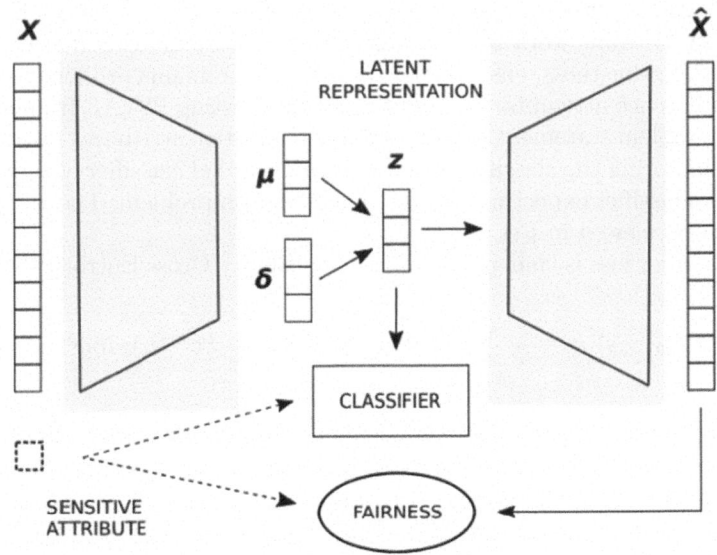

Fig. 2. F2VAE architecture [4].

3.3   Multi-sensitive Attribute Bias Mitigation

Recommender systems play a pivotal role in enhancing user experience by delivering personalized content, products, or services. However, existing models often overlook the intricate nature of user preferences and the potential influence of multiple sensitive attributes. Recommender systems, like F2VAE [4], have shaped how we get personalized recommendations. F2VAE uses a binary classifier in a Variational Autoencoder (VAE) set up to handle sensitive attribute inference. Despite its positive results, there is room to enhance this model, especially considering the evolving nature of user preferences and the need for a more detailed representation of diverse user traits.

F2VAE's binary classifier simplifies user preferences, predicting a single sensitive attribute like a user's country. We propose moving from this binary approach to a more advanced multi-label classifier. The first motivation is to boost the system's sensitivity to user diversity. While F2VAE tailors recommendations well, the binary classifier might oversimplify user preferences. By adopting a multi-label classifier, the improved model intends to capture multiple user-sensitive attributes at once, offering a richer understanding of user profiles. Additionally, the shift to a multi-label classifier aims to enhance personalization beyond binary categories. Concerning fairness and inclusivity, the move to a multi-label classifier is significant. F2VAE's binary approach may unintentionally introduce biases and fail to represent user diversity adequately. The proposed multi-label framework aligns with inclusivity and fairness principles, considering multiple sensitive attributes simultaneously for a more equitable user experience. To strengthen the model against potential threats and protect user privacy, adversarial training within the VAE framework is maintained. The multi-label classifier contributes to adversarial robustness, ensuring the latent space remains resilient to sensitive attribute inference attempts. We build upon the existing F2VAE framework and define the problem statement as using multiple sensitive attributes instead of just one and refactoring the classifier to a deeper multi-label classifier as opposed to a single-layer classifier expecting a binary classification problem. The architectural change can be viewed in Fig. 3.

The classifier loss is changed from Eq. 7 to Binary Cross Entropy loss (BCE):

$$BCEloss = C_\theta(z, s) = -\frac{1}{N} \sum_{i \in U_s}^{N} s_i \cdot log(Pred_i(z_i)) + (1 - s_i) \cdot log(1 - Pred_i(z_i))$$

$$(12)$$

where s_i corresponds to the one-hot encoded ground truth vector of the sensitive attributes related to a user, $log(Pred_i)$ as log probabilities of the predicted classes identified from the latent representation z. $Pred_i(z)$ acts as the output probability function for the classifier defined as the sigmoid $\sigma(z)$ function. The output shape of the function $Pred_i(z)$ matches that of the ground truth vector. Each element from the sigmoid output can be considered active since it is a multi-label classification problem and many sensitive classes can be attributed to a single user. In a multi-label classification problem like ours, the BCE loss treats each label as a mini-binary classification problem where the ground truth

```
MultiVAE(
   (encoder): Encoder(
      (dropout): Dropout(p=0.5, inplace=False)
      (linear_1): Linear(in_features=632663,
out_features=2000, bias=True)
      (linear_2): Linear(in_features=2000, out_features=400,
bias=True)
      (tanh): Tanh()
   )
   (decoder): Decoder(
      (linear_1): Linear(in_features=200, out_features=2000,
bias=True)
      (linear_2): Linear(in_features=2000,
out_features=632663, bias=True)
      (tanh): Tanh()
   )
   (multi_label_classifier): MultiLabelClassifier(
      (fc1): Linear(in_features=200, out_features=128,
bias=True)
      (relu1): ReLU()
      (fc2): Linear(in_features=128, out_features=64,
bias=True)
      (relu2): ReLU()
      (fc3): Linear(in_features=64, out_features=19,
bias=True)
      (sigmoid): Sigmoid()
   )
)
```

```
MultiVAE(
   (encoder): Encoder(
      (dropout): Dropout(p=0.5, inplace=False)
      (linear_1): Linear(in_features=632663,
out_features=2000, bias=True)
      (linear_2): Linear(in_features=2000, out_features=400,
bias=True)
      (tanh): Tanh()
   )
   (decoder): Decoder(
      (linear_1): Linear(in_features=200, out_features=2000,
bias=True)
      (linear_2): Linear(in_features=2000,
out_features=632663, bias=True)
      (tanh): Tanh()
   )
   (classify): Linear(in_features=200, out_features=16,
bias=True)
)
```

Fig. 3. (Left) Proposed solution with sensitive attributes Country & Sex. (Right) F2VAE with sensitive attribute Country.

can be either 0 or 1. Both the positive and negative label errors are summed up and the total loss is averaged on the size of the ground truth vector. The negative sign cancels out the negative values observed after applying the log function to probabilities.

The transition from cross-entropy loss in the existing implementation to binary cross-entropy loss in the proposed multi-label classification marks a substantial improvement in several aspects. Unlike traditional multi-class classification, binary cross-entropy loss excels in handling scenarios where each instance can belong to multiple classes simultaneously. This is particularly advantageous in our context of predicting multiple user-sensitive attributes concurrently. From a fairness perspective, the multi-label approach significantly contributes to mitigating bias by considering the simultaneous influence of multiple sensitive attributes. In contrast to the previous binary classifier focusing on a single aspect, the deeper multi-label classifier offers a more holistic view, reducing the risk of perpetuating biases associated with any single attribute. This shift aligns with the intention to foster fair and unbiased recommendations, promoting a recommender system that better respects the intricacies of individual user profiles. Also, having the ability to classify multiple sensitive attributes associated with a single user and maximizing the error should help mitigate a wider range of group biases and increase long-term fairness.

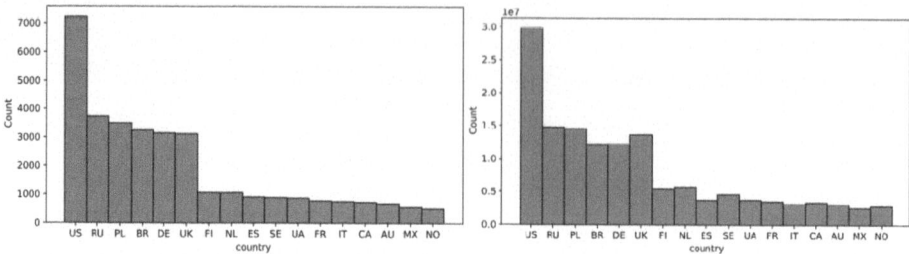

Fig. 4. (Left) Number of users associated with each country in a subset. (Right) The number of interactions associated with each country in the same subset [4].

4   Experimental Evaluation

We used the Last.fm 2B (LFM-2b) dataset [11] to evaluate our proposed multi-sensitive attribute incorporated model and compare the results to the default F2VAE model. The source codes and datasets are available online[1].

Figure 4 shows the user count and listening event distribution in a subset of the total data. It is quite visible that most listening events originate from a single country and this trend carries over to the whole dataset. To prepare the dataset for training, the initial step involved filtering out incomplete user records and those with insufficient listening history. Duplicate listening events were then removed, and minority countries were consolidated. The resulting dataset, consisting of 30,073 users and 632,663 unique tracks, was divided into query and ground truth subsets for model evaluation. The model's recommendations were compared to the ground truth to assess performance using established metrics.

4.1   Evaluation Metrics

We use several quality and fairness measures to evaluate the results of our proposed improvement versus the existing implementations. Namely, Normalized Discounted Cumulative Gain at k (NDCG@k), Precision@k, Recall@k, and Unfairness at k (UFAIR@k). The UFAIR@k metric is a measure of disparity in the precision of top k recommendations across different user groups within a recommender system. It is calculated by summing the absolute differences between Precision@k (PREC@K) values for all pairs of user groups. Groups are formed by grouping the recommendations by sensitive attributes. The sensitive attributes used for grouping can be the ones being isolated during the training process to quantify the unfairness. The unfairness value can be further normalized by the number of comparisons as follows:

$$UFAIR@k = \frac{1}{n \cdot m} \sum_{n=1}^{|C|} \sum_{m=1}^{|C|} |\frac{1}{|C_n|} \sum_{u \in C_n} PREC@k(u) - \frac{1}{|C_m|} \sum_{u \in C_m} PREC@k(u)| \tag{13}$$

---

[1] Github: https://github.com/Git-Uzair/thesis.

where C contains all the groups formed by the sensitive attributes in the dataset selected for testing and $C_{n/m}$ is the subset of users associated with that group.

4.2   Results

This section discusses the results obtained after training the three models, namely, the default F2VAE model, our proposed implementation trained to mitigate only the country attribute, and our implementation trained to mitigate bias from both country and sex attributes. All the hyperparameters were left unchanged as obtained from the default F2VAE model to only record how the results change by adding our proposed changes.

UFAIR@k. Table 1 states unfairness results for the 3 models, F2VAE, Multi-attribute model trained to mitigate bias from the attribute country, and then trained to mitigate bias from the attribute country and sex combined, with the recommendations being grouped by their sensitive attribute, country, and then country and sex to calculate unfairness. The best results are in bold and the worst are underlined. We can see a decrease in unfairness from 12% to 30% based on the number of top k values chosen to calculate unfairness between the default model and our implementations. It is to be considered which model is being used for evaluation since one of our implementations uses two sensitive attributes to mitigate unfairness. Our model exhibits superior performance compared to the original F2VAE implementations for several reasons. Firstly, the deeper architecture of our adversarial classifier enables the VAE to more effectively mitigate the influence of attributes on recommendations. Secondly, the classifier's capability to handle multiple sensitive attributes contributes to mitigating additional biases, as observed in the results of our model trained on multiple attributes. Despite being trained to mitigate the effects of only one attribute, both our implementations perform well when recommendations are grouped by country and sex, suggesting that addressing the country attribute already covers a significant portion of unfairness in the data due to its higher variance in user grouping compared to country and sex combined.

Table 1. UFAIR@K results for recommendations.

	UFAIR@1 (country)	UFAIR@10 (country)	UFAIR@20 (country)	UFAIR@1 (country, sex)	UFAIR@10 (country, sex)	UFAIR@20 (country, sex)
F2VAE	0.0149	0.0140	0.0140	0.0170	0.0122	0.0123
Multi-Attr Trained to mitigate (to mitigate)	0.0960	0.0126	0.0121	0.0040	0.0131	0.0116
Multi-Attr Trained to mitigate (country, sex)	**0.00955**	**0.0125**	**0.0119**	**0.0020**	**0.0100**	**0.0114**

Table 2. PREC@K results for recommendations.

	PRECR@1 (country)	PREC@10 (country)	PREC@20 (country)	PREC@1 (country,sex)	PREC@10 (country,sex)	PREC@20 (country,sex)
F2VAE	0.453	0.420	0.397	0.453	0.419	0.395
Multi-Attr Trained to mitigate (to mitigate)	0.465	0.438	0.412	0.454	0.430	0.403
Multi-Attr Trained to mitigate (country, sex)	**0.483**	**0.438**	**0.414**	**0.470**	**0.435**	**0.410**

PREC@k. Table 2 states precision results for the 3 models, F2VAE, Multi-attribute model trained to mitigate bias from the attribute country, and then trained to mitigate bias from the attribute country and sex combined for recommendations grouped by their sensitive attribute, country, and then country and sex calculate precision. The best results are in bold and the worst are underlined. We see anywhere from around 4% to 6% increase in overall precision across different comparison parameters. Our implementation with a higher number of sensitive attributes added for mitigation outperforms the default implementation and our implementation with a single sensitive attribute.

NDCG@100. We use the NCDG@100 evaluation metric to examine the ranking quality of our recommenders. We set k to 100 to have a better widespread idea of our ranking quality. We have noticed improvement in fairness and the accuracy of our recommender system but it does not always mean a quality improvement. A fairness gain can also sometimes affect ranking quality negatively. The exact numbers for NCDG@100 can be referred to in Table 3. The best results are in bold and the worst are underlined.

Table 3. Recall@100 and NDCG@100 results for all models.

	F2VAE	Multi-Attr (country)	Multi-Attr (country, sex)
Recall@100	0.307	0.314	**0.320**
NDCG@100	0.327	0.335	**0.340**

Recall@100. By focusing on the top 100 recommendations, this metric provides a practical evaluation of the system's effectiveness in capturing a significant number of pertinent items within a reasonably sized recommendation list. We get a similar graph to our previous NDCG@100 calculations and our implementations outperform the existing one by giving a 2.5% to 4% performance gain. The exact numbers can be referred to in Table 3.

NDCG and Recall are only evaluated with the top 100 recommendations to get an overall idea of the quality of recommendations throughout the models.

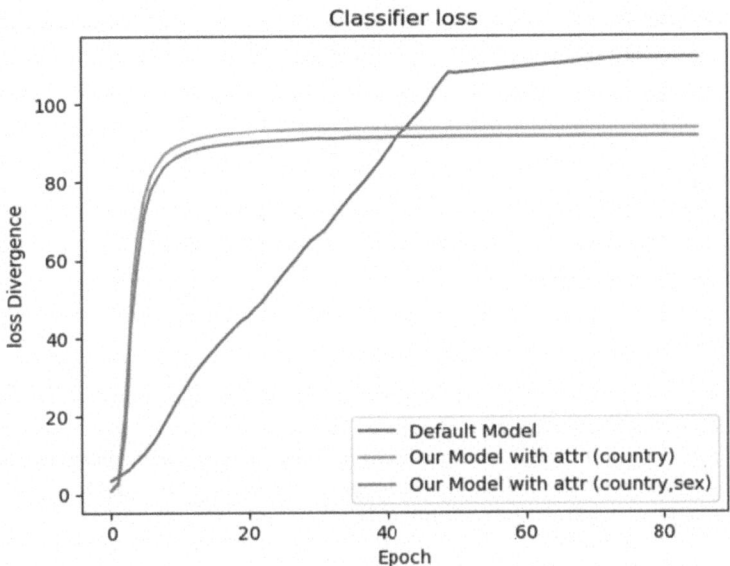

Fig. 5. Classifier loss across models.

Classifier Loss. The change in the classifier loss with the new classifier can be seen in Fig. 5. Our multi-label implementation converges faster than the default implementations which can be attributed to multiple reasons. One is the fact that the new classifier is deeper than the default implementation. The other reason for early convergence of the adversarial loss for the classifier can suggest that the VAE quickly adapts its latent variable representation to better reduce the influence of these sensitive attributes. The model may more effectively find a compromise in the latent space that minimizes the ability of the classifier to predict these attributes.

5   Conclusions

We evaluated the impact of adding a deeper multi-label classifier to mitigate bias and unfairness using multiple sensitive attributes from the dataset as opposed to one. The results obtained conclude an increase in fairness and ranking quality according to all the metrics being evaluated. By using a deeper classifier tuned to predict multiple sensitive attributes, the model is exposed to more information about the users. The additional information helps the model capture more nuanced patterns and preferences leading to better recommendations. We noticed the largest amount of improvement when training the classifier model with country and sex labels as there were significant variations in the users belonging to a sex within a given country. Predicting a wider range of sensitive attributes also acts as a form of regularization which helps prevent overfitting and improves generalization. We used a deeper type of network for the classifier and we saw a faster convergence of the loss in that case. The faster convergence in the multi-label case may suggest that the joint optimization is leading to a more effective balance between these competing objectives. In our future work, to increase trust we will exploit explanations or interpretations of the inner workings of the system. Specifically, we aim to design methods to (i) provide explanations capable of conveying fairness aspects, (ii) audit the system to examine fairness and discover latent sources of bias via explanations, (iii) semi-automatically intervene to ensure system fairness, and (iv) interact with users to facilitate these scenarios.

References

1. Blei, D.M., Kucukelbir, A., McAuliffe, J.D.: Variational inference: a review for statisticians. arXiv:1601.00670 (2016)
2. Borges, R., Stefanidis, K.: Enhancing long term fairness in recommendations with variational autoencoders. In: Proceedings of the 11th International Conference on Management of Digital EcoSystems (MEDES), pp. 95–102 (2019)
3. Borges, R., Stefanidis, K.: On mitigating popularity bias in recommendations via variational autoencoders. In: Proceedings of the 36th ACM/SIGAPP Symposium on Applied Computing (SAC), pp. 1383–1389 (2021)
4. Borges, R., Stefanidis, K.: F2VAE: a framework for mitigating user unfairness in recommendation systems. In: Proceedings of the 37th ACM/SIGAPP Symposium on Applied Computing (SAC), pp. 1391–1398 (2022)
5. Borges, R., Stefanidis, K.: Feature-blind fairness in collaborative filtering recommender systems. Knowl. Inf. Syst. **64**(4), 943–962 (2022)
6. Chouldechova, A., Roth, A.: A snapshot of the frontiers of fairness in machine learning. Commun. ACM **63**(5), 82–89 (2020)
7. Edwards, H., Storkey, A.J.: Censoring representations with an adversary. In: Proceedings of the 4th International Conference on Learning Representations (ICLR) (2016)
8. Gupta, K., Raghuprasad, M.Y., Kumar, P.: A hybrid variational autoencoder for collaborative filtering. arXiv:1808.01006 (2018)
9. Kingma, D.P., Welling, M.: Auto-encoding variational bayes. In: Proceedings of the 2nd International Conference on Learning Representations (ICLR) (2014)
10. Li, X., She, J.: Collaborative variational autoencoder for recommender systems. In: Proceedings of the 23rd ACM SIGKDD International Conference on Knowledge Discovery & Data Mining (KDD), pp. 305–314 (2017)
11. Melchiorre, A.B., Rekabsaz, N., Parada-Cabaleiro, E., Brandl, S., Lesota, O., Schedl, M.: Investigating gender fairness of recommendation algorithms in the music domain. Inf. Process. Manag. **58**(5), 102666 (2021)
12. Pitoura, E., Stefanidis, K., Koutrika, G.: Fairness in rankings and recommendations: an overview. VLDB J. **31**(3), 431–458 (2022)
13. Rumelhart, D.E., Hinton, G.E., Williams, R.J.: Learning internal representations by error propagation. In: Rumelhart, D.E., Mcclelland, J.L. (eds.) Parallel Distributed Processing, vol. 1, pp. 318–362. MIT Press (1986)
14. Wang, Y., Ma, W., Zhang, M., Liu, Y., Ma, S.: A survey on the fairness of recommender systems. ACM Trans. Inf. Syst. **41**(3), 52:1–52:43 (2023)
15. Zhu, Z., Hu, X., Caverlee, J.: Fairness-aware tensor-based recommendation. In: Proceedings of the 27th ACM International Conference on Information & Knowledge Management (CIKM), pp. 1153–1162 (2018)

Indexing and Event Detection

UpLIF: An Updatable Self-tuning Learned Index Framework

Alireza Heidari$^{(\boxtimes)}$ ⓘ, Amirhossein Ahmadi ⓘ, and Wei Zhang ⓘ

Huawei Cloud, Vancouver, Canada
{alireza.heidarikhazaei,amirhossein.ahmadi,wei.zhang6}@huawei.com

Abstract. The emergence of learned indexes has caused a paradigm shift in our perception of indexing by considering indexes as predictive models that estimate keys' positions within a data set, resulting in notable improvements in key search efficiency and index size reduction; however, a significant challenge inherent in learned index modeling is its constrained support for update operations, necessitated by the requirement for a fixed distribution of records. Previous studies have proposed various approaches to address this issue with the drawback of high overhead due to multiple model retraining. In this paper, we present UpLIF, an adaptive self-tuning learned index that adjusts the model to accommodate incoming updates, predicts the distribution of updates for performance improvement, and optimizes its index structure using reinforcement learning. We also introduce the concept of balanced model adjustment, which determines the model's inherent properties (i.e. bias and variance), enabling the integration of these factors into the existing index model without the need for retraining with new data. Our comprehensive experiments show that the system surpasses state-of-the-art indexing solutions (both traditional and ML-based), achieving an increase in throughput of up to 3.12× with 1000× less memory usage.

Keywords: Learned Index · Indexing · Database · Machine Learning · Reinforcement Learning · Data Management

1 Introduction

Context. Machine Learning has been progressively adopted to improve index systems, referred to as *learned indexes*, due to its swift predictive abilities and capacity to encapsulate the distribution patterns of records [5,6]. The recursive model index (RMI) [9] proposed an immutable ML model that, given a key, identifies the range where the corresponding data is located. This approach rivals or even suppresses classical data indexing systems (such as B+Tree) in search performance. In this modeling, the position of each record is estimated by calculating the value of the given key in a monotonic function which can be approximated based on a Cumulative Distributed Function (CDF) [4,12,15]. However, the drawback of learned index modeling is that the distribution of records must remain fixed, which poses challenges for operations that update the key domain

© The Author(s), under exclusive license to Springer Nature Switzerland AG 2025
R. Chbeir et al. (Eds.): IDEAS 2024, LNCS 15511, pp. 345–362, 2025.
https://doi.org/10.1007/978-3-031-83472-1_23

(such as `Insert` and `Delete`). Since the integration over the CDF domain must always sum to *one*, any records' insertion or deletion, necessitates a non-local update to the current learned model.

To overcome the challenge of an updatable learned index, three main approaches have been proposed [16]: *(i) Delta-buffer, (ii) In-place, (iii) Hybrid structures*. The *delta-buffer* solutions involve storing incoming records in a buffer when multiple data records are assigned to the same position to postpone the required updates to the current learned model. However, when the buffer size exceeds a certain threshold, buffer pruning occurs and the data merge into the model, as suggested by LIPP [18]. In contrast, the *in-place* approaches reserve empty placeholders in the original key domain for unseen records. The problem with these solutions, such as Alex [2] is that once the predicted offset by the learned model is filled, subsequent updates are placed in the next available slots regardless of key sequence, leading to extensive search in key lookup. Finally, the *hybrid* approach combines placeholders and buffers for entries that do not fit in the key domain, aiming to strike a balance between lookup efficiency and update speed for different system workloads. DILI [11], a state-of-the-art hybrid solution, employs a tree structure that facilitates level-by-level lookups using the models and buffered data stored in the nodes. The buffered data influence the tree height, leading to an increase in the tree height as more updates are processed.

Challenges. The existing solutions for updatable learned indexes exhibit the following limitations [17]: *(1) Ignoring the Training Cost*. The primary issue with earlier methods is the need to modify the learned model by keeping updates or constructing several models. Retraining in the simplest scenario (that is, using a linear regression model) requires $O(N)$, and a cascading effect through their update structures costs $O(NlogN)$. *(2) Constrained to Specific Incoming Update Distribution*. These approaches strongly consider the specific spread of the incoming data and existing keys in the key domain. When they encounter a discrepancy in the incoming distribution, their throughput drops drastically. *(3) Lack of Generality*. There is no method to mitigate the effects of various algorithms on the learned index. These methods consistently use the same training algorithm, such as linear regression, throughout the system's lifetime.

Approach. In this paper, we present *UpLIF*, an adaptive self-tuning learned index framework, that *adjusts the learned index model on the incoming updates to postpone the model retraining and utilizes the distribution of the updates for its model adjustment*. UpLIF uses the hybrid approach and introduces **B**alanced **M**odel **A**djustment **T**ree (*BMAT*) for its delta-buffer. BMAT captures incoming updates to the base model and calculates a linear adjustment for the learned index model approximation error at query time. UpLIF also learns the distribution of the incoming updates in the runtime to add in-place placeholders in the key domain.

UpLIF also introduces a reinforcement learning-based optimization to adjust the BMAT structure to ensure high performance and minimal memory usage. This agent assesses performance metrics, including the height of BMAT, and takes the following actions to improve the structure of BMAT: *(1)* training on a data subset to lower the height of BMAT, or *(2)* transitioning between different types of BMAT that are more suitable according to the existing workload and data volume.

UpLIF shows notable enhancements over a wide range of indexing techniques: our approach can increase performance by up to $3.12\times$ compared to other indexing methods (both traditional and machine learning-based) while requiring up to $1000\times$ less memory for its index structure. UpLIF also addresses the following technical challenges: *(1)* **Balancing**. Learning a linear model with C features and N training data requires $O(CN)$. UpLIF delays retraining to keep updated operating cost low by employing BMAT. However, balancing in BMAT can be disrupted by incoming update traffic. To address this, UpLIF uses logarithmic operations to rebalance and eliminate the effects of update distribution change. *(2)* **Movement**. Inserting a key in the middle of a list involves dividing the list into two sublists at the insertion point. This process requires relocating one of the sub-lists to a different memory location, leading to a complexity of $O(N)$. UpLIF reduces the relocation numbers by introducing a gap size. It also expands the relocation range by placing placeholders by forecasting the upcoming updates on the relocation gap based on the incoming update distribution. *(3)* **Self-Tuning**. Update operations lead to an increase in the BMAT height, thus increasing the search cost. UpLIF autonomously identifies the optimal balance point to maintain system performance. It decides to reduce tree height based on different *performance signals* by solving an optimization problem in which the cost function aims to maximize system throughput while simultaneously minimizing memory usage. *(4)* **Generic**. UpLIF has generic modeling for modification of the index tree that is compatible with any kind of key space (e.g., numerical, string) and can work with any learned index model type.

Contributions. The main contributions of this work are as follows: *(1) Developing a modular architecture independent of the underlying learned index model* (Sect. 2). This framework accepts any input type for which a comparison measure can be defined within that input space. *(2) Presenting BMAT, a balanced tree structure for storing data needed for model adjustment* (Sect. 3). This structure controls the index adjustment inherent properties (i.e., bias and variance). It captures updates to the base model at query time and uses a linear correction for the learned index model approximation error. *(3) Developing an optimization agent adaptively tunes the framework to maintain the performance high and index size low* (Sect. 4). This agent employs reinforcement learning to identify the BMAT states require tuning and selects the most suitable tuning action to optimize the overall index system. *(4) Evaluating UpLIF against various state-of-the-art indexing systems* (Sect. 5). UpLIF exhibits higher performance among all tested systems through our extensive experiments while maintaining significantly lower index sizes.

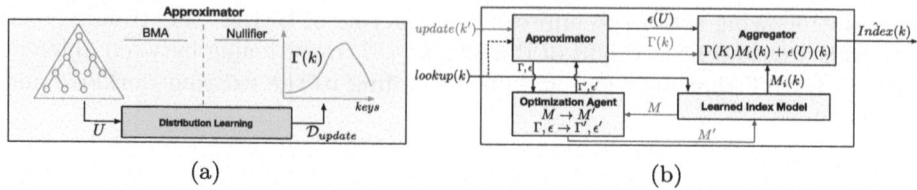

Fig. 1. (a) The approximator module demonstrates that trains the distribution $\mathcal{D}_{update}$ from the incoming updates U. (b) Overview of the UpLIF's modules.

2   Framework Overview

In this section, we explain the constitutional modules of the UpLIF architecture (Fig. 1b). UpLIF begins by initializing the model $M_0 \in \mathbb{M}$ from the sorted array *keys*, where $\mathbb{M}$ is a hypothesis space. Then it takes lookup or update queries with respect to a key k. Once the system is initialized, it accepts queries that look up or modify a specific key, k. To answer these queries, UpLIF uses the following four core modules:

Module 1: Learned Index Model. This module is initialized with the model M_0 learned on the initial sorted keys. The only requirement for choosing a learned index algorithm is that it must maintain the data in a sorted order. This is due to: *(i)* making the data in order causes the inefficiency in the range queries, and *(ii)* the operation of the Approximator module relies on the data being sorted. Consequently, methods such as Alex [2], which alter the key sequences, cannot be used as the learned index base model in our framework. In our evaluations (Sect. 5), we use RadixSpline [8] as the underlying learned index model.

We introduce the concept of Balanced Model Adjustment for the Approximator to accept the incoming updates and control the index model bias and variance for index model correction in a logarithmic time. The model correction that occurs in the approximation module makes the retraining unnecessary to answer queries (Sects. 3.2 and 3.3).

3   Model Adjustment Overview

In this section, we explain our technique to adjust the index model based on the incoming updates to postpone retraining. First, we introduce an index model approximation that is able to solve the model adjustment without requiring model retraining. Then, we present the Balanced Model Adjustment Tree (BMAT) and Nullifier that help to solve the approximation problem.

3.1   Preliminaries

In this section, we provide the necessary definitions to formally state the problem. A learned index model $M(.)$ is a type of index system that uses machine

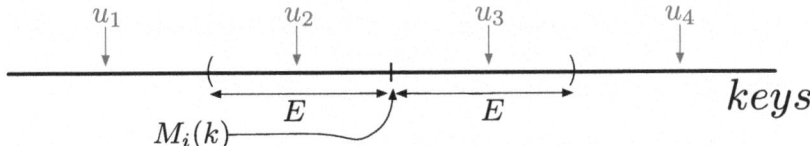

Fig. 2. The learned index model M_i, characterized by a nonzero error E, maps four ranges over the key space. Incoming updates, treated as a random variable, affect this range in four distinct ways: u_1 uniformly shifts all elements without changing the range size; u_2 expands the range on the left by shifting $M_i(k)$ rightward; u_3 enlarges the range on the right; and u_4 neither alters the error range size nor moves any internal components.

learning algorithms to predict the range of the index $M(key) \pm E$ based on the provided *key*. To propose our framework for updating the learned index model, we must introduce the fundamental idea of updating the domain of a monotonically increasing function. Update queries (`INSERT` and `DELETE` operations) can alter the domain of $M(.)$, leading to a residual error (as shown in Fig. 2). By keeping track of these errors, we can apply them to the model output, as if the model were re-trained on the entire updated data. The Definition 1 puts this in a formal linguistic way.

Definition 1. *Let D_0 be the initial data and U_0 be the first batch of update queries. Let $M_0 = M(D_0)$ be the model trained in D_0 and $M_1 = M(D_0 \cup U_0)$ be the model trained in the updated data $D_1 = D_0 \cup U_0$. Let $\epsilon_0(.)$ be an error approximator for the first batch of updates. Our goal is $M(D_0 \cup U_0) \approx M(D_0) + \epsilon_0(U_0)$ or $M_1 = M_0 + \epsilon_0(U_0)$. In general, $M_{i+1} = M_i + \epsilon_i(U_i)$, where i denotes the batch number of updates and $\epsilon_i(U_i)$ is the error of the corresponding batch.*

This modeling suggests that if we can identify the set of incoming updates for the set U_i, we can use this to anticipate modifications to the model that has been trained on the updated data. To properly define this model, certain preliminary steps must be taken. Modeling $\epsilon_i(U_i)$ suggests a parallel way to describe index calculation, as the two parts of the updated model do not depend on the result of the other part.

Definition 2. *Suppose that $f(x)$ be a continuous and monotonic increasing function, $g(x)$ called $t(x)$-expanded range of $f(x)$ iff $g'(x) = t(x)f'(x)$, which shows the relation between derivatives of $g(x)$ and $f(x)$. In the most basic scenario, when $t(x)$ is constant, the range of g increases linearly. For instance, $f(x) = 4x$ is a 2-expanded version of $g(x) = 2x$, meaning that for every increment in x, the function f occupies double the range compared to g.*

Definition 3. *$scalier_s$ is an operator that takes a function f as input and scales it point-wise at the domain intersection of f and s, in other words:*

$$scalier_s(f) = \begin{cases} s(x)f(x) & x \in Domain(f) \cap Domain(s) \\ f(x) & otherwise. \end{cases} \tag{1}$$

Definition 4. *Nullify is an operator that operates on a list of keys L. Introduces gaps between the elements of the list based on a given maximum distance d_{MAX} and a distribution $\mathcal{D}$. In essence, this operator can be seen as a scalar transformation $scaler_{\mathcal{D}}$ on the index function that produces the list L.*

3.2 Index Model Approximation

Upon receiving an update, the index function is modified so that all keys that exceed the updated value increase by one unit, while the index function remains unchanged for values less than the lead update received. Following updates U, each element k in the original dataset D is adjusted by the count of updates x in U where $x < k$. Initially, the index model is denoted as M. Once the comprehensive set of possible updates $\mathcal{U}$, drawn from the distribution $\mathcal{D}_{update}$, is applied, this index model evolves into a modified version $M_{\mathcal{U}}$, symbolized as $M^\star$. The goal of the optimization problem is to identify a model M' that reduces the following cost function by considering a subset of updates $U \subseteq \mathcal{U}$:

$$\mathcal{L}_{M'} = \sum_{k \in D \cup U} \|M^\star(k) - M'(k)\| \tag{2}$$

therefore, our optimization problem can be defined as;

$$\min_{M' \in \mathbb{M}} \mathcal{L}_{M'}$$
$$\text{s.t.} \quad \mathbb{E}_{u \sim \mathcal{D}_{update}} [M'] = \mathbb{E}_{u \sim \mathcal{D}_{update}} [M^\star] \tag{3}$$
$$\mathbb{P}\left[|M'(k) - M'(u)| < \frac{1}{2}\right] \leq \tau \quad \forall u \sim \mathcal{D}_{update}, k \in D \cup U$$

Where $\mathbb{M}$ denotes the hypothesis space of the models and τ is a parameter chosen by the user to specify the size of the gap in the optimization solution. It is crucial to understand that determining the smallest τ renders the problem NP-Hard, requiring an exhaustive search across the whole $\mathbb{M}$ to identify the model that optimally predicts the gap for the unseen data $u \in \mathcal{D}_{update}$. It is essential to recognize that errors are inherent in every index model. Adding a bias value maintains the distance between keys constant, yet scaling a function alters the spacing between values within the function's range, thereby adjusting the estimator's variance. In system index estimation, it is crucial to prioritize an unbiased estimator despite the risk of higher variance, as increased variance results in a wider search range during the final stages, whereas bias compromises the reliability of the outcome. Searches typically progress from the center towards the boundaries, following normal distribution curves. Given that the likelihood of locating the target value in the distribution's tail is low, we strive to keep the optimization problem (Eq. 3) unbiased. This may lead to increased system variance; however, with lengthy tails, such variance is less harmful compared to the bias. It also considers the distribution of incoming updates; it is crucial to consider the interval linked to the most probable incoming updates, as indicated by the second optimization condition.

Optimization Relaxation. Addressing the optimization problem 3 in its general form is exceedingly challenging, thus we introduce a relaxation for the function shape M', representing it as a transformation from the original index function M. Various methods exist to transform a function, with the most straightforward which is the linear transformation, which involves using a coefficient and an additive function. A function ascends or descends along the vertical axis through the addition or subtraction of a constant, referred to as a bias. Furthermore, multiplying by numbers greater than one causes the vertical axis values to expand, whereas multiplication by numbers less than one results in a contraction of these values. These dynamics are encapsulated in linear approximation by the line's slope and the bias.

In this study, to simplify the optimization challenge described for Eq. 3, we introduce the class $\mathfrak{C}$ for $\mathbb{M}$ as linear combinations of the index function, expressed by $M'(k) = \Gamma(k)M(k) + r(k)$. Here, $\Gamma(k)$ acts as the Scalier (Definition 3) for $M(k)$, and $r(k)$ accounts for the influence of bias error (Definition 1). Also from Definition 2, M' is approximately $\frac{d\Gamma(k)}{dk}$-expanded range of M. This linear structure allows us to assign specific roles to each segment of $\Gamma(k)$ and $r(k)$, thereby breaking down the optimization into two separate phases:

Phase 1. During this phase, the optimizer adjusts $r(k)$ to ensure the model is unbiased concerning the existing data in $D \cup U$, using the current model $M(k)$ and scalier $\Gamma(k)$.

Phase 2. In this stage, the optimizer modifies $\Gamma(k)$ to ensure the model remains unbiased towards the unseen data within distribution $\mathcal{D}_{update}$ based on the existing model $M(k)$ and bias function $r(k)$. Initially, given the unknown nature of $\mathcal{D}_{update}$, a uniform distribution is assumed. This is later adapted to an observation likelihood as system knowledge increases, thereby generating additional vacant areas in crucial domains where updates are more likely to occur, creating optimal spaces for future system updates.

While *Phase 2* operates independently on the incoming updates and estimates $\mathcal{D}_{update}$, *Phase 1* involves a step function with $|U|$ steps, each occurring at a unique position within the domain of keys. Constructing this function is feasible, yet updating the index function with each new update is necessary as the system may receive lookup queries. Therefore, a rapid method to execute these phases is essential. In the following section, we introduce a submodular framework that can approximately address the optimization challenge by altering local properties.

3.3   Balanced Model Adjustment Tree

In this section, we introduce **B**alanced **M**odel **A**djustment **T**ree (**BMAT**), a tree-shaped structure provides a deterministic approximation based on the distribution of the incoming update $\mathcal{D}_{update}$. By traversing BMAT, UpLIF determines the parameters needed for the adjustment of the model (that is, $\Gamma(k)$,

$r(k)$, and *hyperparameters*). Therefore, it guarantees that we achieve the performance of the learned index model with an additional logarithmic cost to identify the required adjustment. We implement the concept of BMAT using two types of balanced trees: *(1) Red-Black Tree (RBMAT)*, and *(2) B+Tree (B+MAT)*. Each of these types of BMAT is suitable for different sizes of datasets and operations, as we discuss in Sect. 4.2 (see Fig. 4).

BMAT acts as a delta-buffer structure construct in a subset $U^G \subseteq U$ that cannot be accommodated in the index structure. It offers a time complexity of $O(\log |U^G|)$ and a space complexity of $O(|U^G|)$. Each node in BMAT holds a fixed amount of data, and the structure is balanced to maintain optimal tree height and prevent performance decline. There are two possibilities for a key lookup: *(1)* If the key k is present in U that could not fit in a position of the index system, then the value of the query is already stored in a node, and *(2)* If k is not found in U, then the number of children on the left side of each node on the subtree of the target node can determine the update bias value $r(k)$, which is indeed an effective method for *Phase 1*. This deterministic structure introduces a forecast for the upcoming updates to address *Phase 2* and allocates new empty slots for future updates, thus expanding the scope of the last-mile search.

For the key k update, a lookup occurs first, if the key is in the BMAT, the value is updated. If the key in the index structure has an empty place, its value is inserted in the empty place. Otherwise, the node is broken into three segments. UpLIF employs a hyperparameter K, which denotes the number of elements following the key index k, known as *aux*. The pair k and *aux* constitute the middle segment. This segment is then expanded through Nullifier (Sect. 3.4). Then, the key k and *aux* (a key K element after the key k) are added to the BMAT as nodes and point to the left, middle, and right segments with their updated bias values. Figure 3 shows an example of the update operations in BMAT.

3.4 Update Placeholder (*Nullifier*)

When managing data accompanied by a workload $\mathcal{D}_{keys}$, it is essential to have an in-place system that distributes space between them to accommodate incoming updates based on the approximation of the updates. This is the role of *Nullifier* (Fig. 1a), which operates on a distribution $\mathcal{D}_{keys}$ in *domain*, a value d_{MAX} as the maximum gap, and input data $D = [k_1, k_2, \ldots, k_N]$ drawn from $\mathcal{D}_{keys}$, and then produces a $\mathcal{D}_{update}$-expanded (see Definition 2) of $D \sim \mathcal{D}_{keys}$. This involves calculating the number of gaps between keys k_i and k_j for $i < j$,

$$
GapSize(k_i, k_j) = \left\lceil \frac{d_{MAX} \cdot \int_{k_i}^{k_j} \mathcal{D}_{update}(x)dx}{\int_{k_1}^{k_N} \mathcal{D}_{update}(x)dx} \right\rceil \tag{4}
$$

To approximate the distribution $\mathcal{D}_{update}$ with incoming data U, we use a Gaussian Mixture Model (GMM). Nullifier creates gaps in D by selecting all successive elements and using Eq. 4. For simplicity, when $j = i + 1$, it is denoted by $\Gamma(k) = GapSize(k_{prev}, k)$. Subsequently, according to Definition 4, produces a

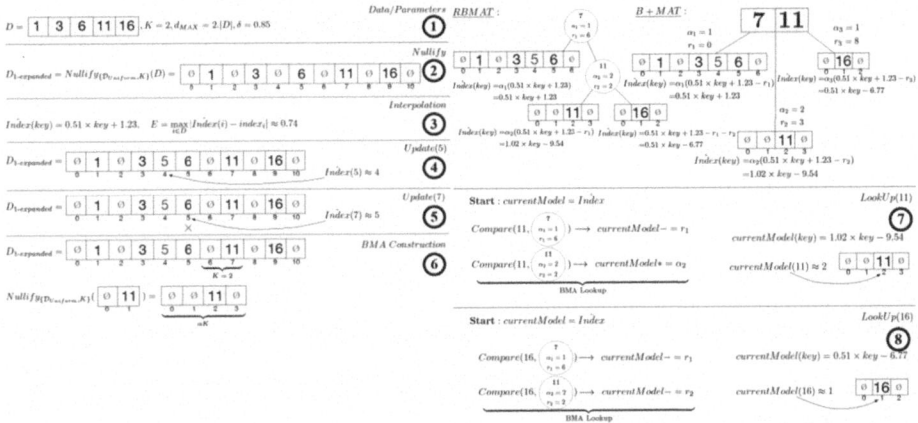

Fig. 3. UpLIF overview on update and lookup operations. *(1)* Initially, UpLIF creates a placeholder structure in the key domain and constructs an index model on top of it. *(2)* An update with $key = 5$ arrives, and it can be placed in the empty placeholder. *(3)* An update with $key = 7$ arrives that makes conflict with $key = 6$ using the current model. UpLIF divides the key domain into three segments and adjusts the previous model for each segment without retraining. It also adds placeholders with $\alpha = 2$ in the middle segment (Sect. 3.4). Finally, two nodes are added to BMAT which can be a red-black tree (left) or a B+Tree (right).

series of NULL values for each $k \in key$. The updated index for each k is calculated as $GapSize(k_1, k) \times Index(k)$, with $\Gamma(k)$ vacant slots preceding it and $\Gamma(k_{next})$ vacant slots following it.

Note that the denominator is constant and only needs to be computed once. Then, we create a new vector using the values in D and the calculated gap between its components, which we designated as NULL. To calculate the expansion of the key k, we determine the distance between the first key and k, as well as the number of elements less than k. This is equal to $\sum_{k':k'<k} \Gamma(k') + |\{k' \in D : k' < k\}|$. Instead of utilizing this formula inefficiently, which involves multiplying the index function for scaling, we opt to average these gaps and utilize a constant value instead, $\alpha = \frac{\sum_{k \in D} \Gamma(k)}{N}$.

4 Adaptive System Tuning

In this section, we present the optimization technique we use in UpLIF to self-tune its structure to maintain high performance and low memory usage. First, we cover the performance measures that influence the performance of UpLIF and index memory size. Then, we show the possible system tuning that can be applied to the index structure. Then, we present our designed reinforcement learning-based optimization agent that considers the performance measures and tuning actions and adjusts UpLIF's index structure with the goal of: *(1)* Increasing the system throughput and *(2)* Decreasing the index memory footprint.

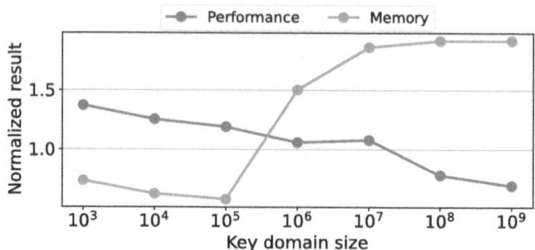

Fig. 4. Comparison between various BMAT types. The plots show the performance and memory consumption of RBMAT normalized to B+MAT.

4.1   Performance Measures

Below are the performance measures that affect overall system performance and memory usage. Our optimizer must consider these metrics for system tuning.

Balanced Structure Height. The most crucial performance measure is the BMAT height. When the height of the index tree increases relative to the size of the processor caches, cache misses increase and cause a performance drop for BMAT traversal. By incorporating this measure into UpLIF, we can mitigate the decline in performance and achieve better overall system performance.

Granularity Measure. This performance measure is the size of the effective segment range when creating a new node for BMAT. If the size gets smaller, the workload is skewed into a small range of keys increasing the system latency for update operations.

Error Scaling. This factor evaluates the influence of $\Gamma(k)$ on the variation error of the model. As discussed in Sect. 3.2, the scaling factor $\alpha > 1$ can increase the last-mile search, which affects the system query performance for lookup operations.

Number of the Models. This performance measure keeps track of the number of distinct models stored on BMAT. The greater number of active models in the system shows higher memory usage for the indexing.

4.2   System Tuning Actions

UpLIF system tuning can be performed in two different ways: *(1)* Transitioning between various BMAT types and *(2)* Reducing the BMAT height by retraining on the subset of keys. These adjustments in the BMAT structure influence the overall performance and memory consumption of UpLIF.

Transition Between BMAT Types. We develop BMAT based on two different balanced trees: Red-Black Tree (RBMAT) and B+Tree (B+MAT). RBMAT is a binary tree and differs from B+MAT in that keys and pointers are clustered in memory; therefore, we obtain efficient cache behavior on both disk and in

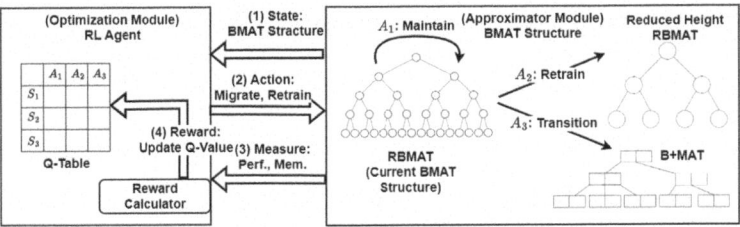

Fig. 5. RL Agent Overview.

memory [3]. However, for small data sizes, RBMAT is more efficient, as it does not have lookup overhead at each node, and while the BMAT traversal path can be kept in the cache, RBMAT performs better. In terms of memory, BMAT has the overhead of empty places in each node and RBMAT has the overhead of pointers to the right, left, and parent nodes for each node. RBMAT has a smaller size compared to B+MAT if the B+MAT nodes are empty. However, if the B+MAT nodes are mostly filled, RBMAT gets higher memory consumption to keep incoming updates. Figure 4 shows the comparison between the performance and memory consumption of RBMAT and B+MAT on a read-heavy workload (see Sect. 5). The results show RBMAT data normalized against B+MAT data. In our setup, we observed that until 100K keys, RBMAT acts quite better than B+MAT as it has higher throughput and lower index size. However, after this point, B+MAT performs better and is the ideal solution for the BMAT type.

Retraining on the Subset of Data. Another type of system tuning on BMAT is retraining on the subset of the data to reduce BMAT height. The growth in tree height due to our model approximation mechanism necessitates a periodic reverse process to prune the BMAT structure, maintaining the performance of our framework. We introduced a greedy approach to select the minimum number of nodes in both RBMAT and B+MAT to have minimal overhead on performance and decrease the height.

4.3   Reinforcement Learning-Based Optimization

RL Agent. We introduce an RL agent to meet the dual objectives of improving system throughput and minimizing memory size. Figure 5 shows the overview of the RL agent. This agent facilitates the discovery of optimal operational strategies through continuous interactions within a well-defined environment, described by a state space $\mathcal{S}$, an action space $\mathcal{A}$, and a reward function R.

States. We consider the state space $\mathcal{S}$ to encapsulate critical system metrics and configurations. Each state $s \in \mathcal{S}$ is a vector $(\mathcal{S}_1, \mathcal{S}_2, \mathcal{S}_3, \mathcal{S}_4, \mathcal{S}_5)$. The four initial state parameters ($\mathcal{S}_1$ to $\mathcal{S}_4$) correspond to the four performance measures outlined in Sect. 4.1, while the final parameter specifies the BMAT type. $\mathcal{S}_1$: Integer representing the current height of the tree structure. $\mathcal{S}_2$: Integer representing the minimum granularity of the data coverage by the models at the

Algorithm 1. System Tuning Agent Algorithm

1: $Q \leftarrow$ initializeQTable()
2: $\alpha, \gamma, \epsilon \leftarrow$ initializeVariables()
3: **while** true **do**
4:     Observe state s_t
5:     $A_t \leftarrow$ getAvailableActions(s_t)
6:     **if** generateRandNumber() $< \epsilon$ **then**
7:         $a_t \leftarrow$ getRandomAction(A_t)
8:     **else**
9:         $a_t \leftarrow \arg\max_{a \in A_t} Q(s_t, a)$
10:    **end if**
11:    tuneSystem(a_t)
12:    **for** $k \leftarrow 1$ to N **do**
13:        fetchRunOperation()
14:    **end for**
15:    Observe state s_{t+1}
16:    $P_{t+1}, M_{t+1} \leftarrow$ measureThroughputMemory()
17:    $R_{t+1} \leftarrow$ calculateReward(P_{t+1}, M_{t+1})
18:    $a' \leftarrow \arg\max_{a \in \text{Actions}} Q(s_{t+1}, a)$
19:    $Q(s_t, a_t) \leftarrow (1 - \alpha)Q(s_t, a_t) + \alpha(R_{t+1} + \gamma Q(s_{t+1}, a'))$
20:    $\epsilon \leftarrow$ updateEpsilon(ϵ)
21:    $s_t \leftarrow s_{t+1}$
22: **end while**

---

leaf nodes. $\mathcal{S}_3$: Real number reflecting the proportional error variation from a baseline, indicative of the stability of the model. $\mathcal{S}_4$: Integer counts the total number of models active within the system and describes its operational scale. $\mathcal{S}_5$: Binary variable indicating the type of tree structure, with 0 for RBMAT and 1 for B+MAT.

Actions. The action space $\mathcal{A}$ has discrete actions designed to optimize system configuration based on the actions we defined in Sect. 4.2. $\mathcal{A}_1$: Maintain current BMAT structure. $\mathcal{A}_2$: Trigger retraining of index models on specific BMAT's branches. $\mathcal{A}_3$: Transition to another BMAT structure (RBMAT to B+MAT, B+MAT to RBMAT).

Reward. The reward function $R : \mathcal{S} \times \mathcal{A} \to \mathbb{R}$ is designed to balance throughput and memory usage:

$$R(s,a) = \eta \cdot \frac{\text{Throughput}(s,a)}{\text{Max System Throughput}} - (1 - \eta) \cdot \frac{\text{Memory Usage}(s,a)}{\text{Total Memory}},$$

where η is the coefficient that weighs the relative importance of throughput and memory efficiency.

RL Algorithm. Algorithm 1 shows the steps of the RL agent for learning the best tuning actions at each state using Q-Learning.

5 Evaluations

In this section, we first outline the experimental setup employed to evaluate UpLIF. Subsequently, we present UpLIF's performance across various metrics and compare it with leading indexing solutions. We demonstrate that UpLIF surpasses both traditional and learned indexes in terms of performance and memory efficiency.

5.1 Experimental Settings

Environment. We implement UpLIF in C++ and compile it with GCC 9.0.1. We perform our evaluation on an Ubuntu 20.04 Linux machine with AMD Ryzen ThreadRipper Pro 5995WX 64-core 2.7G Hz CPU and 256 GB DDR4 RAM. All experiments were performed with RadixSpline serving as the underlying learned index model, configured with a spline degree of 128.

Datasets. We conduct all experiments using 8-byte keys sourced from a dataset and 8-byte values generated randomly. We evaluated UpLIF on the three following datasets from the SOSD benchmark [7]: *(1) FB* [13]: 200M Facebook user ids, *(2) WikiTS* [14]: 190M unique request integer timestamps of log entries of the Wikipedia website, and *(3) Logn:* 200M unique values sampled from a log-normal distribution with heavy tail with $\mu = 0$ and $\sigma = 1$.

Baselines. We compare UpLIF's results with the following methods to have comparisons with all the updatable learned index solutions as well as the classical indexing method: *(1) B+Tree:* This baseline represents a standard B+Tree implementation, implemented in STX B+Tree [1]. *(2) Alex* [2]: A state-of-the-art learned index method that utilizes an in-place approach for handling updates. *(3)* LIPP [10]: A state-of-the-art leaned-index using the delta-buffer method for updates. *(4) DILI* [11]: A hybrid learned index model that combines different indexing techniques to achieve efficient updates.

Workloads. The primary measure used to assess UpLIF is the average throughput. Throughput is evaluated across four different workloads: *(1) Read-Only:* a workload focused on read-only operations, *(2) Read-Heavy:* a workload with a high proportion of reads (90%) and a small percentage of inserts (10%), *(3) Write-Heavy:* a workload involving 50% reads and 50% inserts, and *(4) Write-Only:* a workload dedicated solely to write operations.

To begin with, an index is initialized with 100 million keys for a given dataset. Each workload is executed for 60 s and the remaining keys are inserted during this time. The reported throughput indicates the number of operations (i.e., inserts and reads) completed within that timeframe, averaged over 10 runs to account for variability.

RL Training. We pre-train the RL agent for each workload and use the trained agent in the Optimization Module (Fig. 2) to tune the system based on the system condition. In our evaluations, the RL agent only exploits the calculated Q-Table.

Table 1. Throughput: Comparisons with state-of-the-art methods

Workload	Dataset	UpLIF	B+Tree	Alex	LIPP	DILI	Average	Max
		Million Operations Per Second					Improvement	Improvement
Read-Only (0% write rate)	WikiTS	**5.4**	2.0	4.1	5.2	**5.4**	50.5%	2.68x
	Logn	**6.3**	1.9	**6.2**	6.0	**6.3**	56.1%	**3.18x**
	Facebook	**4.5**	2.0	2.3	4.3	4.4	55.5%	2.23x
Read-Heavy (10% write rate)	WikiTS	**4.2**	1.5	2.2	3.9	4.1	**69.3%**	**2.78x**
	Logn	**4.1**	1.5	3.2	2.9	3.8	61.2%	2.71x
	Facebook	**2.6**	1.3	1.1	2.3	2.5	62.8%	2.28x
Write-Heavy (50% write rate)	WikiTS	**3.4**	1.3	1.9	2.8	2.1	73.8%	2.61x
	Logn	**3.9**	1.3	3.2	2.9	2.9	76.2%	**3.00x**
	Facebook	**3.9**	1.3	1.1	2.1	2.4	**82.2%**	2.48x
Write-Only (100% write rate)	WikiTS	**2.8**	1.3	1.4	2.2	2.4	66.2%	2.14x
	Logn	**3.1**	1.2	2.8	2.2	2.7	57.6%	**2.36x**
	Facebook	**2.5**	1.2	1.2	1.7	1.9	**68.1%**	2.04x
Distribution Shift (50% write rate)	WikiTS	**2.7**	1.3	1.7	2.2	1.8	54.8%	2.08x
	Logn	**3.1**	1.3	2.8	2.3	2.2	**59.4%**	**2.49x**
	Facebook	**2.1**	1.3	1.0	1.6	1.4	59.0%	1.93x

5.2   Performance on the Workloads

Table 1 shows the performance of UpLIF in terms of throughput compared to the other baselines in different workloads and datasets. Our results indicate that as the update rate increases in workload, the gap in throughput improvement between UpLIF and other learned index models widens. This can be attributed to UpLIF's optimization for updates through model adjustment with minimal retraining (in the Approximator module) as well as the index structure tuning (in the Optimization Agent module). As the number of updates increases, other methods such as DILI require more frequent retraining and node conflict handling, resulting in lower throughput. In contrast, UpLIF's efficient update handling allows it to maintain higher throughput even with a high update rate, further highlighting its superiority in handling dynamic workloads.

Read-Only Workloads. In the Read-Only workload scenario, we train UpLIF on the entire dataset without any updates. In this case, UpLIF's performance is identical to the underlying model it utilizes (i.e., RadixSpline). UpLIF consistently achieves more than 2× throughput compared to B+Tree in all datasets. However, compared to the other learned index frameworks, UpLIF's throughput is comparable to DILI since no updates have been applied to UpLIF in this particular workload. In subsequent workloads where the number of updates increases, we observe that UpLIF consistently outperforms all other learned index models, resulting in higher throughput.

Read-Heavy Workloads. For Read-Heavy workloads, UpLIF begins to lead all baselines in performance as there are a small number of updates in this workload type. UpLIF achieves up to 2.78× throughput compared to B+Tree on the Logn Dataset. UpLIF also outperforms all other learned index models and achieves more than 60% throughput on average. Although DILI has less performance than UpLIF, it performs relatively well as the number of updates is low, and its structure is not influenced by the updates.

Write-Heavy Workloads. In the Write-Heavy workload, UpLIF exhibits its best performance due to its model adjustment and structure tuning by retraining and switching between BMAT trees. Compared to B+Tree, UpLIF shows up to 3× higher throughput, and compared to the state-of-the-art updatable learned indexes, it achieves up to 82% higher performance. Specifically, considering the Facebook dataset, UpLIF achieves 3.9 million operations per second, while B+Tree, Alex, LIPP, and DILI are far behind with 1.3, 1.1, 2.1, and 2.4 million ops/sec.

Write-Only Workloads. The Write-Only workload is the most challenging since the index structure is being influenced the most. However, the tuning and approximation techniques of the UpLIF structure mitigate this impact and outperform all other benchmarks. UpLIF achieves performance levels up to 2.49× greater than B+Tree. Furthermore, it delivers an average throughput increase of up to 59.4% in all three datasets compared to the learned indexes.

5.3   Distribution Shift

To evaluate UpLIF and the distribution shift baselines, we initialize the keys with the first 100 million smallest keys for each dataset and then use the write-heavy workload to read and write the remaining data. The last row of Table 1 shows the robustness of UpLIF to changes in the data distribution. UpLIF maintains up to 2.49× and 1.76× higher throughput compared to B+Tree and DILI. DILI cannot work well on distribution shift, as its structure is built based on the dataset characteristics, and inserting keys from different distributions causes more conflicts in its leaf structures which need more node adjustment.

5.4   Range Query Performance

On the range query workload, UpLIF shows significantly lower response times compared to other methods in all data sets (see Fig. 6a). This is attributed to UpLIF's optimization for range queries by maintaining sorted data at the leaf nodes. UpLIF outperforms DILI by a considerable margin ranging from 55% to 70%. The performance gap between UpLIF and DILI becomes more pronounced due to DILI's less efficient handling of range queries, as its model tree necessitates frequent retraversals, resulting in slower processing and longer response times. Similarly, other methods such as B+Tree, Alex, and LIPP also exhibit higher response times compared to UpLIF. For instance, B+Tree and Alex consistently demonstrate slower performance by notable margins, while LIPP exhibits the highest response times among all the tested methods. These comparisons underscore UpLIF's superior efficiency and reliability in efficiently processing range queries, establishing it as the most effective solution across all the evaluated datasets.

5.5   Index Memory Consumption

We compare the memory usage of the index structure in UpLIF under a write-heavy workload across different datasets and compare it with other baselines. Figure 6b shows that UpLIF always achieves significantly lower memory consumption compared to the other methods, with a notable difference of up to 1000 times lower memory usage compared to DILI. This advantage is primarily due to UpLIF's ability to determine the update distribution and strategically place placeholders to efficiently collect updates, thereby minimizing the need to add nodes to BMAT. In contrast, DILI and LIPP exhibit the highest memory consumption, since these models create new leaf nodes and empty slots in the entry array whenever conflicts occur. Although Alex utilizes a similar in-place placeholder approach as UpLIF, it still results in slightly higher memory consumption due to its inefficiency in accurately placing the placeholders based on the update distribution.

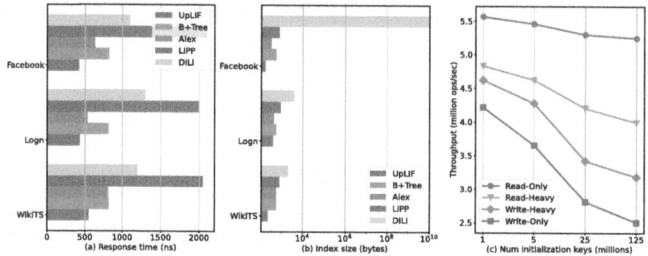

Fig. 6. (a) Range query performance (b) Index memory sizes (c) Throughput with different numbers of initialization data

5.6   Scalability

Figure 6c shows the UpLIF's throughput scalability on different scales of initialization keys, ranging from 1 million to 125 million using the WikiTS dataset. The diagram organizes the data according to the four workloads described earlier. As observed, in the Read-Only and Read-Heavy scenarios, the system maintains a high throughput, demonstrating its efficiency in handling large volumes of read operations. The throughput decreases slightly but remains robust even as the initialization keys increase. In Write-Heavy and Write-Only workloads, initially, there is a noticeable decline in throughput as the volume of initial data expands. However, once a specific threshold of initial keys is reached, the BMAT begins to retrain and prune the tree, which stabilizes the system's throughput.

6   Conclusion

UpLIF introduces an efficient and robust updatable learned index framework that overcomes the limitations of traditional indexes. It incorporates the Balanced Model Adjustment Tree and uses a hybrid approach with placeholders and delta buffers to manage updates without frequent retraining. It also incorporates an RL-based system tuning to optimize its index structure. Our extensive testing shows that UpLIF exceeds other advanced indexing methods in various workloads and maintains high performance and low memory usage even with distribution changes and increases in the data scale.

References

1. Bingmann, T.: STX B+tree C++ (2024). https://github.com/bingmann/stx-btree/
2. Ding, J., et al.: ALEX: an updatable adaptive learned index. In: Proceedings of the ACM International Conference on Management of Data (SIGMOD), pp. 969–984. SIGMOD (2020)
3. Heidari, A., Ahmadi, A., Zhi, Z., Zhang, W.: MetaHive: a cache-optimized metadata management for heterogeneous key-value stores. In: Proceedings of the VLDB Workshop on Cloud Databases (CloudDB) (2024)
4. Heidari, A., Kushagra, S., Ilyas, I.F.: On sampling from data with duplicate records. arXiv:2008.10549 (2020)
5. Heidari, A., McGrath, J., Ilyas, I.F., Rekatsinas, T.: HoloDetect: few-shot learning for error detection. In: Proceedings of the ACM International Conference on Management of Data (SIGMOD), pp. 829–846 (2019)
6. Heidari, A., Michalopoulos, G., Ilyas, I.F., Rekatsinas, T.: Record fusion via inference and data augmentation. ACM/JMS J. Data Sci. $\mathbf{1}$(1), 1–23 (2024)
7. Kipf, A., et al.: SOSD: a benchmark for learned indexes. In: Proceedings of the NeurIPS Workshop on Machine Learning for Systems (2019)
8. Kipf, A., et al.: RadixSpline: a single-pass learned index. In: Proceedings of the 3rd SIGMOD International Workshop on Exploiting Artificial Intelligence Techniques for Data Management (aiDM) (2020)
9. Kraska, T., et al.: The case for learned index structures. In: Proceedings of the ACM International Conference on Management of Data (SIGMOD), pp. 489–504 (2018)
10. Lan, H., Bao, Z., Culpepper, J.S., Borovica-Gajic, R.: Updatable learned indexes meet disk-resident DBMS: from evaluations to design choices. In: Proceedings of the ACM on Management of Data, vol. 1, no. 2 (2023)
11. Li, P., et al.: DILI: a distribution-driven learned index. Proc. VLDB Endow. $\mathbf{16}$(9), 2212–2224 (2023)
12. Livshits, E., Heidari, A., Ilyas, I.F., Kimelfeld, B.: Approximate denial constraints. arXiv:2005.08540 (2020)
13. Marcus, R., Kipf, A., van Renen, A.: Searching on sorted data: fb_200m_uint64.xz (2019). https://doi.org/10.7910/DVN/JGVF9A/Y54SI9
14. Marcus, R., Kipf, A., van Renen, A.: Searching on sorted data: wiki_ts_200m_uint64.zst (2019). https://doi.org/10.7910/DVN/JGVF9A/SVN8PI
15. Sabek, I., Kraska, T.: The case for learned in-memory joins. Proc. VLDB Endow. $\mathbf{16}$(7), 1749–1762 (2023)

16. Sun, Z., Zhou, X., Li, G.: Learned index: a comprehensive experimental evaluation. Proc. VLDB Endow. **16**(8), 1992–2004 (2023)
17. Wongkham, C., et al.: Are updatable learned indexes ready? Proc. VLDB Endow. **15**(11), 3004–3017 (2022)
18. Wu, J., et al.: Updatable learned index with precise positions. Proc. VLDB Endow. **14**(8), 1276–1288 (2021)

Scheduling of Intermittent Query Processing

Saranya Chandrasekaran[1,2]([✉]) [iD] and S. Sudarshan[2] [iD]

[1] U R Rao Satellite Centre, ISRO, Bengaluru, India
[2] IIT Bombay, Mumbai, India
{saranyac,sudarsha}@cse.iitb.ac.in

Abstract. There are many application queries on windows defined over
a stream of tuples that must be processed within specified deadlines
which are after the window end. Stream processing is usually done either
on a tuple-by-tuple basis or in micro-batches. Processing queries over
large windows using stream processing engines can be very inefficient
since there is often a significant overhead per tuple or micro-batch. Conversely, processing all tuples at the end of the window may result in
missed deadlines, and idling of system resources before the window end.
We present scheduling schemes for queries on large windows, using large
batches, and using priority schemes based on query deadlines and slack
time. Our scheduling scheme handles multiple concurrent queries without any prior knowledge of the future query requirements. The proposed
scheduling algorithms have been implemented as a custom scheduler,
on top of Apache Spark. Our performance study with TPC-H queries
shows that our approach of processing can achieve significant computation time reduction compared to naively using Spark Streaming and can
also handle stringent deadline cases efficiently.

Keywords: Stream processing · Intermittent query processing

1 Introduction

Many applications carry out analysis on data streams and require results within
a specified deadline, i.e. in real time. Stream Processing Engines (SPEs) are
widely used for doing such real time analytics. These systems are characterized
by high input data rates and usually run a large number of concurrent queries.
Stream Processing Engines usually do tuple-by-tuple processing or processing
in micro batches. However, doing computation eagerly is not needed for many
applications, which perform aggregation on large windows.

Our work was motivated by a leading E-Commerce site in India where analysis was performed on data collected over the day and the results must be made
available at some time in the morning of the following day; they wished to move
the analysis window from daily to every few hours to support faster responses.
The same queries were run on successive windows, i.e. they are recurring queries;
as per Zhang et al. [17] 60% of queries in Microsoft SCOPE are recurring in

© The Author(s), under exclusive license to Springer Nature Switzerland AG 2025
R. Chbeir et al. (Eds.): IDEAS 2024, LNCS 15511, pp. 363–376, 2025.
https://doi.org/10.1007/978-3-031-83472-1_24

nature. Also, Wang et al. [13] describe Grosbeak, a data warehouse implemented at Alibaba, to handle similar requirements where daily analysis queries must be processed within certain deadlines.

Stream processing engines such as Apache Spark and Apache Flink, update the aggregate as and when a new tuple arrives or when a micro batch of tuples arrives. This method of processing eagerly can lead to significant overheads. Since the results of queries are needed only at the deadline, tuples can be processed in larger batches. In the example considered, if the deadline of the query is say 2 h after the end of the day, one option is to start computation after the end of the window. However, in general, there may not be sufficient time to process the entire data in the gap between the end of the window and the query deadline. In such cases, tuples may be collected for a duration of say one hour, and then processed, and partial aggregates can be finally aggregated at the end. Such batched computation helps in not only reducing the overall computation cost but also in meeting the deadline.

The problem addressed in this paper is that of finding an appropriate batch execution schedule that meets the required query deadlines while minimizing the cost. Here, cost refers to the total time required to process the query. Tang et al. [8] introduce the concept of intermittent query processing, where parts of the query are executed on parts of the input at intermediate points, and the intermediate results are combined at the end to get the final result. Shang et al. [7], Tang et al.[10] propose query optimization by utilizing the query slackness. However, [7,8,10] do not consider query deadlines. Wang et al.[13] propose incremental computation over the available data. Though Wang et al. [13] discuss scheduling, they do not provide further details on how the schedule is generated.

We consider the problem of batching and scheduling of multiple queries, each with its deadline, which runs in a time-shared manner. The system may be dynamic and queries may be added at any time. Our contributions are:

1. We first address the problems of finding batch sizes that keep the computation cost within a predefined multiple of the minimum possible cost (when run as a single batch).
2. We then consider the problem of scheduling the batches of each query based on its input availability and deadlines.
3. We consider both fixed and varying input rates for tuple arrival.
4. We have implemented the scheduling schemes on a custom query scheduler module, built on top of Apache Spark.
5. Our performance study carried out under different scenarios on TPC-H data/queries demonstrate that our optimizations provide significant benefits in terms of reducing cost while meeting the query deadlines.

The special case of batching for the single query case is described in the full version of this paper [1]. The rest of this paper is organized as follows. Section 2 gives the problem description and explains the factors that affect query scheduling. Techniques for batching and scheduling are described in Sect. 3. Section 4 gives an overview of the related work. Implementation aspects are described in Sect. 5. Experimental results are presented in Sect. 6. Section 7 summarizes the conclusions and future work.

2   Problem Description

In this section, we describe the problem specifications and the factors that impact the scheduling of queries with intermittent query processing.

Input data arrives as a stream, and queries analyze the data over a certain time duration and the output is expected within a deadline. The system is assumed to be a soft real time system, where missing the deadline reduces the utility of the results. Our techniques endeavor to complete query execution within the deadline, provided it is feasible.

Query scheduling depends on the ability to model the query execution time. Since the queries are recurring in nature, the time cost model can be derived from historical data. We model the input data rate for computing when batches will be ready for processing, but in Sect. 3.3 we also consider situations where arrival rates may vary from the predicted rate.

Stream processing systems usually allow multiple queries to be processed simultaneously. In this paper, we assume that queries are independent of each other. We assume that queries compute aggregates on windows, and that queries can be computed in an incremental fashion: more specifically, we assume that partial aggregates can be computed on parts (batches) of a stream, and the partial results can be combined later to get the final result. For example, the query to determine the total purchases of each item can be computed by computing partial aggregates for each batch, and later combining the partial aggregates to get the final aggregate values. While aggregation of partial aggregates can also be done intermittently to reduce the final aggregation cost, in this paper, we restrict ourselves to strategies where partial aggregates are combined only in a single final aggregation step.

We assume that each query runs on one input stream and can join the stream with multiple stored or static relations. Extensions to handle some cases of joins between multiple streams are discussed in Sect. 5. We also assume that only tuples that are available at the start of the execution of a given batch are processed in that batch.

We assume that queries run on a time-shared system, where only one query runs at a time. The algorithm proposed in this paper can also be used for a cluster environment, where the same query will be executed in all the cluster nodes, but extensions to allow dividing of resources between queries are part of future work.

The parameters of a query that affect scheduling decisions and relevant notations are given in Table 1. The tuple input rate is assumed to be constant initially, but extended later to handle variable input rates.

The computation cost depends on the batch size and number of batches. A simple linear cost model combines a per tuple processing cost, and a per-batch overhead cost, as follows:

$$Cost = (NumTuples * TupleProcCost) + (NumBatches * OverheadCost)$$

Since actual computation costs may be non-linear, we use a piecewise linear model as an approximation. The model is learnt from actual query execution. Details of cost modelling as applied to TPC-H queries are explained in Sect. 5.

Table 1. Query Parameters

Notation	Description
queryID	Unique Identifier for the query
windStartTime	Time at which tuple arrival starts
windEndTime	Time at which tuple arrival stops
$deadline_Q$	Time by which the query processing must be completed
inputStream	Denotes the query input stream
inputRate	Rate at which tuples arrive for inputStream
numTupleTotal	Total number of tuples to be processed
minCompCost	Time required for processing all the tuples as a single batch
slackTime	The maximum time beyond which the processing cannot be delayed without missing the deadline

3   Scheduling in Dynamic Scenario

In a data analytics system, there may be multiple queries running with the same or different deadlines. Queries may or may not use the same input data stream, and queries may be added or removed from the system arbitrarily. The input rates for each stream and the total number of tuples in the window may also vary.

In this section, we consider such a dynamic scenario. The methodology for determining the batch size is explained in Sect. 3.1. Section 3.2 explains the scheduling scheme with a fixed arrival rate, while Sect. 3.3 extends the scheduling scheme to handle variable input rates.

3.1   Determining Batch Size

The Minimum Computation Cost, minCompCost, which is the time required to process all the tuples as a single batch, can be computed from the cost model for the query. The slack time for the query if computed as a single batch after the window end can be computed as:

$$slackTime = deadline_Q - windEndTime - minCompCost$$

If the slack time is positive, the query can be scheduled after window end but no later than:

$$schStartTime = deadline_Q - minCompCost$$

If the slack time is negative, the query processing cannot be delayed until the end of the window. Instead, the query has to be processed in multiple batches, starting before the window end time.

Analytical queries perform aggregations of data in each window. Processing tuples of one window in multiple batches results in partial aggregation being done on each batch. Hence once all the batches have been processed, the intermediate aggregation results need to be aggregated to get the final aggregation results; we call this step the *final aggregation step*.

In the static case where the arrival rates are fixed, it is possible to break up the input into different batch sizes, and schedule batch execution and final aggregation in such a way that costs are minimized while deadlines are met; details are in the full version of this paper [1]. However, such an approach may delay computation even if the system is idle.

In the dynamic scenario, the scheduler does not have a priori knowledge about future queries. Delaying query execution for the appropriate batch size may result in avoidable missing of deadlines, since new queries may be added to the system at any time. To handle the dynamic scenario, our approach is to process queries intermittently, i.e. whenever the number of tuples available for processing exceeds some minimum batch size. The scheduling of batches is done keeping query deadlines in mind.

The minimum batch size referred to as MinBatchSize, is determined based on the Resource Slack Factor δ_{RSF}. The goal is to pick a minimum batch size such that the overall computation cost is not increased by more than a factor δ_{RSF}. Let N denote the total number of tuples, and $minCompCost_{BatchSize=x}$ denote the computation cost for processing with a batch size of x tuples. Note that the lowest cost is obtained with $x = N$, i.e. processing all the tuples in a single batch. Then MinBatchSize is set to the smallest batch size x such that:

$$minCompCost_{BatchSize=x} \leq \delta_{RSF} * minCompCost_{BatchSize=N}.$$

The parameter δ_{RSF} can be set based on the system utilization. If the system is lightly loaded then a larger δ_{RSF} can be used, allowing for smaller batches. Extensions to automatically adjust δ_{RSF} based on the system load are part of future work.

3.2 Scheduling Using Minimum Batch Size

Once the MinBatchSize is determined, queries can be processed using scheduling techniques such as Least Laxity First (LLF), Earliest Deadline First (EDF), Shortest Job First (SJF), or Round Robin (RR). We assume the system is non-preemptive while processing a single batch of tuples.

For any new queries added to the system, the MinBatchSize is determined as explained earlier, with an additional requirement that the time for processing a batch must be at most some value, denoted as C_{max}. Since the scheduler is non-preemptive, C_{max} ensures that the system can start processing any new query with a delay of not more than C_{max} in case the query has very low slack time. The value of C_{max} has to be decided based on the application latency requirements.

We now consider scheduling with LLF. A query batch can be scheduled at a point in time if the number of tuples available at that time is greater than or equal to MinBatchSize. For each schedulable query i in the list of current queries, $qList$, with batch size x, its SlackTime or laxity at the current point in time is determined as $deadline_{Q_i} - currentTime - CompCost_{Q_{i(batchSize=x)}}$. The query with the least laxity is given the highest priority and its batch is scheduled for execution.

Once a batch has been processed both CPU and memory are released. The intermediate results of the batch are stored on disk. This is unlike streaming data systems which typically retain intermediate results in memory. If all tuples of a query have been processed, then the final aggregation is done and the query is removed from q_{list}.

EDF, SJF, and RR based scheduling can be implemented with small variations of the LLF implementation. A discussion on the *schedulability* of a given set of jobs, i.e. feasibility of execution of the jobs within the specified deadlines, may be found in the full version of our paper [1] but is omitted here for lack of space.

3.3   Handling Variable Input Rate

So far we have assumed that the input data rate and the total number of tuples in the window are both predictable, i.e., known ahead of time. In practice these can vary, and handling these uncertainties is explained in this section. Scheduling using LLF is explained below. Scheduling using EDF, SJF and RR approaches can be done similarly.

Consider the scenario where the total number of tuples is fixed, but the input rate varies. Here, after MinBatchSize determination, the expected time point at which MinBatchSize will be ready as per the input rate is also estimated. A query batch is considered schedulable if either the input has reached MinBatchSize or the time point has crossed the estimated time for availability of MinBatchSize tuples; in the latter case, a query batch is schedulable even if there are fewer tuples than MinBatchSize. Schedulable queries are sorted based on their slack time, and the query with the least slack time is scheduled for processing. If the actual input rate is faster than or equal to the predicted model then processing will be triggered as and when the required batch size is ready. If the actual input rate is slower, then processing gets triggered based on the estimated input available time, thereby trying to meet the deadline by processing the available tuples instead of waiting for the MinBatchSize readiness. Processing using the available tuples reduces the risk of missing the query deadline.

For dynamic systems where both the input rate and the total number of tuples can vary, we can estimate the expected total number of tuples in the window using any appropriate estimator, which can take into account the actual input rate. Laxity is then computed based on the updated estimates for the total number of tuples in the window. Then the query with the least estimated slack time is processed.

4   Related Work

Many stream processing engines run on the YARN infrastructure. Vavilapalli et al. [11] describe scheduling schemes supported by YARN such as FIFO, Capacity, and Fair, but none of these schemes considers deadlines. Stream processing engines such as Apache Spark and Flink process tuples eagerly with some fixed minimum batch size. However, they too do not consider deadlines. Tuning of batch size is done by some stream processing systems to achieve a balance between throughput and latency, but not in a deadline aware manner, unlike our work. Ye et al. [15] propose an optimization technique for Spark Streaming configurations without considering deadlines.

Tang et al.[8], Shang et al. [7] point out that the slack period available in queries can be utilized to reduce resource consumption. Tang et al. [8] propose intermittent processing of queries which is triggered at some time interval or based on the number of tuples accumulated. Shang et al. [7] have built a database system namely CrocodileDB which processes queries intermittently based on user inputted frequency. Tang et al. [9] define a new metric, Incrementability, to denote the amount by which a query supports incremental operations. However, none of these papers considers query deadlines.

Grosbeak [13] schedules analysis jobs in non peak hours based on the history of resource utilization. The job is processed in batches which is similar to our approach, but details on scheduling are not discussed. Wang et al. [14] discuss optimization of intermittent query processing, but unlike our case, they do not consider absolute deadlines or scheduling. Zhang et al. [16] show that join queries processed in a lazy manner can perform better than eager processing, but deadlines are not considered.

In hard real time systems where the incoming tuple has to be processed within a certain time to make critical decisions, each tuple is modeled with a deadline. This is different from our problem statement where all tuples in a query have to be processed within a common deadline. Ou et al. [6] propose Tick scheduling where a tick denotes a set of tuples that have the same deadline, but they do not consider the minimization of computation cost by batching.

Scheduling in real time systems has been widely explored and some of the prominent algorithms are EDF, LLF, etc. While EDF and LLF scheduling only aim at completing the query within its deadline, our approach reduces the overall computational cost of each query by processing queries in batches. Other deadline aware scheduling algorithms (see for example the survey [4]) do not consider batching.

In a cluster environment, choosing the optimal resources (e.g. nodes) to minimize cost while meeting deadlines, is considered in [2,3,5,12]. In all these approaches, once the resources are allocated, either Tuple-by-tuple processing or micro batch processing is used.

To the best of our knowledge, our work is the first of its kind which combines batching and scheduling to honor deadlines while minimizing the cost.

5   Implementation Details

The scheduling schemes proposed in this paper have been implemented by building a Custom Scheduler over the Apache Spark architecture. Our scheduling algorithms are agnostic to the underlying stream processing engine. The Custom Scheduler consists of a Query Repository, Schedule Optimizer and Query Scheduler components.

The Query scheduler runs periodically, whenever a query batch completes execution, or if the system is idle, it rechecks query batch readiness periodically.

Since batch sizes are chosen to ensure that no batch takes more than C_{max} time for execution, the scheduler will be invoked within a maximum interval of C_{max} from the previous invocation. The scheduler first checks if any new query has been submitted, and if so it invokes the Schedule Optimizer to compute the MinBatchSize for the query, which does so using the cost model for the query, along with the chosen δ_{RSL}, and C_{max}. The scheduler then selects the queries whose batches are ready for processing and determines the query to be processed based on the chosen scheduling strategy.

When the Query scheduler schedules a query for execution, the Schedule Optimizer invokes the appropriate query operations. Query Repository contains the actual query operations which are to be carried out for each query. As the current implementation uses Spark, it consists of the spark operations which are executed for each batch and the ones which are executed as part of the final aggregation. If any other stream processing framework is used then its corresponding query operations can be implemented in the Query Repository component of our Custom Scheduler. As each batch is processed the intermediate results are stored in a file.

The Schedule Optimizer also keeps track of the batches processed and invokes the final aggregation once all batches of a query have been processed.

For handling queries with joins the following strategy is adopted. For the stream to static join, as the static data does not change, each batch is joined against the static data to get the join results. Typically join conditions in queries ensure that matching tuples from different streams will have timestamp values that are the same or within some bound. For simplicity, our implementation assumes that the corresponding tuples from two streams are available in the same batch. For example, with the TPC-H schema, Orders and their associated Lineitem tuples are assumed to be in the same batch. This assumption can be relaxed, but implementation details depend on the stream-processing application used.

To derive the time cost model, each query is executed individually to measure the execution time at different input batch sizes. Based on observed execution times, a piece-wise linear cost model was arrived at. Similarly, a piece-wise linear model varying on the number of batches was designed to fit the final aggregation cost. As described in [17], most production environments run the same set of queries repeatedly and hence we can build the cost model for queries when they are first executed on the system, and use the cost model when the same query is executed again.

Though file-based input is widely used in many applications as it enables easy information exchange, streaming data platforms such as Kafka are a commonly used alternative. We explored input from a Kafka system by creating two Kafka topics namely Orders and Lineitem with 36 partitions to support parallel processing. We read from Kafka using both the stream and batch approaches. It was observed that the cost incurred using Kafka is at least 3 times more than using file-based inputs. Also, we observed that between Kakfa streaming and batching, streaming incurs considerably more cost. However, processing in batch mode significantly reduces the cost incurred compared to stream processing, whether we use files as input or the Kafka platform. For our performance studies in Sect. 6, we used file-based inputs to avoid the overheads encountered in reading data from Kafka.

The Custom scheduler does not add any significant time overhead to the overall query processing as the time taken is in the order of milliseconds. Determination of MinBatchSize is done only once for each query.

6 Performance Evaluation

In this section, we present the performance evaluation for our scheduling strategies. Queries were run on a Spark cluster deployed in a standalone mode having 2 Intel Xeon Silver 4116 Processors (2.10 GHz) with 250 GB of RAM. Spark context was configured with 48 cores and 20 GB of memory.

As explained in Sect. 4, there is no prior work that does both batching and scheduling. Hence, in this section, we compare our methods against the standard Spark Streaming and the traditional scheduling algorithms to assess the effect of batching and scheduling respectively. We also further evaluate our method with stringent deadline cases along with variable input rates.

We use a modified version of the TPC-H Dataset of 25 GB. To simulate the input data stream, a timestamp has been added to each record in the relations Orders and Lineitem which are considered as streaming relations. The other relations are considered as static relations. The input stream consists of 4500 files inputted at the rate of 1 file of Orders and 1 file of Lineitem per second. Our study considers a subset of 9 of the TPC-H queries (including queries with stream-to-stream joins, i.e. between Orders and Lineitem, and stream-to-static-relation joins), along with 4 custom queries shown in Table 2.

6.1 Comparison of Custom Scheduler Against Spark Streaming

We first consider the cost reduction obtained due to batching with our approach as well as using Spark Streaming. The cost of execution of a query refers to the

Table 2. Custom Queries

QueryID	Query
CQ1	SELECT count(*) as totalOrders FROM orders
CQ2	SELECT count(*) as totalOrders, orderPriority FROM orders GROUP BY orderPriority
CQ3	SELECT count(*) as totalItems, suppKey FROM lineItems GROUP BY suppKey
CQ4	SELECT count(*) as totalItems, partKey FROM lineItems GROUP BY partKey

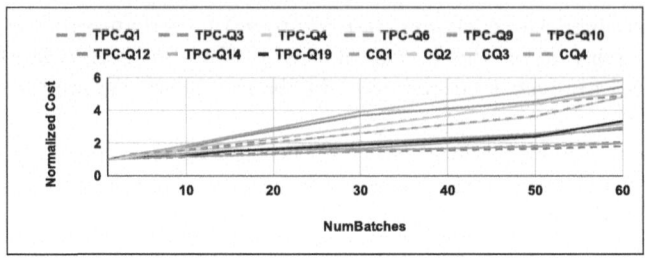

Fig. 1. Cost versus Number of Batches

sum of the query execution time of all the batches and the final aggregation cost. We study the effect of the increase in computation cost as the number of batches increases. The batch size of x in our experiments refers to the number of files processed as part of a single batch. Since there are 4500 files, the number of batches is $4500/x$. All 13 queries were evaluated for different batch sizes. For each of the queries, the minimum cost required for processing it in a single batch is taken as the baseline. The cost incurred when processed with different batch sizes has been normalized w.r.t. this baseline and shown in Fig. 1. It can be observed that the more the number of batches, the more the overall computation cost.

To compare our approach against Spark Streaming, the queries were processed using a Streaming job in Apache Spark with the default i.e. immediate and different batch intervals of 5, 10, 30 and 40 min, with the window aggregation duration of 4500 s. In addition, experiments were done in a one-shot mode where all files were processed in one go. Figure 2 shows the cost incurred for each query, for each batch interval, normalized to the cost of computation in a single batch. It can be observed that the computation cost decreases as the batch interval increases. The least computation cost incurred by Spark Streaming is with the one shot mode of processing. Among all the queries, TPC-Q14 (data labels marked in the figure) has the least normalized computation cost

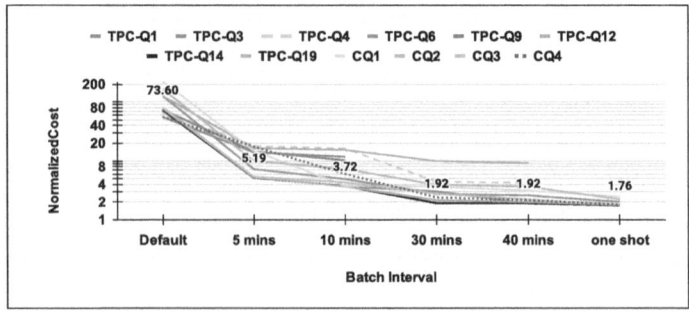

Fig. 2. Normalized Cost (log scale) of Batch processing versus Spark Streaming

with Spark Streaming, which is 1.76 times more than the cost incurred when all tuples are processed in a single batch using our approach.

With Spark Streaming, TPC-Q3, TPC-Q4, TPC-Q9, TPC-Q10 and TPC-Q12 failed for one shot computation and for runs with batch intervals of 30 and 40 min. All these failed queries have join on Orders and Lineitem. Spark Streaming keeps the data in memory for doing the join operations. Thus, as the batch interval increases in Spark Streaming, the amount of data to be stored in memory increases. These cases failed even with the increased memory of 45GB. In contrast with our approach all queries executed successfully with Spark context of 20 GB memory.

Since big data systems need to run multiple queries, the following experiment was carried out where all the TPC-H and the Custom queries were run simultaneously. In our approach, concurrent queries use time-sharing, with one batch of one query executing at a time. Spark Streaming could not support concurrent execution since it ran out of memory. Hence, multiple runs were used, where each run streamed the data to a subset of the queries. The cost incurred in each of the runs was summed up to get the total computation cost. Spark Streaming experiments were done using the default and 10-min batch intervals; for larger intervals, queries TPC-Q3, TPC-Q4, TPC-Q9, TPC-Q12 failed.

Spark streaming costs are compared against the total computation cost incurred using our dynamic mode of scheduling with 50% δ_{RSF} factor using LLF. The cost incurred by Spark Streaming for default and 10-min batch intervals were, respectively, 60 and 12 times the cost using our approach. Thus our approach of batching performs much better than running Spark Streaming for large window operations for multiple queries simultaneously.

6.2   Evaluation of Custom Scheduler for Different Deadlines

Next, we ran experiments to evaluate the performance of the Custom Scheduler with respect to meeting deadlines. For the dynamic scenario, all the TPC-H and the custom queries were considered simultaneously with δ_{RSF} factor of 50% and C_{max} of 30 s. All queries were set with the same window start time and window end time. We chose an arbitrary sequence of queries and set their deadlines such that each query ran as a single batch. The deadline of the first query is set as C_{max} plus the time required for processing all tuples in a single batch starting at the window end. Deadlines of other queries are set as the time required for processing all tuples in a single batch starting at the previous query's deadline. We refer to this set of deadlines as 1D. Further, cases with reduced deadlines were generated, where the deadline was set to window end time plus 0.8, 0.6, 0.4, 0.2, and 0.1 times the assigned gap from the window end time to the deadline for 1D case.

Experiments were carried out for all the above cases using EDF, LLF, SJF and RR. SJF failed for 0.2D and 0.1D and RR failed for 0.4D, 0.2D and 0.1D as some of the queries could not meet their deadlines. EDF and LLF passed all cases except for 0.1D as there is no feasible solution for 0.1D with $\delta_{RSF} = 50\%$. The fact that SJF and RR failed on multiple cases shows that scheduling strategies based on deadline or slack time are essential to meet query deadlines.

6.3 Comparison with EDF and LLF Without Minimum Batch Size

To demonstrate the importance of having a minimum batch size, we ran experiments using EDF and LLF approaches without a minimum batch size, queries are considered schedulable with all available tuples. Since tuples are input in units of files, which in our experiments consist of around 9300 tuples, 1 file could also be viewed as the minimum batch size. With this approach, EDF and LLF failed to meet deadlines for the 0.4D case, while with our approach of computing minimum batch size with $\delta_{RSF} = 50\%$, deadlines were met down to 0.2D, with both EDF and LLF. Also, even for runs where deadlines were met since the queries with the earliest deadline/least laxity were scheduled more frequently with the small batch sizes, the costs incurred by EDF and LLF were 10 and 7 times more compared to the case where minimum batch size was set with $\delta_{RSF} = 50\%$. Thus our methodology minimizes cost while meeting the deadlines.

6.4 Evaluation of Custom Scheduler With Variable Input Rates

Next, we carried out experiments to assess the impact of variable input rates on the scheduling, where the scheduler cannot predict the arrival rate. Figure 3 shows the data (in units of number of files) that are received at different points in time. FR denotes the fixed rate of arrival case while VR1 to VR4 shows variable rates of input. While both VR1 and VR2 are faster compared to FR, VR2 contains bursty input. Both VR3 and VR4 are slower than FR, and some tuples arrive after the window end of FR.

Results for the case of 0.1D deadlines, with $\delta_{RSF} = 100\%$ are shown in Fig. 4a. With both EDF and LLF the scheduler could complete all queries within their deadlines for all cases as in Fig. 4a. SJF and RR completed all queries for VR1 and VR2 but failed for FR, VR3 and VR4 as some queries missed their deadlines. Similarly, experiments were carried out with variable input rates for 0.2D, $\delta_{RSF} = 50\%$ and the results are shown in Fig. 4b. EDF, LLF and SJF passed all cases while RR failed for VR3 and VR4.

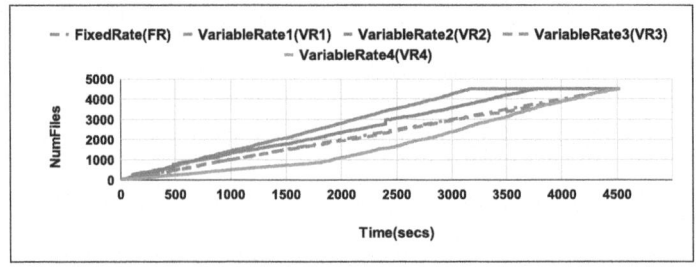

Fig. 3. Variable Input Rate For Multi Query Scenario

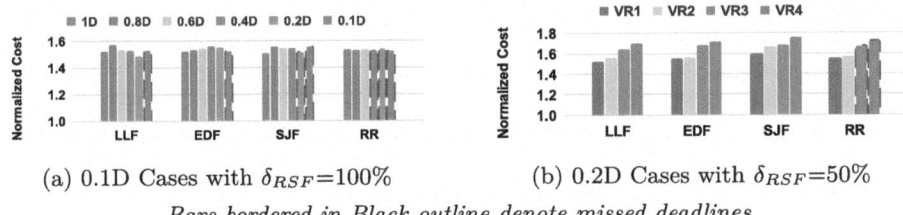

(a) 0.1D Cases with δ_{RSF}=100%   (b) 0.2D Cases with δ_{RSF}=50%

Bars bordered in Black outline denote missed deadlines

Fig. 4. TPC-H, Custom Queries in Dynamic Scenario with Variable Input Rates

When δ_{RSF} is increased from 50% to 100%, the overall computation cost increases due to a reduction in the MinBatchSize. This can be observed in the Figs. 4a and 4b where the normalized computation cost is around 1.5 and 2.0 for δ_{RSF} of 50% and 100% respectively. Also for slower input profiles(VR3 and VR4), as the total number of batches is more the normalized computation cost is more compared to the cost incurred for the other input profile cases.

Thus, processing on very small batches may not lead to benefits. Further, it may be noted that partial aggregation within a batch is beneficial only if the aggregation result is significantly smaller than the input batch size, which requires the batch to contain on average multiple tuples for each group.

The results show that our scheduling algorithms complete the benchmark queries within their deadlines while keeping the overall computation cost not more than δ_{RSF} fraction compared to the computation cost of processing all tuples in a single batch. The results confirm that EDF and LLF perform better in meeting the deadlines compared to SJF and RR for both fixed and variable input rates. Also, our approach handles stringent deadlines better than just using EDF and LLF with a default batch size that is set without considering δ_{RSF}.

7   Conclusion and Future Work

We have presented techniques for determining appropriate batch sizes and scheduling of multiple queries under a dynamic environment, while handling the uncertainties in the input rate. The results presented in the performance section show that our methods perform better in terms of reducing the overall computation cost while meeting deadlines compared to Spark streaming as well as EDF and LLF with a default small batch size.

There are several directions for future work. Our scheduling techniques can be extended for a cluster setup where resources can be added dynamically to complete queries within the deadline. The cost model proposed in this paper can be correlated to the monetary value that would be required for resource allocation. Our current scheduling model runs one batch of one query at a time across all available resources. This can be extended to support the simultaneous execution of different queries on different subsets of nodes in the cluster.

References

1. Chandrasekaran, S., Sudarshan, S.: Scheduling of intermittent query processing. arXiv:2306.06678 (2024)
2. Cheng, D., et al.: Adaptive scheduling of parallel jobs in spark streaming. In: Proceedings of the IEEE Conference on Computer Communications (INFOCOM), pp. 1–9 (2017)
3. Dimopoulos, S., Krintz, C., Wolski, R.: Justice: a deadline-aware, fair-share resource allocator for implementing multi-analytics. In: Proceedings of the IEEE International Conference on Cluster Computing (CLUSTER), pp. 233–244 (2017)
4. Hedayati, S., et al.: MapReduce scheduling algorithms in Hadoop: a systematic study. J. Cloud Comput. **12**(1), 143 (2023)
5. Islam, M., Karunasekera, S., Buyya, R.: dSpark: deadline-based resource allocation for big data applications in Apache Spark. In: Proceedings of the 13th IEEE International Conference on e-Science (e-Science), pp. 89–98 (2017)
6. Ou, Z., et al.: Tick scheduling: a deadline based optimal task scheduling approach for real-time data stream systems. In: Proceedings of the 6th International Conference in Web-Age Information Management (WAIM), pp. 725–730 (2005)
7. Shang, Z., et al.: CrocodileDB: efficient database execution through intelligent deferment. In: Proceedings of the 10th Conference on Innovative Data Systems Research (CIDR) (2020)
8. Tang, D., et al.: Intermittent query processing. Proc. VLDB Endow. **12**(11), 1427–1441 (2019)
9. Tang, D., et al.: Thrifty query execution via incrementability. In: Proceedings of the ACM International Conference on Management of Data (SIGMOD), pp. 1241–1256 (2020)
10. Tang, D., et al.: CrocodileDB in action: resource-efficient query execution by exploiting time slackness. Proc. VLDB Endow. **13**(12), 2937–2940 (2020)
11. Vavilapalli, V.K., et al.: Apache Hadoop YARN: yet another resource negotiator. In: Proceedings of the 4th Annual Symposium on Cloud Computing (SOCC) (2013)
12. Wang, G., Xu, J., Liu, R., Huang, S.: A hard real-time scheduler for Spark on YARN. In: Proceedings of the 18th IEEE/ACM International Symposium on Cluster, Cloud and Grid Computing (CCGRID), pp. 645–652 (2018)
13. Wang, Z., et al.: Grosbeak: a data warehouse supporting resource-aware incremental computing. In: Proceedings of the ACM International Conference on Management of Data (SIGMOD), pp. 2797–2800 (2020)
14. Wang, Z., et al.: Tempura: a general cost-based optimizer framework for incremental data processing. VLDB J. **32**(6), 1315–1342 (2023)
15. Ye, Q., Liu, W., Wu, C.Q.: Nostop: a novel configuration optimization scheme for Spark Streaming. In: Proceedings of the 50th International Conference on Parallel Processing (ICPP), pp. 1–10 (2021)
16. Zhang, S., et al.: Parallelizing intra-window join on multicores: an experimental study. In: Proceedings of the ACM International Conference on Management of Data (SIGMOD), pp. 2089–2101 (2021)
17. Zhang, W., et al.: Deploying a steered query optimizer in production at Microsoft. In: Proceedings of the International Conference on Management of Data (SIGMOD), pp. 2299–2311 (2022)

Monitoring Environmental Factors for Optimal Health and Welfare of Broiler Chickens Using Sensor Technology in Southwest Nigeria

Saidu Oseni[1(✉)] ⓘ, Hameed Bashiru[1] ⓘ, Rasheed Lawal[1] ⓘ, Ayobami Ajayi[1] ⓘ,
Olamide Akintaro[2] ⓘ, and Kamran Munir[3] ⓘ

[1] Obafemi Awolowo University, Ile-Ife 220005, Nigeria
soseni@oauife.edu.ng
[2] Taro Agric Farm, Ile-Ife, Osun, Nigeria
[3] University of the West of England, Bristol, UK

Abstract. Multiple factors affect the health and welfare of broiler chickens under the deep litter system. These factors include daily variations in temperature, humidity, noxious gases (such as ammonia, hydrogen sulfide, and nitrous oxide), particulate matter, and volatile organic compounds. When these factors exceed threshold levels, broiler chickens experience distress, leading to slower growth rates and higher mortality. Digital technologies, through the deployment of sensors and smart devices, enable concurrent monitoring and smart tracking of these factors to maintain comfort zones for the healthy growth of broiler chickens. However, in tropical climates, the use of sensors and smart devices to monitor these factors in broiler houses is rare. This study presents preliminary data on the monitoring of key health and welfare parameters by tracking ambient factors in a commercial broiler enterprise. The methodology involved deploying sensor nodes on a commercial broiler farm to facilitate data capture and transmission from sensors to dedicated databases. Specific variables monitored included temperature, humidity, average and peak noise levels, light intensity, particulate matter, volatile organic compounds, and gases (such as ammonia, hydrogen sulfide, and nitrous oxide). Results indicated a consistent trend of threshold values being exceeded for temperature, humidity, ammonia, and particulate matter. Correlation coefficients were calculated to establish the magnitude and direction of interrelationships among these variables. Overall, the results highlight the need for mitigation strategies to maintain optimal comfort zones, safeguarding the health and welfare of broiler chickens, which in turn positively impacts the profitability of the enterprise.

Keywords: IoT sensors · environmental parameters · broiler chickens · welfare and health

1 Introduction

Health and welfare of broiler chickens are affected and influenced by multiple factors including environmental parameters such as ammonia, among others [1, 2]. For all these factors, comfort zones and threshold values have been established for broilers. For ammonia for instance, levels above 25 PPM [1] reduces weight gain, feed conversion and liveability, lowering profit margins. When threshold levels are exceeded for particulate matter, the liver, kidney and respiratory system of broilers are impaired, leading to morbidity and mortality of chickens [2]. Similar, albeit negative trends have been reported for hydrogen sulfide, nitrous oxide and volatile organic compounds.

From the foregoing, it is obvious that tracking and monitoring of these ambient variables and gases have a bearing on key health, welfare and survival parameters of broiler chickens. Consequently, a dire need to monitor all these factors is critical to the profitability of broiler enterprises. Machine learning is commonly employed to create this desired forecasting system, where the model typically incorporates a neural network. Here, the pre-stress environment is used to train the network and prediction involves classifying weather data points based on if they constitute a stressful event. This provides the motivation for this paper as it present a database engineered system; namely, a recurrent neural network (RNN) model for forecasting stressful conditions in the pens of broiler chickens.

Furthermore, the deployment of digital technologies via sensors and smart devices for the capture and transmission of these variables will help to ensure that critical levels are not exceeded [3]. This study presents preliminary data for the monitoring of key health and welfare parameters of chickens by tracking ambient factors in a commercial broiler enterprise.

2   Materials and Methods

2.1   Site Description

The study was conducted at Taro Agric Farm, a commercial farm located at Kurundun village, Erefe, Ife East Local Government Area, Ile-Ife, Osun State, Nigeria. The Farm is located on Longitude 7.474028 and Latitude 4.539396. The Farm has a 4000-capacity broiler chicken production facility and covers a land expanse of 5 acres. The farm has a motorised borehole as the source of water supply and uses manually operated feeders and drinkers.

2.2   Experimental Birds and Management

The broiler chickens were raised in batches. A batch comprises 2000 chickens. The birds were raised in a deep litter system. Two weeks before arrival of the chicks, the deep litter pens were thoroughly washed and disinfected with IZAL and fumigated with Formalin. Prior to the arrival of the chicks, wood shavings, obtained from local sawmills, were spread as litter materials. Lighting and heating equipment were put in place and used for chicks' brooding. The chicks were brooded for 2 weeks. Water and feed were provided to birds *ad-libitum* throughout the brooding period. All the routine medications and vaccinations were administered as at when due. The birds were raised for a period of 6 weeks to attain a body weight of 2 kg or higher, prior to marketing.

2.3 Data Collection

The devices detailed in this article are designed to monitor environmental parameters using low-cost hardware and the connectivity framework of the Internet of Things (IoT). These devices, along with their associated applications, enable real-time recording and tracking of temperature, relative humidity, ammonia and hydrogen sulfide concentrations, luminosity, average and peak noise levels, nitrous oxide concentrations, particulate matter and volatile organic compounds. This allows poultry farmers to quickly analyse data and make informed decisions regarding health, welfare and survival of chickens. A detailed list of sensors and variables captured by each sensor is presented in Table 1.

Table 1. List of sensors and variables captured by each sensor

Sensor Code	Sensor Name	Description
SHT45	Adafruit Sensirion SHT45 Precision Temperature & Humidity sensor	Measures ambient temperature and relative humidity
MICS6814	3-in-1 Gas Sensor Breakout	Measures concentrations of carbon monoxide, nitrogen dioxide and ammonia
968–003	Hydrogen sulfide sensor	Measures concentration of hydrogen sulfide
SGP40	Adafruit SGP40 Air Quality Sensor Breakout	detects and quantifies various gas concentrations in the air
ESP32	Adafruit ESP32 board	A microcontroller widely used in IoT applications for communication and real-time data processing

2.4 System Layout

The block diagram showing the key components of the sensors used is shown in Fig. 1. The sensors measure various environmental parameters (temperature, relative humidity, air quality, and gas concentrations) and send the data to the microcontroller board. The microcontroller processes the raw data, performing initial filtering and formatting before being processed for transmission to the cloud server, which is then received by the cloud server for storage. The IoT server manages communication between the devices and ensures data integrity and security. The processed data is then visualized using analytics tools, allowing the monitoring of environmental conditions in real time, and aiding in quick decision making.

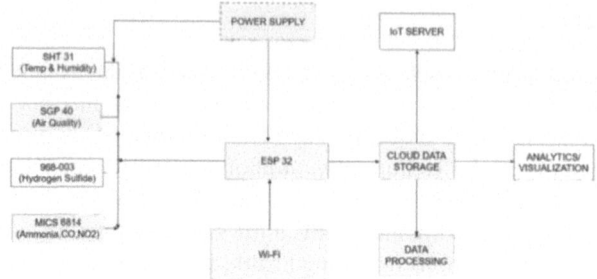

Fig. 1. Key Components of the Sensors Used

Fig. 2. (left) A coupled sensor node, (right) A coupled sensor node strategically placed in broiler pen for data capture

2.5   Installation of Nodes

The sensor components are coupled together with the ESP32 microcontroller in a node (as shown in left of Fig. 2) and is placed strategically in the pen house (as shown in right of Fig. 2) to be able to capture ambient readings seamlessly. Three sensor nodes are placed in three separate broiler house compartments on the farm.

2.6   Solar and Network Installations

The three pillars of the IoT setup in a broiler production unit include power source, network or internet connection, and the sensor nodes for capturing data. A 2.5Kva solar inverter system, which included 2 batteries, was installed at the farm to provide constant power supply to the farm, and to also power the sensor nodes for collecting data. A network booster with an antenna as well as a 5G Router was used to make internet connection available on the farm.

2.7   Summary of Data

A total of 42,302 data points collected from three sensor nodes over the months of April, May, and June 2024 were utilized for this study. A limitation of the dataset used in this study is that it does not cover a more extended period, such as a full year. This limited timeframe was due to the availability of data at the time of the study, as new data continues to be generated daily. It should be noted that Node 3 did not begin operations until May because it had not yet been manufactured and deployed in April. The data was collected daily from all nodes at an interval of 5 min, making an approximate of 288 data points collected per day. A summary of the records is presented in Table 2.

Table 2. Summary of records used in the study

Number of records	April	May	June
Node 1	7512	8103	3635
Node 2	7759	8069	3154
Node 3	NA	338	3732
TOTAL	15271	16510	10521

2.8 Data Analysis

The data collected from the IoT sensors at Taro Agric Farm were managed using Microsoft Excel, where charts were plotted for visualization. The correlation (CORR) procedure of SAS was employed to analyse the data and generate correlation coefficients among the various parameters. Data were further summarized using descriptive statistics such as means, standard deviation, minimum and maximum.

3 Results

3.1 Daily Fluctuations in Environmental Factors from April to June, 2024 at Taro Agric Farm

Figure 3 shows the maximum temperature fluctuations from April to June 2024 at Taro Agric Farm. Maximum temperature fluctuated between 29 °C and 39 °C from April to June. April and May had more consistent temperatures, mostly oscillating around the 35 °C distress threshold, indicating periods of potential distress for chickens. June showed a slight decrease, with temperatures ranging from 29 °C to 36 °C, occasionally dipping below the 35 °C distress threshold but remaining above the 30 °C stress threshold, indicating a generally stressful but less distressful environment. April experienced temperatures mainly between 34 °C and 39 °C, with most values exceeding the distress threshold of 35 °C. The month of May showed similar trends, with fluctuations from 30 °C to 38 °C. In June, temperatures generally ranged from 29 °C to 36 °C, slightly lower than the previous months but still frequently surpassing the distress threshold. These consistent exceedances indicate persistent thermal stress and potential distress conditions for chickens during the period.

Figure 4 presents the maximum relative humidity fluctuations from April to June at Taro Agric Farm. Relative humidity levels ranged from 80% to 100% between April

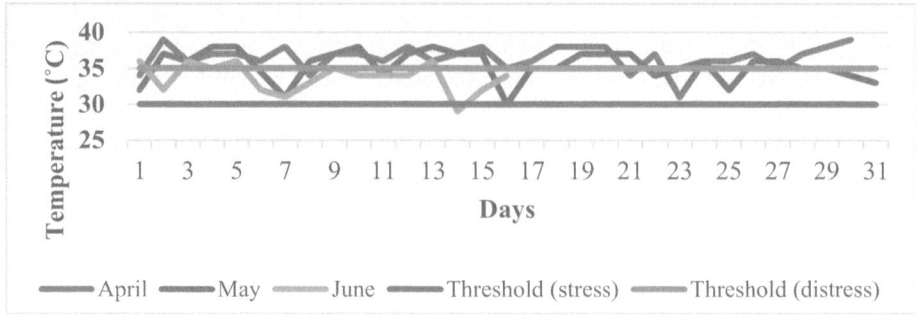

Fig. 3. Daily maximum temperature fluctuations from April to June 2024 at Taro Agric Farm

and June. Across these months, the maximum threshold level (70%) was consistently exceeded daily. The peak maximum humidity level was reached in June while the lowest maximum relative humidity was obtained in April. Overall, this indicated that the broiler chicken production environment was persistently beyond the optimal relative humidity threshold.

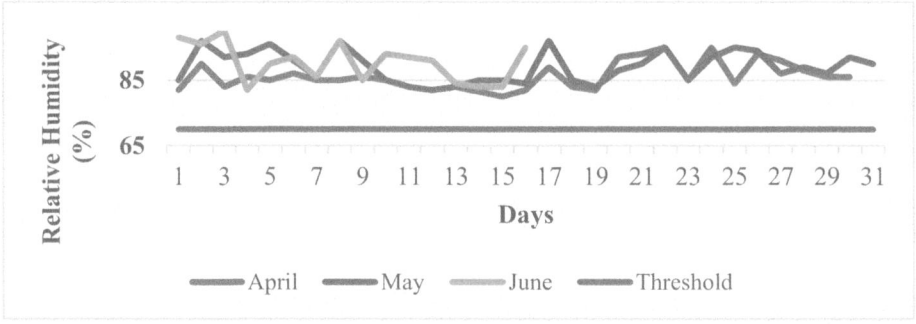

Fig. 4. Daily maximum relative humidity fluctuations from April to June 2024 at Taro Agric Farm

Figure 5 presents daily maximum temperature-humidity index (THI) fluctuations from April to June 2024 at Taro Agric Farm. Maximum daily THI values consistently exceeded threshold in April and May. In June, THI values were lower even though the values were mostly higher than the threshold. In April, maximum THI peaks of 99.74, 98.5, 97.01 and 98.7 were observed at the 2nd, 20th, 26th and 30th day. Similarly, THI peaks of 99.46, 96.86, 96.38 and 96.34 were observed in the 5th, 10th, 13th and 21st day, respectively.

Figure 6 presents the maximum hydrogen sulfide fluctuations from April to June 2024 at Taro Agric Farm. Hydrogen sulfide levels across these months ranged from 0 PPM to 53 PPM. These values frequently exceeded the thresholds for stress (2 PPM) and distress (6 PPM), especially in April and June which indicated periods of potential distress for broiler chickens. In April, daily maximum hydrogen sulfide levels were consistently higher than the thresholds and multiple peaks were observed at days 12, 27

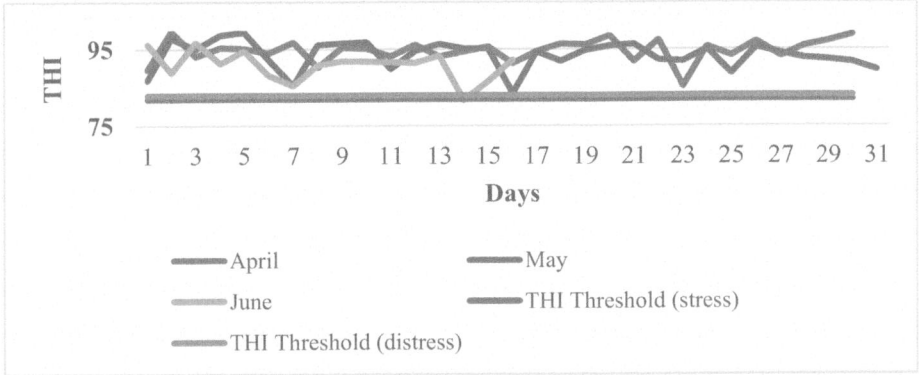

Fig. 5. Daily maximum temperature-humidity index (THI) fluctuations from April to June 2024 at Taro Agric Farm

and 29. Similarly, daily maximum hydrogen sulfide levels were also at higher threshold levels and peaks were observed at the 5th, 7th and 15th day. In May, maximum daily hydrogen sulfide levels were generally lower, although threshold levels were mostly exceeded apart from the 7th and 27th -31st day when hydrogen sulfide levels were lower than the thresholds. These maximum daily hydrogen sulfide levels suggest significant air quality issues, potentially harmful to broiler chickens and could possibly predispose them to chronic respiratory diseases.

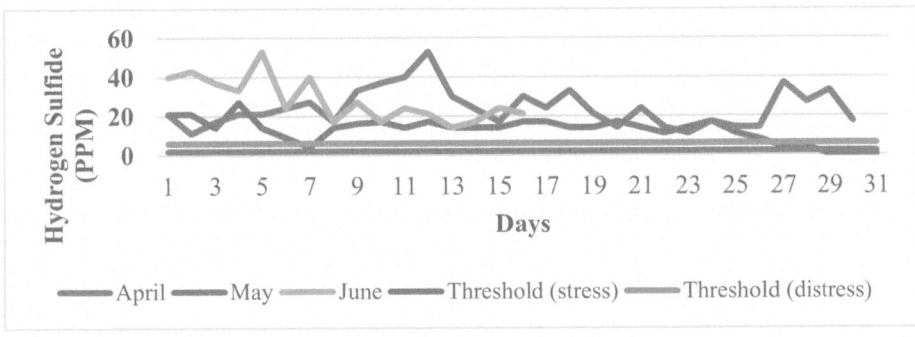

Fig. 6. Daily maximum hydrogen sulfide fluctuations from April to June 2024 at Taro Agric Farm

Figure 7 shows the daily maximum ammonia fluctuations from April to June 2024 at Taro Agric Farm. Daily maximum ammonia levels fluctuated widely, with April and June showing extreme peaks up to 205 PPM. Daily maximum ammonia levels typically exceeded the threshold levels for stress (25 PPM) and distress (50 PPM), especially in April and June. Peaks for daily maximum ammonia levels in April, May and June were observed in the 2nd, 15th and 8th day, respectively. These levels are well above the 25 PPM stress and 50 PPM distress thresholds, indicating persistent and severe ammonia pollution. The exceedance of these thresholds, especially in April and June, indicates potential ammonia-related stress conditions at the farm.

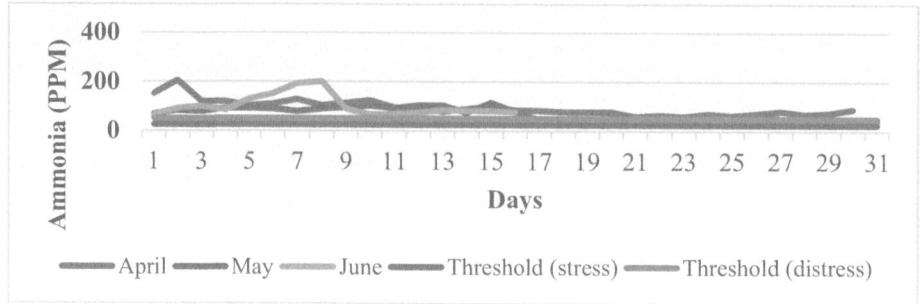

Fig. 7. Daily maximum ammonia fluctuations from April to June at Taro Agric Farm

Figure 8 shows the maximum NO2 fluctuations from April to June 2024 at Taro Agric Farm. There were wide variations in NO2 levels across these months. In April, NO2 levels were consistently above the thresholds (1 PPM and 2PPM for stress and distress, respectively) apart from days 23- 25. Further, NO2 level peaked at 11 PPM at the 10th day of April which was the highest NO2 level recorded. In May and June, lower NO2 levels were recorded but the levels were still mostly above the thresholds.

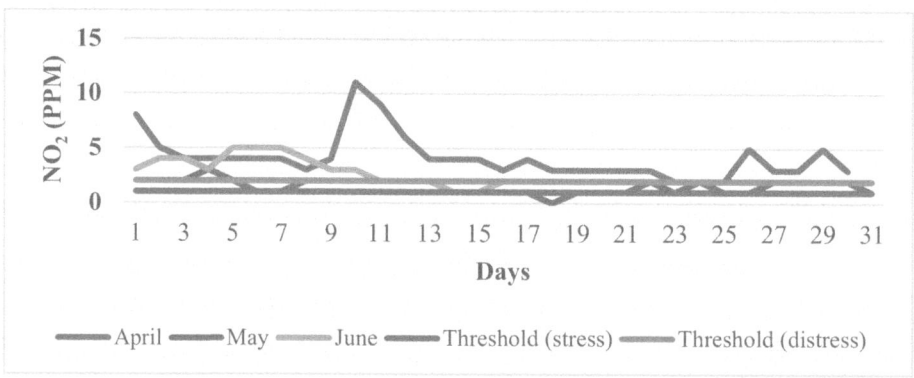

Fig. 8. Daily maximum NO_2 fluctuations from April to June 2024 at Taro Agric Farm

Figure 9 presents the daily maximum particulate matter (2.5 $\mu g/m^3$) fluctuations from April to June, 2024 at Taro Agric Farm. Particulate matter levels across these months were mostly within the threshold levels. However, there were spikes in the level of particulate matter with peaks that are as high as 1000% greater than the threshold levels.

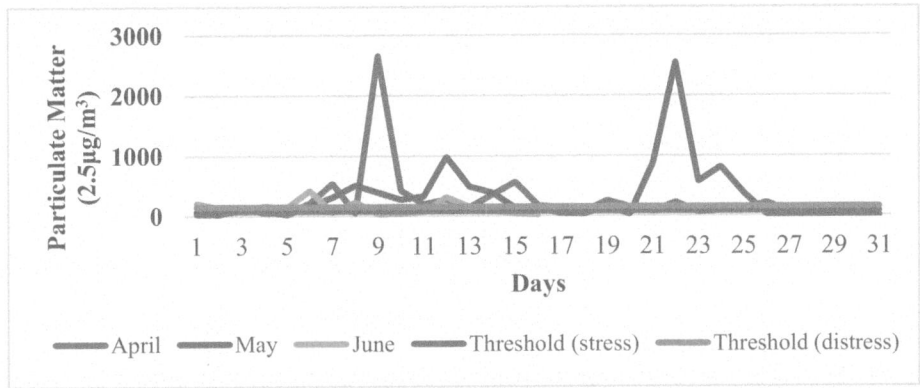

Fig. 9. Daily maximum particulate matter (2.5 µg/m^3) fluctuations from April to June 2024 at Taro Agric Farm.

Figure 10 shows the daily maximum particulate matter (10 µg/m^3) fluctuations from April to June 2024 at Taro Agric Farm. The trends were similar to observable trends for particulate matter (2.5 µg/m^3). PM10 levels also varied widely. Months of April and May showed extreme levels, frequently surpassing the 150 µg/m^3 stress and 300 µg/m^3 distress thresholds while June showed lower values.

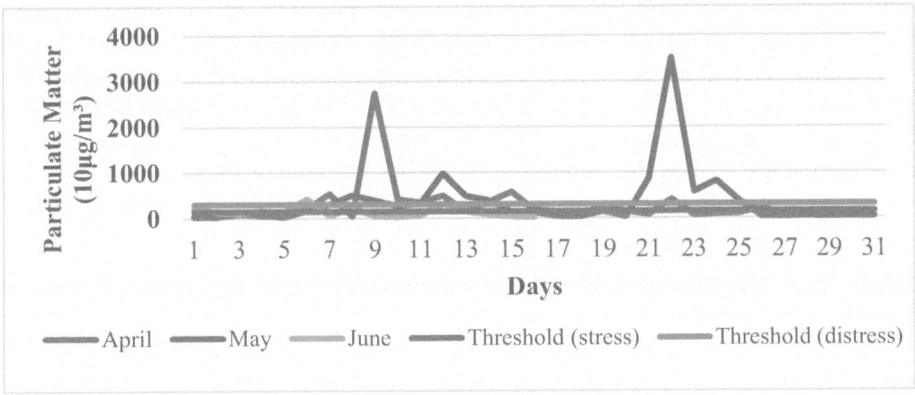

Fig. 10. Daily maximum particulate matter (10 µg/m^3) fluctuations from April to June 2024 at Taro Agric Farm

Figure 11 shows the daily maximum total volatile organic compound (TVOC) fluctuations from April to June 2024 at Taro Agric Farm. Consistently, TVOC values were significantly higher than threshold values across April, May and June. This indicated severe and persistent TVOC pollution in the broiler chicken production environment.

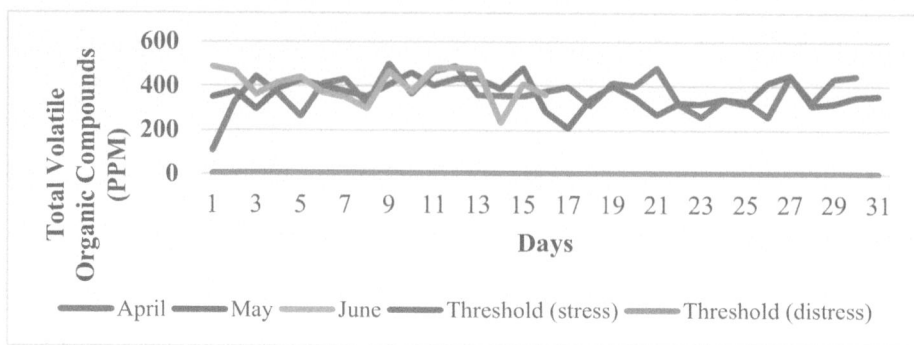

Fig. 11. Maximum total volatile organic compound fluctuations from April to June at Taro Agric Farm

3.2 Descriptive Statistics of Environmental Factors from April to June 2024 at Taro Agric Farm

Table 3 presents descriptive statistics of environmental parameters captured by IoT sensors at Taro Agric Farm. The average temperature recorded was 30.15 °C with a standard deviation of 3.44 °C, indicating moderate overall variability. Temperatures ranged from 22 °C to 39 °C. However, relative humidity ranged widely from 46% to 100%. In addition, the average relative humidity was 72.27%, with a standard deviation of 9.10%. For hydrogen Sulfide, the average level was 0.95 PPM, with a standard deviation of 3.10 PPM, ranging from 0 to 56 PPM. The average ammonia concentration was 48.27 PPM, with a standard deviation of 19.82 PPM, ranging from 13 PPM to 205 PPM. For nitrogen dioxide, the average concentration was 1.28 PPM, with a standard deviation of 1.05 PPM, ranging from 0 to 11 PPM. For particulate matter (PM2.5), the average concentration was 9.80 µg/m^3, with a high standard deviation of 29.30 µg/m^3, while the average concentration for particulate matter (PM10) was 10.34 µg/m^3, with a standard

Table 3. Descriptive statistics of environmental parameters captured by IoT sensors at Taro Agric Farm from April through June, 2024

Variable	n	Mean	Std Dev	Minimum	Maximum
Temperature	42302	30.15	3.44	22.00	39.00
Relative humidity	42302	72.27	9.10	46.00	100.00
Hydrogen sulfide	42302	0.95	3.11	0.00	56.00
Ammonia	42302	48.27	19.82	13.00	205.00
Nitrogen dioxide	42302	1.28	1.054	0.00	11.00
Particulate matter (2.5 µg/m^3)	42302	9.80	29.30	0.00	2657.00
Particulate matter (10 µg/m^3)	42302	10.34	33.58	0.00	3511.00
Volatile organic compounds	42302	109.64	74.52	2.00	500.00

deviation of 33.58 µg/m^3. The average total volatile organic compounds level was 109.64 PPM, with a standard deviation of 74.52 PPM, ranging from 2 PPM to 500 PPM.

Table 4 presents descriptive statistics of environmental parameters captured by IoT sensors at Taro Agric Farm at the coldest period of the day (2 AM). The average temperature at 2AM was lower at 27.16 °C, with less variability (standard deviation of 1.60 °C), ranging from 24 °C to 34 °C. However, the average relative humidity was higher at 78.96%, with less variability (standard deviation of 3.72%), ranging from 57% to 86%. For hydrogen Sulfide, the average level was 0.56 PPM, with a standard deviation of 2.60 PPM, ranging from 0 to 56 PPM. The average ammonia concentration was 44.82 PPM, with a standard deviation of 16.17 PPM, ranging from 22 PPM to 141 PPM. For nitrogen dioxide, the average concentration was 1.06 PPM, with a standard deviation of 0.83 PPM, ranging from 0 to 5 PPM. The average concentration of particulate matter (PM2.5) was 10.63 µg/m^3, with a high standard deviation of 56.43 µg/m^3. Furthermore, the average concentration of particulate matter (PM10) was 11.18 µg/m^3, with a standard deviation of 70.10 µg/m^3. The average total volatile organic compounds level was 61.14 PPM, with a standard deviation of 33.03 PPM, ranging from 11 PPM to 288 PPM.

Table 4. Descriptive statistics of environmental parameters captured by IoT sensors at Taro Agric Farm at the coldest period of the day (2 AM), from April through June, 2024

Variable	n	Mean	Std Dev	Minimum	Maximum
Temperature	1792	27.16	1.60	24.00	34.00
Relative humidity	1792	78.96	3.72	57.00	86.00
Hydrogen sulfide	1792	0.56	2.61	0.00	56.00
Ammonia	1792	44.82	16.17	22.00	141.00
Nitrogen dioxide	1792	1.057	0.83	0.00	5.00
Particulate matter (2.5 µg/m^3)	1792	10.63	56.43	0.00	2191.00
Particulate matter (10 µg/m^3)	1792	11.17	70.10	0.00	2813.00
Volatile organic compounds	1792	61.14	33.03	11.00	288.00

Table 5 presents descriptive statistics of environmental parameters captured by IoT sensors at Taro Agric Farm at the hottest period of the day (2 PM). The average temperature at 2PM was higher at 34.28 °C, with moderate variability (standard deviation of 2.52 °C), ranging from 26 °C to 39 °C. The relative humidity is lower at 60.56%, with moderate variability (standard deviation of 6.76%), ranging from 46% to 82%. The average level of hydrogen sulfide was 1.14 PPM, with a standard deviation of 3.04 PPM, ranging from 0 to 30 PPM. The average ammonia concentration was 52.21 PPM, with a standard deviation of 23.45 PPM, ranging from 23 PPM to 201 PPM. For nitrogen dioxide, the average concentration was 1.38 PPM, with a standard deviation of 1.20 PPM, ranging from 0 to 9 PPM. For particulate matter (PM2.5), the average concentration is 7.59 µg/m^3, with a standard deviation of 7.02 µg/m^3, ranging from 0 to 94 µg/m^3 while for particulate matter (PM10), the average concentration is 8.16 µg/m^3, with a standard deviation of 7.79 µg/m^3, ranging from 0 to 135 µg/m^3. The average total volatile organic

Table 5. Descriptive statistics of environmental parameters captured by IoT sensors at Taro Agric Farm at the hottest period of the day (2 PM)

Variable	n	Mean	Std Dev	Minimum	Maximum
Temperature	1792	34.28	2.52	26.00	39.00
Relative humidity	1792	60.56	6.76	46.00	82.00
Hydrogen sulfide	1792	1.14	3.04	0.00	30.00
Ammonia	1792	52.21	23.45	23.00	201.00
Nitrogen dioxide	1792	1.38	1.20	0.00	9.00
Particulate matter (2.5 $\mu g/m^3$)	1792	7.60	7.02	0.00	94.00
Particulate matter (10 $\mu g/m^3$)	1792	8.16	7.79	0.00	135.00
Volatile organic compounds	1792	170.43	78.46	7.00	432.00

compounds level was 170.43 PPM, with a standard deviation of 78.46 PPM, ranging from 7 PPM to 432 PPM.

3.3   Correlation Coefficients Among Environmental Parameters

Table 6 presents the overall correlation coefficients among environmental parameters captured by IoT sensors at Taro Agric Farm. Overall, a strong and consistent inverse relationship was observed between temperature and relative humidity, with a correlation coefficient of −0.915. This implied that as the temperature increases, the relative humidity significantly decreases, likely due to the air's decreased capacity to hold moisture at higher temperatures. Particulate matter concentrations (PM2 and PM10) were highly correlated with each other (0.988), indicating similar causal or attributions. However, their correlations with other parameters were generally weak, suggesting that particulate matter levels vary independently of temperature, and humidity. Gaseous pollutants showed positive inter-correlations, highlighting a potential commonality in their sources or release mechanisms. For instance, hydrogen sulfide, ammonia, and nitrogen dioxide were all positively correlated with one another, reinforcing the idea that certain farm activities are common contributions to these gases.

Table 6. Overall correlation coefficients among environmental parameters captured by IoT sensors at Taro Agric Farm from April through June, 2024

	Temperature	Relative humidity	Hydrogen sulfide	Ammonia	Nitrogen dioxide	Particulate matter 2	Particulate matter 10	Volatile organic compounds
Temperature	1	−0.92***	0.18***	0.30***	0.29***	−0.002NS	0.003NS	0.55***
Relative humidity		1	−0.16***	−0.22***	−0.20***	0.02*	0.01*	−0.47***

(*continued*)

Table 6. (*continued*)

	Temperature	Relative humidity	Hydrogen sulfide	Ammonia	Nitrogen dioxide	Particulate matter 2	Particulate matter 10	Volatile organic compounds
Hydrogen sulfide		1	0.18***	0.34***	0.09***	0.09***	0.19***	
Ammonia			1	0.63***	0.05***	0.05***	0.20***	
Nitrogen dioxide				1	0.08***	0.08***	0.21***	
Particulate matter 2					1	0.99***	0.07***	
Particulate matter 10						1	0.07***	
Volatile organic compounds							1	

*-$P < 0.05$; ***$P < 0.0001$; $^{NS}P > 0.05$

Table 7 presents the correlation coefficients among environmental parameters captured by IoT sensors at Taro Agric Farm at the coldest period of the day (2 AM). During the coldest period at 2 AM, the correlations among environmental parameters exhibit some shifts compared to the overall trend. The inverse relationship between temperature and relative humidity remains significant but is less strong (-0.636) than the overall trend, indicating that humidity still decreases as temperature rises, albeit less sharply during the night. At 2 AM, the correlation coefficients among these gases were generally positive but weaker compared to the overall trend, suggesting that their production or release may be less influenced by different factors at night. However, correlation coefficients of gaseous pollutants with particulate matter and volatile organic compounds were negligible during the night, suggesting a different dynamic in pollutant behaviour compared to daytime.

Table 8 presents the correlation coefficients among environmental parameters captured by IoT sensors at Taro Agric Farm at the hottest period of the day (2 PM). At the hottest period of the day, 2 PM, the trends among environmental parameters again exhibited distinct characteristics. The inverse relationship between temperature and relative humidity was very pronounced (-0.886), reflecting a significant drop in humidity as temperatures peak. This strong correlation underscores the substantial effect of high temperatures on moisture levels in the air. At 2 PM, the correlation coefficients among hydrogen sulfide, ammonia, and nitrogen dioxide were stronger, indicating that the production or release of these gases was more intense or interconnected during the hottest part of the day. Gaseous pollutants such as hydrogen sulfide, ammonia, and nitrogen dioxide remain positively correlated with each other. However, their correlations with particulate matter and volatile organic compounds were generally weak.

Table 7. Correlation coefficients among environmental parameters captured by IoT sensors at Taro Agric Farm at the coldest period of the day (2 AM) from April through June, 2024

	Temperature	Relative humidity	Hydrogen sulfide	Ammonia	Nitrogen dioxide	Particulate matter 2	Particulate matter 10	Volatile organic compounds
Temperature	1	-0.64^{***}	0.39^{***}	0.41^{***}	0.47^{***}	-0.016^{NS}	-0.02^{NS}	0.15^{***}
Relative humidity		1	-0.50^{***}	-0.29^{***}	-0.40^{***}	-0.017^{NS}	-0.02^{NS}	0.02^{NS}
Hydrogen sulfide			1	0.10^{***}	0.26^{***}	-0.009^{NS}	-0.01^{NS}	0.01^{NS}
Ammonia				1	0.59^{***}	-0.003^{NS}	-0.01^{NS}	0.31^{***}
Nitrogen dioxide					1	-0.009^{NS}	-0.01^{NS}	0.15^{***}
Particulate matter 2						1	0.10^{***}	-0.03^{NS}
Particulate matter 10							1	-0.03^{NS}
Volatile organic compounds								1

***P < 0.0001; ^{NS}P > 0.05

Table 8. Correlation coefficients among environmental parameters captured by IoT sensors at Taro Agric Farm at the hottest period of the day (2 PM) from April through June, 2024

	Temperature	Relative humidity	Hydrogen sulfide	Ammonia	Nitrogen dioxide	Particulate matter 2	Particulate matter 10	Volatile organic compounds
Temperature	1	-0.89^{***}	0.12^{***}	0.25^{***}	0.32^{***}	0.18^{***}	0.17^{***}	0.39^{***}
Relative humidity		1	-0.04^{NS}	-0.09^{***}	-0.16^{***}	-0.08^{**}	-0.07^{*}	0.29^{***}
Hydrogen sulfide			1	0.19^{***}	0.43^{***}	0.20^{***}	0.20^{***}	0.001^{NS}
Ammonia				1	0.63^{***}	0.19^{***}	0.23^{***}	0.19^{***}
Nitrogen dioxide					1	0.31^{***}	0.32^{***}	0.20^{***}
Particulate matter 2						1	0.98^{***}	0.08^{**}
Particulate matter 10							1	0.07^{*}
Volatile organic compounds								1

*-P < 0.05; **P < 0.001; ***P < 0.0001; ^{NS}P > 0.05

4 Discussion

Results showed that Taro Agric Farm experienced persistent environmental stressors such as high temperatures, excessive humidity, poor air quality, and gaseous pollutants, all of which could adversely affect broiler chicken health and productivity [4, 5]. The maximum temperatures at Taro Agric Farm from April to June fluctuated between 29 °C and 39 °C. April and May showed consistent temperatures around 35 °C, indicating potential distress for the broiler chickens. June showed a slight decrease, with temperatures occasionally surpassing the distress threshold, indicating persistent thermal stress and potential distress conditions for the farm. High temperatures can have several adverse effects on broiler chickens, by predisposing them to heat stress [6]. Further, it could also lead to reduced feed intake [7] and stunted growth rate [8], increased mortality and impaired immune function [9]. In addition, broiler chickens may exhibit behavioural changes in response to high temperatures. Proper management practices, such as providing adequate ventilation, cooling systems, and ensuring access to fresh water, are essential to mitigate these adverse effects and maintain the health and productivity of broiler chickens. The maximum relative humidity levels ranged from 80% to 100%, consistently exceeding the optimal threshold of 70%. This indicates a persistently suboptimal environment for broiler chickens, potentially affecting their health and productivity. Maximum daily THI values consistently exceeded thresholds in April and May, indicating a high level of heat stress. In June, THI values were lower but still frequently above the threshold, suggesting persistent heat stress conditions. Hydrogen sulfide levels frequently exceeded the stress (2 PPM) and distress (6 PPM) thresholds, particularly in April and June, indicating periods of potential distress and significant air quality issues that could predispose broiler chickens to respiratory diseases. Particulate matter (PM2.5 and PM10) levels showed spikes significantly higher than threshold levels. These spikes indicate periods of poor air quality, which could adversely affect the health of the broiler chickens. The spike in ammonia levels in early April and June, as shown in Fig. 7, could be attributed to a combination of factors. During these periods, we observed a temporary increase in bird density due to overlapping production cycles, which led to higher waste production and subsequently increased ammonia levels. Additionally, variations in ambient temperature and humidity may have exacerbated the ammonia concentrations.

There is a strong and consistent inverse relationship between temperature and relative humidity (correlation coefficient of -0.915). This implies that as temperature increases, the relative humidity significantly decreases due to the air's decreased capacity to hold moisture at higher temperatures. The concentrations of particulate matter PM2.5 and PM10 are highly correlated with each other (0.988), suggesting similar sources or behaviours. However, their correlations with other parameters are generally weak, indicating that particulate matter levels may vary independently of temperature, and humidity. Specific period analysis showed that at the coldest period of the day (2 AM), the inverse relationship is less strong (-0.636) compared to the overall trend, indicating a less sharp decrease in humidity as temperature rises at night. The correlation coefficients among gases were generally positive but weaker, indicating that their production or release may be less active or influenced by different factors at night. At the hottest period of the day (2 PM), the inverse relationship between temperature and humidity was very pronounced (-0.886), reflecting a significant drop in humidity as temperatures peaked. This underscores the substantial effect of high temperatures on moisture levels in the air. Correlations among hydrogen sulfide, ammonia, and nitrogen dioxide are stronger, indicating more active or interconnected production or release during the hottest part of the day. The correlations among hydrogen sulfide, ammonia, and nitrogen dioxide were stronger during the hottest part of the day, indicating that their production or release was more intense during this period.

5 Future Work

The current study provides preliminary data on the monitoring of key environmental factors affecting broiler chickens. Future research will expand on this by incorporating more advanced sensor technologies and machine learning algorithms to enhance the precision and efficiency of data collection and analysis. Specifically, planned work includes the assessment of the influence of heat stress and gaseous pollutants on broiler performance/welfare under commercial production, evaluation of chicken welfare parameters across production cycles in Southwest Nigeria (with special focus on the combined effect of volatile organic compounds, heat stress and noxious gases), development and implementation of machine learning models to predict stress and distress levels based on real-time environmental data, allowing for proactive management strategies.

6 Conclusions

The findings of this study underscore the critical impact of environmental parameters on the health and welfare of broiler chickens. The deployment of sensor technology for real-time monitoring revealed that threshold levels for temperature, humidity, ammonia, and particulate matter are frequently exceeded, leading to potential distress and health issues for broiler chickens. However, further studies are required to validate these sensors. These insights highlight the necessity for continuous monitoring and the implementation of mitigation strategies to maintain optimal conditions within broiler houses. By doing so, it is possible to enhance the welfare and productivity of broiler chickens, ultimately contributing to the profitability and sustainability of poultry enterprises in tropical climates.

Acknowledgements. The study was funded by Innovate UK (Project Number: 10049067) under the United Kingdom Research and Innovation (UKRI) as part of an African Agri-food Knowledge Transfer Partnership (AAKTP) collaboration between the University of the West of England (UWE), Bristol, United Kingdom; Obafemi Awolowo University (OAU), Ile-Ife, Nigeria; and Taro Agric Consulting (TAC), Ile-Ife, Nigeria.

References

1. Sheikh, I.U., et al.: Ammonia production in the poultry houses and its harmful effects. Int. J. Vet. Sci. Anim. Husbandry 3(4), 30–33 (2018)

2. Bist, R.B., Chai, L.: Advanced strategies for mitigating particulate matter generations in poultry houses. Appl. Sci. **12**(22), 11323 (2022)
3. Ojo, R.O., et al.: Internet of things and machine learning techniques in poultry health and welfare management: a systematic literature review. Comput. Electron. Agric. **200**, 107266 (2022)
4. Akinyemi, F., Adewole, D.: Environmental stress in chickens and the potential effectiveness of dietary vitamin supplementation. Front. Anim. Sci. **2** (2021)
5. Gržinić, G., et al.: Intensive poultry farming: a review of the impact on the environment and human health. Sci. Total. Environ. **858**, 160014 (2023)
6. Apalowo, O.O., Ekunseitan, D.A., Fasina, Y.O.: Impact of heat stress on broiler chicken production. Poultry **3**(2), 107–128 (2024)
7. Onagbesan, O.M., et al.: Alleviating heat stress effects in poultry: updates on methods and mechanisms of actions. Front. Vet. Sci. **10** (2023)
8. Huerta, A., et al.: Resiliency of fast-growing and slow-growing genotypes of broiler chickens submitted to different environmental temperatures: growth performance and meat quality. Poult. Sci. **102**(12), 103158 (2023)
9. Lara, L.B., Rostagno, M.H.: Impact of heat stress on poultry production. Animals **3**(2), 356–369 (2013)

River Levels Affecting Firefighter Interventions: Factor Analysis

Naoufal Sirri(✉)Ⓘ and Christophe GuyeuxⒾ

FEMTO-ST Institute, UMR 6174 CNRS, University of Franche-Comté,
Belfort, France
{naoufal.sirri,christophe.guyeux}@univ-fcomte.fr

Abstract. This research extends the continuum of studies exploring various variable categories and their implications for firefighters. The motivation stems from recognizing that river height, while it may impact the environment, is generally an immediate priority for firefighting interventions. Firefighters are primarily called for emergencies, prompting consideration of how river height, despite its environmental impact, could influence their operations. In this context, our investigation aims to comprehensively assess how river height influences firefighter interventions, with a particular emphasis on their frequency. The study's relevance is underscored by the significant impacts of flooding risks on the environment. Notably, in France, 23% of accidental drownings occurred in rivers, representing a major cause of death according to a public health survey in 2021, highlighting the critical nature of our research. Over nine years, from 2015 to 2024, our methodology encompasses data preparation, thorough analysis, and the application of the predictive XGBoost model, renowned for its speed and resilience to outliers. The iterative training pipeline selects features that improve the RMSE score over a 24-hour horizon, emphasizing the crucial importance of variables related to river height, particularly across all horizons. The main conclusion drawn from this study is that these variables exert an immediate and persistent impact on interventions, suggesting increased relevance for predicting outcomes over the entire duration. This precision in understanding models associated with the presence or absence of river flooding offers a practical approach to anticipate resource management, improve firefighter response times, and contribute to saving lives by mitigating intervention failures during major incidents. This study initiates a comprehensive exploration of variable families to understand the factors influencing firefighting activities.

Keywords: Firefighters intervention · Height of rivers impact · Data mining · Feature selection · XgBoost · Intervention causes

1 Introduction

The influence of river heights on the frequency and nature of firefighter interventions, particularly in critical areas like the environment, is substantial. In France, accidental drownings represent a major cause of death, especially among individuals under 25 years old. An investigation conducted by "Public Health France" revealed that in 2021, 23% of accidental drownings occurred in watercourses, with a higher mortality rate (41%) compared to other locations [1]. However, harnessing river height surveillance data for specific event detection proves to be complex. The intricacies of spatial and temporal variability, coupled with the establishment of detection thresholds, require thorough analysis. This study aims to explore the potential correlation between river heights and firefighter interventions, thereby contributing to the development of more effective surveillance strategies.

In recent years, significant progress has been made in predicting river heights and firefighting interventions, thanks to the advancements in artificial intelligence and its applications. These developments have enabled researchers to enhance the accuracy and efficiency of prediction models, ultimately contributing to improved emergency response and risk management strategies. The study [2] developed a flood model using artificial neural networks (ANN) and geographic information systems (GIS) to simulate flood-prone areas in southern Peninsular Malaysia, with satisfactory results that can aid governments in future planning and infrastructure development. On the other side the study [3] compared the accuracy of three software computing methods artificial neural networks (ANNs), adaptive neuro-fuzzy inference system (ANFIS), coupled wavelet and neural network (WANN), and conventional sediment rating curve (SRC) approach to estimate daily suspended sediment load (SSL) at two gauging stations in the US, finding that WANN was the most accurate model in SSL estimation, with better performance than other models and the SRC method. In contrast, [4] estimated river bedform using Artificial Neural Network (ANN) and Support Vector Machine (SVM) methods, finding that the SVM model with RBF kernel function predicted bedform more accurately than other methods, with higher values of statistical parameters and better performance compared to empirical formulas and ANN. Additionally, [5] developed an efficient AI platform for real-time urban flood forecasting, integrating rainfall hyetographs embedded with uncertainty analyses as well as hydrological and hydraulic modeling, using deep learning for feature extraction and prediction, showing reliable results despite some inconsistencies. Moreover, recently the authors evaluated the use of support vector regression, long short-term memory, and their combination for river height prediction using historical sensor data, with LSTM-SVR showing better performance in capturing rapid transient changes in river levels [6]. [7] implemented data privacy using Differential Privacy (DP), employing a local differential privacy approach, while predicting firefighter interventions in specific locations from statistical estimators. Similarly, [8] applied natural language processing techniques to extract features from weather bulletin texts to predict peak intervention periods caused by rare events. Additionally, [9] demonstrated the feasibility of a continuously updated and optimized database using dedicated feature selection tools for XGBoost. Finally, in the same context as our paper, two studies were conducted to analyze the impact of specific features, namely air quality and solar activity, on firefighter interventions by [10] and [11], respectively.

Our research group tackles this issue by embracing an innovative approach: studying the influence of river heights on firefighter interventions. Leveraging rich historical datasets containing information on both river heights and firefighter interventions, we delve into this relationship. The main objective of this article is to assess the impact of river heights on firefighter interventions, focusing particularly on their frequency. To achieve this, we have formulated a hypothesis and its prediction within the scope of our research:

– Hypothesis: An increase in river heights will exhibit a correlation with a sustained rise in firefighter interventions from the outset.
  Prediction: The presence of elevated water levels is expected to systematically impact the environment, leading to interventions over a prolonged and sustained period.

To ensure a systematic and organized approach to our research, we meticulously developed and adhered to a methodological plan. In Sect. 2, we provide a comprehensive overview of the methods and materials utilized in the experiments, outlining our experimental approach. The research process and obtained results are presented in Sect. 3. Section 4 delves into an in-depth analysis and critical reflection on the findings, addressing the initial research question and emphasizing relevant implications and interpretations. The study concludes in Sect. 5, synthesizing the main conclusions, highlighting significant contributions, and suggesting potential avenues for future research. This methodological framework facilitated a systematic exploration of the correlation between of river heights and firefighting interventions while maintaining a holistic approach throughout our investigation.

2 Methods

2.1 Data Preparation

Data Acquisition. This study utilizes an extensive dataset sourced from the Service d'Incendie et de Secours du Doubs (SDIS 25), France, encompassing 322,197 documented interventions spanning from January 1, 2015, to March 30, 2024. Each intervention entry is meticulously documented, comprising an identification code, precise start and end timestamps, geographical coordinates, intervention type, and response durations. Various contextual features were incorporated to comprehensively delineate incident circumstances. These encompassed meteorological parameters, solar activity metrics, river water levels within the Doubs department (French firefighters are responsible for various rescues, including floods), air quality indices, epidemiological statistics, holiday periods, lunar phases, and additional relevant factors. To prognosticate forthcoming intervention occurrences, a comprehensive dataset was curated by amalgamating firefighter intervention records with supplementary information from diverse sources. The process is detailed as follows:

- The key entries within the dictionary are organized into hourly segments, spanning from "01/01/2015 00:00:00" to "30/03/2024 18:00:00", formatted as "YYYY-MM-DD hh:mm".
- Data regarding the water levels of the initial forty rivers in Doubs were sourced from the governmental service "Hydroreel" and incorporated into the study. The process entailed populating a dictionary with the nearest average measurements for each hourly interval, as referenced in [12].
- To evaluate the influence of solar activity, features including 10 cm radio flux, sunspot count, sunspot area, and X-ray emissions were incorporated, as referenced in [13].
- To evaluate the influence of air quality, features including particulate matter ($PM_{2.5}$, PM_{10}), ozone (O_3), and nitrogen oxides (NO_2) from various air quality monitoring stations nearby, as referenced in [14].
- Data from NASA's VIIRS and MODIS satellites were continuously collected, capturing Earth images with diverse wavelengths and resolutions to analyze fire propagation in specific areas, as documented in [15] and [16].
- The [17] libraries were employed to compute the spatial separation among the Earth, Moon, and Sun. Utilizing Astral [18], various parameters related to the sun and moon, including moon phase and moonrise, were analyzed to assess their influence on natural calamities. Furthermore, sunrise and sunset data were utilized to establish a boolean variable signifying "night" or "day".
- Integration of weekly epidemiological data sourced from the Sentinelles network, encompassing ailments such as chickenpox, influenza, and acute diarrhea, as documented in [19].
- Incorporation of variables associated to French league and Champions League football matches, acknowledged as probable factors impacting interventions, as referenced in [20].
- Integration of temporal details encompassing time, day, day of the week, day of the year, month, and year, alongside information on holidays, academic breaks sourced from [21], and events like Ramadan observances, lockdowns, and curfews.
- Initially, our meteorological data reference was Météo France [22] (the french public meteorological service). However, difficulties arose due to access limitations to remote main stations and a three-hour sampling interval, which affected geographical and temporal accuracy. Despite Météo France bulletins offering information on diverse weather risks, the introduction of MeteoStat [23] was crucial in addressing this constraint. MeteoStat offers forthcoming forecasts, augmenting the functionalities of Météo France.
  - Data retrieved from three meteorological stations in the Doubs department includes a comprehensive array of atmospheric metrics. These encompass atmospheric pressure, cloud cover, barometric trends, temperature, humidity levels, precipitation within the last hour, dew point, precipitation over the past three hours, gust speeds over a specified interval, average wind speed recorded at 10-minute intervals, horizontal visibility, average wind direction tracked every 10 min, and the prevailing climatic conditions, all sourced from Météo France.

- Collected MétéoFrance weather advisories encompassing diverse meteorological hazards such as wind, rainfall, floods, storms, snowfall, freez-
  ing rain, heatwaves, and extreme cold, each categorized with color codes
  (green, orange, red, yellow), augmenting the significance of the meteorological dataset [24].
- The MeteoStat API was used to access climatic variables such as temperature, dew point, precipitation, snowfall, wind speed and direction, pressure, and humidity sourced from openly available meteorological and climate datasets. Temperature data was extracted from an extensive 11×11
  grid network spanning the entirety of the department.

The selected variables, derived from an examination of firefighter interventions, are designed to ascertain their potential influence. While this inclusive
methodology poses the possibility of integrating non-significant variables, it
enables the assessment of correlations between variables related to the heights
of rivers and additional parameters. Encompassing various risks including accidents, fires, and floods, among others, these variables feature prominently in the
dataset of firefighter interventions.

Data Pre-processing. During the data processing phase, linear interpolation
was employed to handle missing values within certain meteorological datasets. To
suit our learning model, two techniques from the Scikit-learn library [25] were utilized. The "StandardScaler" approach was implemented to standardize numerical features, encompassing variables such as year, hour, humidity, dew point,
wind speed and direction, influenza, cloud cover, precipitation, gusts, visibility,
varicella statistics, temperature, acute diarrhea, river levels, and lunar distance.
This method adjusts the distribution of values to achieve a mean of zero and
unit variance. Additionally, we employed the "TargetEncoder" technique [26] to
encode categorical attributes, including the day of the week, year, month, barometric trend, holidays, and events. This method transformed these variables by
replacing each category with the mean of the corresponding target variable. We
retained the original target values (the count of interventions) as discrete entities,
as they better reflect the distribution of interventions.

Data Mining. A comprehensive examination of the dataset proved indispensable in extracting pertinent insights for our study. On average, there were approximately 30,000 interventions recorded per year, with a noticeable upward trend
in intervention frequency over time. Concerning variables associated with river
heights, Table 1 presents key statistical metrics for various river height parameters. It offers a concise summary of the central tendencies, variability, and
observed distribution patterns across different categories of variables within this
domain. Initially, data from 40 rivers within the department were collected, but
only those situated in areas with a significant population density surrounding
the river were retained. All these variables are represented as continuous entities,
predominantly demonstrating a right-skewed distribution. However, two variables, Le_Doubs_de_Mouthe and La_Savoureuse_de_Giromagny, exhibit bimodal

distribution patterns (see Fig. 1). In the context of temporal analysis, these modes may correspond to distinct seasons characterized by varying precipitation levels, implying a connection between river height fluctuations and seasonal weather patterns, geographical features like steep valleys or mountainous terrain, and human interventions such as dam constructions or alterations in river courses.

Table 1. Data analysis of river height variables (cm)

Variable	Mean	Std	IQR	Max	Distinct values
La_Bourbeuse	43.4	47.4	44.1	279.0	17,511
Le_Drugeon	45.5	35.8	46.4	1051.0	19,457
Le_Doubs_de_Mouthe	58.7	20.2	30.0	136.0	12,392
L'Allan	61.6	18.1	15.0	205.9	9,846
Le_Dessoubre	58.5	37.8	39.8	502.0	10,136
La_Savoureuse_de_Giromagny	43.1	18.6	23.1	245.0	11,461
Le_Doubs_canal_de_Besançon	235.7	69.4	70.0	2557.3	10,570
La_Rosemontoise	8.0	26.3	17.1	305.0	10,648
Le_Doubs_de_Labergement	33.9	29.7	32.0	191.7	15,557
La_Savoureuse_de_Belfort	20.9	18.5	23.0	147.9	10,961

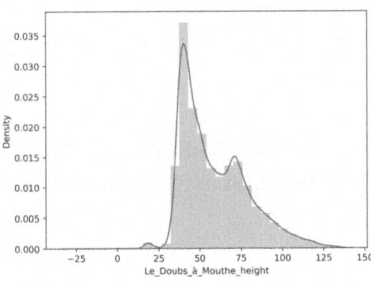   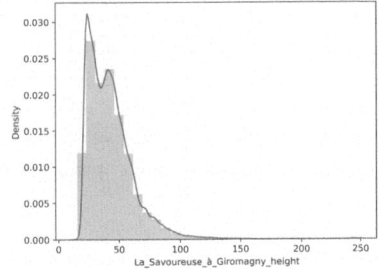

(a) Distribution of Le_Doubs   (b) Distribution of La_Savoureuse

Fig. 1. Analysis of the distribution of Le_Doubs_de_Mouthe and La_Savoureuse_de_Giromagny

Moreover, the examination of these variables through time series analysis demonstrates a notable degree of similarity. To substantiate this observation, we conducted a correlation analysis among these variables, revealing findings that affirm the lack of localized variability while indicating significant correlations. These correlations span from 0.71 to 0.95, suggesting that these variables furnish redundant information for prediction, indicative of a shared influential factor. Consequently, these outcomes pivot our research towards a broader departmental outlook rather than a localized one, underscoring the absence of discernible localized effects.

In concluding our analysis, we sought to assess the influence of river heights on both the overall target variable and specific categories thereof, employing a correlation analysis. The findings indicate a tenuous and statistically insignificant impact on the overarching target variable. However, a modest effect was discernible, particularly in the context of all river heights, demonstrating correlations ranging from 0.21 to 0.27 with a specific category of the target variable, namely floods. These insights offer preliminary glimpses into the constrained role of river heights in shaping intervention patterns. While notable correlations are absent, the prospect of a more intricate relationship underscores our comprehension of the nuanced interplay between river heights and incidents necessitating emergency services intervention.

2.2 Feature and Model Selection

In the past, it was customary to incorporate all available features from the training dataset under the assumption that maximizing information inclusion would yield an optimal model. However, advocating for a constraint on the number of features considered arises from two principal reasons. Firstly, certain variables may display strong interdependencies, while others might contribute minimally to predictive capacity, potentially leading to diminished model generalization or the introduction of redundant information. Secondly, the inclusion of numerous features can substantially escalate computational complexity without commensurate enhancements in model performance [27]. Thus, employing a more restricted feature set holds promise for achieving more efficient outcomes. In this investigation, we employed the 'feature importance' method for feature selection, assigning scores to each variable in the dataset, with higher scores denoting greater relevance [28]. A threshold was established to retain the top 400 most pertinent features. Diverse selection techniques were applied, including:

1. High Variance: preserving characteristics with variances exceeding 0.5.
2. Pearson and Spearman correlation coefficients: filtering out correlations with the target variable whose absolute value equals or exceeds 0.4.
3. Chi-Square Selector: applied the chi-square test to evaluate the association between each feature and the target variable after normalizing the features using the 'Min-Max Scaler' function.
4. Extreme Gradient Boosting (XGBoost) [29]: employed preset hyperparameters (maximum depth = 7, number of estimators = 100000, early stopping after 10 rounds), conducted training, and subsequently calculated or obtained feature importance.
5. Light Gradient Boosting Model (LightGBM) [30]: used specified hyperparameters (learning rate = 0.1, objective function = regression, metric = RMSE,

number of leaves = 2^7, maximum depth =7, number of estimators = 100000, early stopping after 10 rounds), and extracted feature importance as mentioned previously.

The culmination of this selection process yields a refined roster of features, prioritizing those consistently pinpointed by multiple techniques. From the initial pool of 3912 features, roughly 10% were preserved, culminating in a subset of 400 features employed for model training.

In our model selection process, we opt for Extreme Gradient Boosting (XGBoost) [29]. Renowned for its scalability, efficiency, flexibility, and speed, XGBoost is adept at navigating intricate datasets while achieving remarkable predictive accuracy. Its efficacy lies in its integration of gradient-boosting principles, allowing it to iteratively enhance the performance of weak learners. Widely embraced across diverse domains, from finance to healthcare, XGBoost exhibits versatility in addressing regression, classification, and ranking tasks with exceptional proficiency. Its popularity is underscored by its adeptness in handling missing data, feature selection, and adaptation to various data types.

2.3 Approaches Implemented for The Prediction Tool

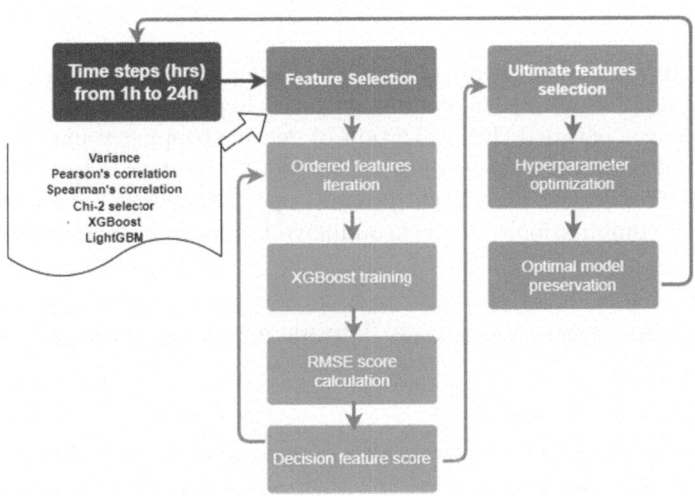

Fig. 2. Training pipeline representation model

To translate our theoretical analysis into practical application, we constructed a robust training and optimization pipeline (see Fig. 2).

1. We initiated numerous training iterations, covering various timeframes ranging from 1 to 24 h, aiming to evaluate the influence of the height of rivers up to 24 h in advance.

2. Subsequently, we performed feature extraction as delineated in the "Feature Selection" section.
3. Iteratively, each selected variable from the preceding phase was incorporated into the XGBoost model for training purposes.
4. In each iteration, we partitioned the dataset into training, testing, and validation subsets. Initially, 20% of the data was allocated for testing, while the remaining 80% was designated for training-validation. Subsequently, we adjusted the allocation to 80% for training and 20% for validation.
5. We conducted training for the XGBoost model, employing preset hyperparameters (see Table 2) and integrating early stopping with "early_stopping_rounds=15". This mechanism was employed to halt training if there was no improvement in the performance on the validation set for 15 consecutive rounds. This decision was influenced by the fixed 100,000 estimators in our model.
6. We calculated prediction scores using the RMSE metric (Eq. 1) and compared them to previous scores. Variables that exhibited improvement were preserved in the input list via a sequential feature selection approach, and the preceding scores were accordingly adjusted.

$$RMSE = \sqrt{\frac{1}{n} \sum_{i=1}^{n} (y_i - \hat{y}_i)^2} \tag{1}$$

7. Finally, we embarked on an ultimate endeavor to fine-tune prediction outcomes by exploring various combinations of hyperparameters (see Table 3). This endeavor entailed the utilization of "early_stopping_rounds=25" alongside a fixed parameter set comprising 100,000 estimators. Through our iterative methodology, we identified optimal hyperparameter configurations, pinpointing a combination of values conducive to enhanced performance.

Table 2. XGBoost default hyperparameters

Hyperparameter	Values
max_depth	7
min_child_weight	1
gamma	0
subsample	0.8
colsample_bytree	0.8
learning_rate	0.1

Table 3. Settings XGBoost hyperparameters

Hyperparameter	XGBoost
max_depth	[2, 14]
min_child_weight	[0, 14]
gamma	[0.0, 0.4]
subsample	[0.6, 0.9]
colsample_bytree	[0.6, 0.9]
learning_rate	[0.01, 0.009]

3 Results

Substantial efforts have been dedicated to assembling, processing, and consolidating a comprehensive dataset, with a keen focus on amalgamating information from diverse sources of past interventions. Special attention was directed towards evaluating the significance of each variable within this framework. As articulated earlier, the essence of this study lies in unearthing pertinent insights regarding the impact of river heights on intervention forecasting. The preceding section delineated a methodology for identifying crucial features. Following the training of our model across a day-long timeframe, we present the outcomes of this feature selection process, honing in on river height-related variables whose inclusion led to enhanced prediction accuracy (see Table 4).

Table 4. Exploring the impact of Feature Selection on height of rivers analysis

Time horizon	Height of rivers	Feature Selection Technique	Rank
1 h	La_Savoureuse_de_Belfort	Variance and XGBoost	38
2 h	Le_Dessoubre	Variance, XGBoost and LightGBM	90
4 h	Le_Drugeon	Variance and LightGBM	120
5 h	Le_Doubs_de_Mouthe	Variance and LightGBM	51
9 h	Le_Dessoubre	Variance, XGBoost and LightGBM	75
10 h		Variance, XGBoost and LightGBM	65
11 h	Le_Doubs_de_Labergement	Variance, XGBoost and LightGBM	45
14 h	La_Savoureuse_de_Belfort	Variance and XGBoost	134
15 h		Variance and XGBoost	78
17 h	Le_Bourbeuse	Variance and LightGBM	111
19 h	La_Rosemontoise	Variance, XGBoost and LightGBM	84
20 h		Variance, XGBoost and LightGBM	53
21 h	L'Allan	Variance and LightGBM	96
22 h	La_Savoureuse_de_Giromagny	Variance and XGBoost	109
23 h	La_Bourbeuse	Variance and LightGBM	122
	Le_Dessoubre	Variance, XGBoost and LightGBM	67
24h	Le_Doubs_canal_de_Besançon	Variance, XGBoost and LightGBM	31

Additionally, we present the classification outcomes yielded by our model, poised for interpretation in the subsequent section. Table 5 delineates the predictive efficacy of the XGBoost model concerning intervention counts, encompassing temporal horizons inclusive of river height variables, with the most notable performances accentuated in bold. Moreover, it outlines the RMSE metrics before and after the incorporation of river height variables, along with the percentage improvement. Figure 3 illustrates the outcomes derived from 300 samples aiming to forecast an atypical intervention count at the 2nd hour. Notably, a

Table 5. Insights into classification outcomes

Time horizon	Height of rivers	RMSE pre-sel	RMSE post-sel	Improvement
1 h	La_Savoureuse_de_Belfort	2.9066	2.8898	0.58%
2 h	Le_Dessoubre	1.6883	**1.6500**	2.32%
4 h	Le_Drugeon	2.3412	2.32ç63	0.64%
5 h	Le_Doubs_de_Mouthe	3.7112	3.6889	0.60%
9 h	Le_Dessoubre	3.8322	3.8299	0.06%
10 h		2.4238	2.3996	1.01%
11 h	Le_Doubs_de_Labergement	3.5223	3.5022	0.57%
14 h	La_Savoureuse_de_Belfort	2.3098	2.3054	0.19%
15 h		3.4083	3.4038	0.13%
17 h	Le_Bourbeuse	3.7851	3.7614	0.63%
19 h	La_Rosemontoise	3.9134	3.8838	0.76%
20 h		2.4863	2.4549	1.28 %
21 h	L'Allan	3.5325	3.5309	0.04%
22 h	La_Savoureuse_de_Giromagny	2.7966	2.6840	4.20 %
23 h	La_Bourbeuse	3.7371	3.7186	0.50%
	Le_Dessoubre	3.9556	3.9335	0.56%
24 h	Le_Doubs_canal_de_Besançon	2.8899	2.6307	**9.84%**

significant enhancement of 9.84% in RMSE is observed by leveraging a singular river height variable at the final hour within this timeframe, in contrast to the inclusion of alternative variables in prior horizons. Figure 4 depicts the precision of predictions, exhibiting a maximum deviation of 0 to 17 errors for the 2nd hour before and after integrating the river height variable, respectively. Given that the XGBoost model generates decimal predictions (e.g., 8.32 interventions), the outcomes have been rounded to the nearest integer (here, 8 interventions) to align with practical applicability.

Table 6. Best hyperparameter for the 2nd hour

Hyperparameter	Values
max_depth	4
min_child_weight	4
gamma	0.2
subsample	0.9
colsample_bytree	0.8
learning_rate	0.1

Table 7. Best hyperparameter for the 24th hour

Hyperparameter	Values
max_depth	6
min_child_weight	2
gamma	0
subsample	0.9
colsample_bytree	0.9
learning_rate	0.02

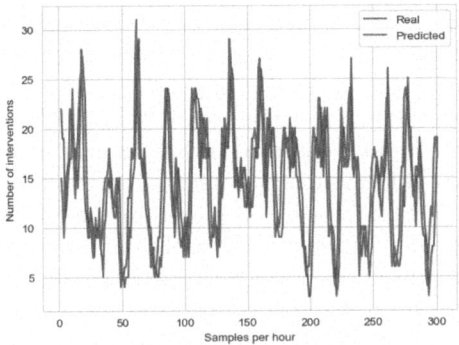

Fig. 3. Prediction for the 2nd hour

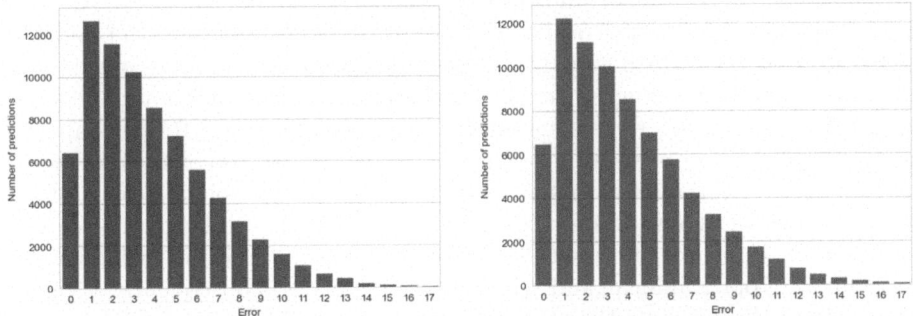

Fig. 4. Insights from predictive modeling. Prediction's error at the 2nd hour: (left) pre-selection, (right) post-selection

Significantly, XGBoost exhibited its most commendable RMSE score of 1.6500 at the 2nd hour, prominently influenced by the inclusion of one of the scrutinized river height variables, particularly that of the "Dessoubre" river. Noteworthy is the consistent enhancement in scores across temporal horizons owing to the recurrent selection of this variable, with certain rivers persistently contributing to score improvement for at least 2 h, exemplified by the "Dessoubre," "Savoureuse," and "Rosemontoise" rivers at the 9th, 14th, and 19th hours, respectively. Subsequently, the XGBoost model underwent comprehensive training across all horizons via a meticulous grid search procedure for parameter tuning. This approach systematically explores diverse attribute combinations to ascertain the optimal solution. Tables 6 and 7 furnish comprehensive insights into the identified and employed hyperparameters respectively for the best RMSE score at the second hour and the greatest improvement percentage at the last hour, ensuring that the model achieves its maximal performance.

4   Discussion

The present study aimed to assess the impact of river water level rise on fire-fighter intervention forecasting over nine years, from 2015 to 2024. Our methodological approach was carefully designed, involving meticulous data preparation and in-depth analysis to inform our experimental decisions. This included feature selection through various statistical techniques and machine learning methods. XGBoost was selected as the predictive model due to its robustness in handling outlier values, a particularly advantageous quality for continuous river height variables (see Table 1). Our methodology involved the implementation of an iterative training pipeline, in which features that improved the RMSE score over a 24-hour horizon were systematically selected (see Fig. 2). The results, presented in Table 5, highlight the importance of variables associated with river height, particularly from the outset and during prolonged perennial periods. Notably, at the 2nd hour, the presence of the "Dessoubre" river contributed to a promising RMSE score of 1.65, which was able to continue improving the score in extended horizons, sometimes taking up to 2 h. Furthermore, at the final hour, the selection of the "Besancon" river resulted in the most substantial RMSE improvement at 9.84%. This observation suggests a persistent and progressive impact of river water level variables across all periods, highlighting their increased utility for both short and long-term forecasts. Thus, our findings indicate that certain river-related variables exert a notable influence across all horizons, owing to various temporal, topographical, and atmospheric influences. Consequently, it can be inferred from these results that there may be other variables producing delayed effects or other factors on river behavior influencing firefighter interventions. Weather conditions such as humidity and precipitation vary seasonally, while human activities such as dam construction, river diversions, and irrigation can also influence the distribution of river heights. Moreover the existence of complex topography, such as steep valleys or mountainous regions, where interactions between watercourses can lead to significant local variations in river height, resulting in varied consequences at different times of the day. This may explain why the impact becomes apparent from the outset and persists for extended hours. All these factors lead us to conclude that the potential impact of rising water levels on the environment is not negligible, its effects can have dangerous consequences, systematically triggering firefighter intervention.

The inquiry aims to comprehend why there aren't enough selected variables belonging to the river height family that could enhance prediction. For instance, considering the improved prediction outcome of 1.65 after selecting the "Dessoubre" river height at the 2nd hour, it becomes evident that other features from the same category simply provide redundant information and cannot be chosen as they will not contribute to enhancing the prediction outcome. Moreover, as depicted in Fig. 5, it becomes apparent, after calculating the correlation of variables in different categories with the river height feature, that variables belonging to the same family exhibit a high correlation, ranking them high, while also observing a correlation with other variables not belonging to the same family, such as precipitation and humidity, which is present in all horizons.

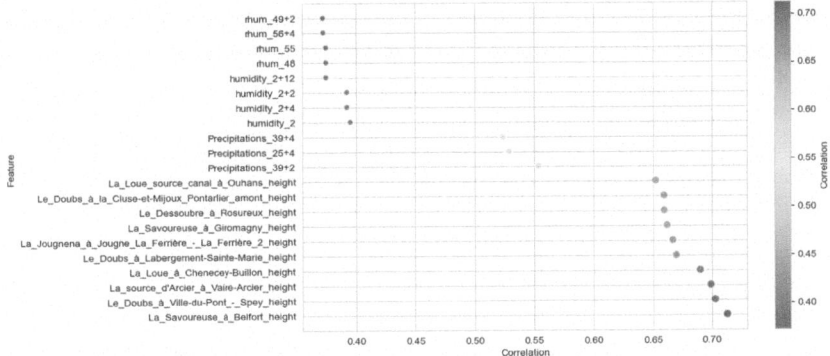

Fig. 5. Correlation with the river "Le_Dessoubre" at the 2nd hour

The logical explanation for this result is an indirect causal correlation, implying that heavy precipitation could have an indirect influence on rising water levels. To delve deeper, a brief detailed analysis of the forecast improvement after selecting a variable belonging to the river height family, as shown in Fig. 6, reveals a consistently higher average improvement in forecasts during the summer. This strongly advocates for the inclusion of variables from this family during this season to achieve promising prediction outcomes.

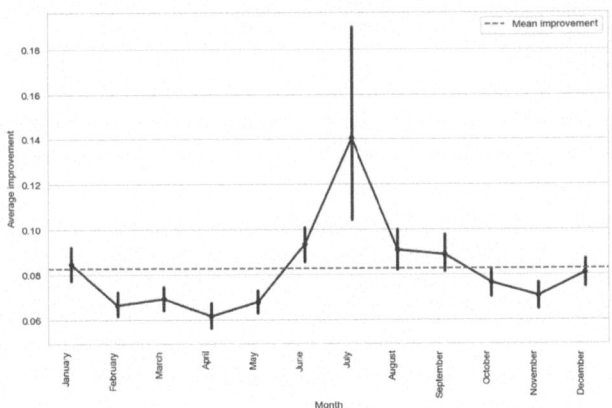

Fig. 6. Improvement statistics per month

A detailed examination, as depicted in Fig. 4, unveils a reduction in prediction errors, thus validating our findings. Specifically, the figure highlights a decrease in errors, notably an increase in 0 errors and a decrease in 1 and 2 errors, following the inclusion of the "Dessoubre" river height variable compared to errors observed before its inclusion. This study deepens our understanding of how river height influences firefighter interventions, presenting novel insights.

However, certain limitations, such as the implementation of other boosting models or machine learning methods for comparison purposes, as well as the exclusive use of grid search for hyperparameter optimization, have tempered our results. Future avenues will involve exploring additional variables on other rivers, studying the combined impact of river height and meteorology, employing alternative models, and adopting more sophisticated sequential optimization approaches, such as Bayesian optimization.

5   Conclusion

This study represents a deeper dive within a series of articles aimed at investigating various categories of variables and their implications for firefighting interventions. Focusing on forecasting the number of future interventions, a critical aspect for firefighting services globally, this article explores the influence of river heights on emergency service predictions. Utilizing an extensive nine-year dataset from the Fire and Rescue Service (SDIS 25) in the Doubs region, France, the research examines how the predictive model strives to identify significant trends associated with river water level variables during distinct periods of the day. The findings suggest that variables in this category may have a sustained impact from the outset for an extended duration. For instance, the influence of the "Dessoubre" river from the 2nd hour until the 23rd hour underscores the increased utility of forecasts for prolonged time intervals. While this approach proves valuable for resource management and optimizing response times, ongoing enhancements are imperative for seamless integration into firefighting decision-making processes.

Continuing our research efforts, we will investigate the nuanced impact of variable categories (epidemiology, alert bulletins, events, etc.) on forecasts across various time horizons. We aim to extract valuable insights from these domains and identify the most influential variables based on intervention types and forecasting timeframes. By leveraging this information, we intend to refine our predictions, especially for short-term and long-term scenarios, by selecting models and their hyperparameters that align with the increased relevance of specific variables. Additionally, we will focus on integrating geolocated variables to fine-tune forecasts based on factors such as population density, forest area, etc. Our primary goal is to enhance our predictive capabilities, enabling the implementation of operational strategies that proactively address the demand for firefighting services.

References

1. Public   Health   France.   https://www.santepubliquefrance.fr/maladies-et-traumatismes/traumatismes/noyade/documents/rapport-synthese/surveillance-epidemiologique-des-noyades.-resultats-de-l-enquete-noyades-2021.   Accessed   10 Oct 2024

2. Kia, M.B., et al.: An artificial neural network model for flood simulation using GIS: Johor River Basin, Malaysia. Environ. Earth Sci. **67**, 251–264 (2012)
3. Olyaie, E., Banejad, H., Chau, K.W., Melesse, A.M.: A comparison of various artificial intelligence approaches performance for estimating suspended sediment load of river systems: a case study in United States. Environ. Monit. Assess. **187**, 1–22 (2015)
4. Javadi, F., Ahmadi, M.M., Qaderi, K.: Estimation of river bedform dimension using artificial neural network (ANN) and support vector machine (SVM). J. Agric. Sci. Technol. **17**, 859–868 (2015)
5. Chang, D.-L., et al.: Artificial intelligence methodologies applied to prompt pluvial flood estimation and prediction. Water **12**, 3552 (2020)
6. Borwarnginn, P., Haga, J.H., Kusakunniran, W.: Predicting river water height using deep learning-based features. ICT Express **8**, 588–594 (2022)
7. Arcolezi, H.H., et al.: Forecasting the number of firefighter interventions per region with local-differential-privacy-based data. Comput. Secur. **96**, 101888 (2020)
8. Cerna, S., Guyeux, C., Laiymani, D.: The usefulness of NLP techniques for predicting peaks in firefighter interventions due to rare events. Neural Comput. Appl. **34**(12), 10117–10132 (2022)
9. Guyeux, C., Makhoul, A., Bahi, J.M.: How to build an optimal and operational knowledge base to predict firefighters' interventions. In: Proceedings of SAI Intelligent Systems Conference (IntelliSys), vol. 1, pp. 558–572 (2022)
10. Sirri, N., Guyeux, C.: Air quality impact on firefighter interventions: factors analysis. In: Proceedings of the 7th International Conference on Big Data & Internet of Things (2024)
11. Sirri, N., Guyeux, C.: Solar activity impact on firefighter interventions: factors analysis. In: Proceedings of the 5th International Conference on Deep Learning & Applications, pp. 107–122 (2024)
12. Ministry of Ecological Transition. http://www.hydro.eaufrance.fr/. Accessed 10 Oct 2024
13. NASA. https://www.swpc.noaa.gov/. Accessed 10 Oct 2024
14. ATMO-BFC. https://www.atmo-bfc.org/accueil. Accessed 10 Oct 2024
15. NASA. https://lance.modaps.eosdis.nasa.gov/viirs/. Accessed 10 Oct 2024
16. MODIS. https://lance.modaps.eosdis.nasa.gov/modis/. Accessed 10 Oct 2024
17. Skyfield. https://github.com/skyfielders/python-skyfield. Accessed 10 Oct 2024
18. Astral. https://pypi.org/project/astral/0.5/. Accessed 10 Oct 2024
19. The Sentinel Network. https://www.sentiweb.fr/?page=table. Accessed 10 Oct 2024
20. Soccer. https://www.footendirect.com/. Accessed 10 Oct 2024
21. Ministry of National Education. http://www.education.gouv.fr/pid25058/le-calendrier-scolaire.html. Accessed 10 Oct 2024
22. Météo-France. https://www.ecologie.gouv.fr/. Accessed 10 Oct 2024
23. Meteo-Stat. https://pypi.org/project/meteostat/. Accessed 10 Oct 2024
24. Vigilance-France. https://vigilance.meteofrance.fr/fr. Accessed 10 Oct 2024
25. Pedregosa, F., et al.: Scikit-learn: machine learning in python. J. Mach. Learn. Res. **12**, 2825–2830 (2011)
26. Target Encoder. https://scikit-learn.org/stable/modules/generated/sklearn.preprocessing.TargetEncoder.html. Accessed 10 Oct 2024
27. Garreta, R., Moncecchi, G.: Learning scikit-learn: machine learning in Python: experience the benefits of machine learning techniques by applying them to real-world problems using Python and the open source scikit-learn library. Packt Publishing (2013)

28. Zien, A., Krämer, N., Sonnenburg, S., Rätsch, G.: The feature importance ranking measure. In: Proceedings of the European Conference on Machine Learning & Knowledge Discovery in Databases (ECML/PKDD), Part II, pp. 694–709 (2009)
29. Chen, T., Guestrin, C.: Xgboost: a scalable tree boosting system. In: Proceedings of the 22nd ACM SIGKDD International Conference on Knowledge Discovery & Data Mining, pp. 785–794 (2016)
30. Ke, G., et al.: Lightgbm: a highly efficient gradient boosting decision tree. In: Proceedings of the 31st Conference on Neural Information Processing Systems (NIPS) (2017)

Correction to: Made to Measure: Towards Approximability of Query Evaluation Engines

Daniel Flachs[iD] and Guido Moerkotte[iD]

Correction to:
Chapter 16 in: R. Chbeir et al. (Eds.):
Database Engineered Applications, **LNCS 15511,**
https://doi.org/10.1007/978-3-031-83472-1_16

The original version of the book was inadvertently published with an error related to the text part and alignment in Tables 2,3, and 4. It has been corrected.

The updated version of this chapter can be found at
https://doi.org/10.1007/978-3-031-83472-1_16

Author Index

R. Chbeir et al. (Eds.): IDEAS 2024, LNCS 15511, pp. 411–412, 2025.
https://doi.org/10.1007/978-3-031-83472-1

The manufacturer's authorised representative in the EU is Springer
Nature Customer Service Centre GmbH, Europaplatz 3, 69115 Heidelberg,
Germany. If you have any concerns regarding our products, please
contact ProductSafety@springernature.com

Printed and bound by CPI Group (UK) Ltd, Croydon, CR0 4YY
29/04/2026
02099551-0004